Omics in Plant Disease Resistance

Current Issues in Molecular Biology. Volume 19

Edited by:

Vijai Bhadauria

Crop Development Centre/Dept. of Plant Sciences
51 Campus Drive
University of Saskatchewan
Saskatoon, SK S7N 5A8 Canada

Caister Academic Press

www.caister.com

ISBN (paperback): 978-1-910190-35-7
ISBN (ebook): 978-1-910190-36-4

Contents

Current publications of interest

Microalgae
Current Research and Applications
Edited by: MN Tsaloglou
152 pp, January 2016
Book: ISBN 978-1-910190-27-2 £129/$259
Ebook: ISBN 978-1-910190-28-9 £129/$259
The latest research and newest approaches to the study of microalgae.

Bacteria-Plant Interactions
Advanced Research and Future Trends
Edited by: J Murillo, BA Vinatzer, RW Jackson, et al.
x + 228 pp, March 2015
Book: ISBN 978-1-908230-58-4 £159/$319
Ebook: ISBN 978-1-910190-00-5 £159/$319
"an up-to-date overview" (Ringgold)

Microarrays
Current Technology, Innovations and Applications
Edited by: Z He
x + 246 pp, August 2014
Book: ISBN 978-1-908230-49-2 £159/$319
Ebook: ISBN 978-1-908230-59-1 £159/$319
"a valuable and useful source ... recommended" (Biotechnol. Agron. Soc. Environ.)

Metagenomics of the Microbial Nitrogen Cycle
Theory, Methods and Applications
Edited by: D Marco
xiv + 268 pp, September 2014
Book: ISBN 978-1-908230-48-5 £159/$319
Ebook: ISBN 978-1-908230-60-7 £159/$319
"a strong overview" (Ringgold)

Proteomics
Targeted Technology, Innovations and Applications
Edited by: M Fuentes, J LaBaer
x + 186 pp, September 2014
Book: ISBN 978-1-908230-46-1 £159/$319
Ebook: ISBN 978-1-908230-62-1 £159/$319
"many excellent chapters" (Doodys)

Applied RNAi
From Fundamental Research to Therapeutic Applications
Edited by: P Arbuthnot, MS Weinberg
x + 252 pp, June 2014
Book: ISBN 978-1-908230-43-0 £159/$319
Ebook: ISBN 978-1-908230-67-6 £159/$319
"Essential reading" (Biotechnol Agron Soc Environ); "recommended" (Fungal Diversity); "an excellent resource" (Doodys)

Next-generation Sequencing
Current Technologies and Applications
Edited by: J Xu
xii + 160 pp, March 2014
Book: ISBN 978-1-908230-33-1 £120/$240
Ebook: ISBN 978-1-908230-95-9 £120/$240
"written in an accessible style" (Zentralblatt Math); "recommend this book to all investigators" (ChemMedChem)

Omics in Soil Science
Edited by: P Nannipieri, G Pietramellara, G Renella
x + 198 pp, January 2014
Book: ISBN 978-1-908230-32-4 £159/$319
Ebook: ISBN 978-1-908230-94-2 £159/$319
"a recommended reference" (Biotechnol. Agrom. Soc. Environ.); "a must for Soil scientists" (Fungal Diversity)

Genome Analysis
Current Procedures and Applications
Edited by: MS Poptsova
xiv + 374 pp, January 2014
Book: ISBN 978-1-908230-29-4 £159/$319
Ebook: ISBN 978-1-908230-68-3 £159/$319
An up-to-date and comprehensive overview of next-generation sequencing data analysis, highlighting problems and limitations, applications and developing trends in various fields of genome research.

RNA Editing
Current Research and Future Trends
Edited by: S Maas
viii + 240 pp, June 2013
Book: ISBN 978-1-908230-23-2 £159/$319
Ebook: ISBN 978-1-908230-88-1 £159/$319
"an essential book" (Doodys)

Real-Time PCR
Advanced Technologies and Applications
Edited by: NA Saunders, MA Lee
viii + 284 pp, July 2013
Book: ISBN 978-1-908230-22-5 £159/$319
Ebook: ISBN 978-1-908230-87-4 £159/$319
"an invaluable reference" (Doodys); "wide range of real time PCR technologies" (Food Sci Technol Abs); "I was impressed by this text" Aus J Med Sci

Bionanotechnology
Biological Self-assembly and its Applications
Edited by: BHA Rehm
x + 310 pp, February 2013
Book: ISBN 978-1-908230-16-4 £159/$319
Ebook: ISBN 978-1-908230-81-2 £159/$319
"the most striking and successful approaches" Book News

OMICS in Plant Disease Resistance

Vijai Bhadauria

Crop Development Center / Department of Plant Sciences, University of Saskatchewan, Saskatoon, Saskatchewan S7N 5A8 Canada
E mail: vijai.bhadauria@usask.ca
Tel: (306) 866-8380

Abstract
The term OMICS, which look into the global profiling and analysis of various cellular molecules, has gained new heights with the advancement of next-generation sequencing and mass spectrometry technologies. It has broader implication in genetic improvement of crops for resistance against various diseases of economic significance. This focus issue entitled **OMICS in Plant Disease Resistance** highlights the implication of OMICS (genomics, transcriptomics, proteomics and metabolomics) in agricultural research.

Genomics and post-genomics era OMICS technologies have broader implication in genetic improvement of crops for disease resistance (Bhadauria et al., 2010a, 2010b, 2014). This focus issue summarizes the technological innovations in the area of OMICS for enhancing genetic resistance of crops to various pests, such as fungi, oomycetes and insects (Figure 1). Below are some of the highlights of the focus issue on **OMICS in plant disease resistance.**

Wild germplasm is being utilized to enhance genetic resistance in lentil (*Lens culinaris*) against fungal diseases using next generation sequencing-based genotyping and comparative genomics approaches (Bhadauria et al.). Such approaches allow pinpointing and mapping the genomic regions in wild species (*Lens ervoides*) conferring resistance against diseases like anthracnose (*Colletotrichum lentis*), Ascochyta blight (*Ascochyta lentis*) and Stemphylium blight (*Stemphylium botryosum*). Similar approaches in pests could lead to identification of effectors (avirulence proteins; effectoromics) that can then be used in fishing out pathogen recognition specificities encoded by NBS-LRR class of resistance (*R*) genes in wild germplasm. The *R* genes can be introgressed in cultivars through marker-aided backcrossing to enhance genetic performance against fungal diseases.

Plants respond to pathogen attacks by deploying antimicrobial metabolites, such as phytoanticipins (like glucosinolates that are converted to toxic molecules after pathogen recognition), phytoalexins and altered volatile organic compounds. Liquid and gas chromatography coupled with mass spectrometry allow the global

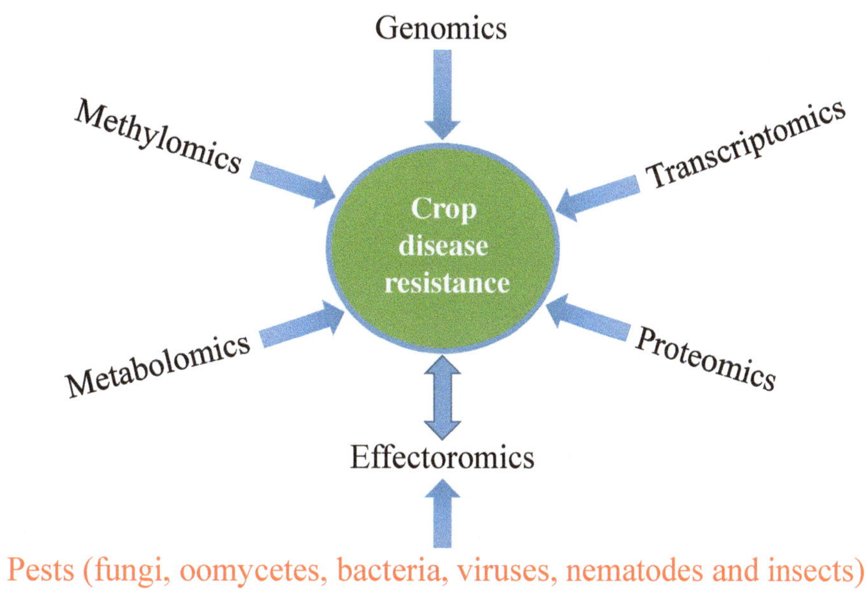

Figure 1. Omics in plant disease resistance.

identification of metabolites (Arbona and Gómez-Cadenas). Nitric oxide (NO) is considered as a signaling molecule and produced during biotic and abiotic stresses. Pérez et al., describe the role of cDNA-amplification fragment length polymorphism and RNA-Seq to unravel NO-based plant disease resistance and catalogue NO-responsive genes. Rodríguez's group at Misión Biológica de Galicia (MBG-CSIC), Spain (Francisco et al.) describes the advancement in OMICS (genomics, transcriptomics, proteomics and metabolomics) with the aim of identifying key components of brassica resistance against clubroot (*Plasmodiophora brassicae*), blackleg (*Leptosphaeria maculans*), etc. Cui et al., provide a comprehensive overview on interaction at transcriptome (RNA-Seq) and proteome levels between rice and planthopper- borne viruses, such as rice stripe virus (Tenuivirus), rice dwarf virus (Reoviridae), rice gall dwarf virus (Reoviridae), rice ragged stunt virus (Reoviridae), rice yellow stunt virus (Rhabdoviridae), rice grassy stunt virus (Tenuivirus), rice black-streaked dwarf virus (Reoviridae) and southern rice black-streaked dwarf virus (Reoviridae). Escobar's group (Cabrera et al.) describes the power of OMICS in identifying susceptibility and resistance factors in host to root-knot nematodes *Meloidogyne* spp. Burra et al., highlight technological advancements and pitfalls of transcriptomics and proteomics to unravel the molecular basis of complex plant-oomycete interaction and shed light on oomycete effector biology. Gomez-Casati's group at Universidad Nacional de Rosario, Argentina discussed the implication of OMICS in generating pathogen-resistant crops. Grandellis and colleagues summarize various gel and non-gel based protein separation methods coupled mass spectrometry to decipher plant defense against pathogens.

One of the best researched pathosystems (plant-fungal pathogen) is Rice-*Magnaporthe oryzae* not only because of economic significance but also genetic and molecular tractability of both host and pathogen. Sharma's group at ICAR-National Research Centre on Plant Biotechnology, India discuss transcriptomics-based oscillating interactome especially regulation of rice resistance genes and *M. oryzae* effectors, including avirulence proteins. Chen et al., discussed proteomics-based approaches to decipher molecular mechanisms underlying rice defense against broad spectrum of pathogens, such as viruses, bacteria and fungi, and in rice lesion mimic mutants.

Acknowledgements

I would like to acknowledge Lucia Popescu at Department of Soil Science, University of Saskatchewan, Saskatchewan S7N 5A8 Canada for assisting in compilation of the focus issue- OMICS in plant disease resistance.

References

Bhadauria, V., and Banniza, S. (2014). What lies ahead in post-genomics era: A perspective on genetic improvement of crops for fungal disease resistance. Plant Signal. Behav. *9*, e28503.

Bhadauria, V. (2010). Rice blast fungus transcriptomics. Arch. Phytopathol. Plant Protect. *43(8)*, 723-727.

Bhadauria, V., and Peng, Y.L. (2010). Rice blast fungus proteomics. Arch. Phytopathol. Plant Protect. *43(2)*, 149-153.

Other Publications of Interest

Xu, J.-P. (2014). Next-generation Sequencing: Current Technologies and Applications. Caister Academic Press, U.K. ISBN: 978-1-908230-33-1

He, Z. (2014). Microarrays: Current Technology, Innovations and Applications. Caister Academic Press, U.K. ISBN: 978-1-908230-49-2

Poptsova, M.S. (2014). Genome Analysis: Current Procedures and Applications. Caister Academic Press, U.K. ISBN: 978-1-908230-29-4

Fuentes, M. and LaBaer, J. (2014). Proteomics: Targeted Technology, Innovations and Applications. Caister Academic Press, U.K. ISBN: 978-1-908230-46-1

Murillo, J., Vinatzer, B.A., Jackson, R.W. and Arnold, D.L. (2015). Bacteria-Plant Interactions: Advanced Research and Future Trends. Caister Academic Press, U.K. ISBN: 978-1-908230-58-4

Caranta, C., Aranda, M.A., Tepfer, M. and Lopez-Moya, J.J. (2011). Recent Advances in Plant Virology. Caister Academic Press, U.K. ISBN: 978-1-904455-75-2

Nannipieri, P., Pietramellara, G. and Renella, G. (2014). Omics in Soil Science. Caister Academic Press, U.K. ISBN: 978-1-908230-32-4

Herold, K.E. and Rasooly, A. (2009). Lab-on-a-Chip Technology: Fabrication and Microfluidics. Caister Academic Press, U.K. ISBN: 978-1-904455-46-2

Wild Help for Enhancing Genetic Resistance in Lentil Against Fungal Diseases

Vijai Bhadauria§, Melissa M.L. Wong§, Kirstin E. Bett and Sabine Banniza*

Crop Development Centre / Department of Plant Sciences, University of Saskatchewan, Saskatoon, Saskatchewan S7N 5A8 Canada
§Contributed equally to this work
*Corresponding author. E-mail: sabine.banniza@usask.ca

Abstract
Lentil (*Lens culinaris*) is one of the cool season grain legume crops and an important source of dietary proteins and fibre. Fungal diseases are main constraints to lentil production and account for significant yield and quality losses. Lentil has a narrow genetic base presumably due to a bottleneck during domestication and as a result, any resistance to fungal diseases in the cultivated genepool is gradually eroded and overcome by pathogens. New sources of resistance have been identified in wild lentil (*Lens ervoides*). This article provides an overview of harnessing resistance potential of wild germplasm to enhance genetic resistance in lentil cultivars using next-generation sequencing-based genotyping, comparative genomics and marker-assisted selection breeding.

In agricultural systems, crops are constantly exposed to a variety of pests, such as viroids, viruses, phytoplasma, bacteria, fungi, oomycetes, nematodes and insect herbivores. In response, plants have developed an amazing capacity for recognizing and effectively responding to pest attacks. This capacity is known as innate immunity, which relies on the recognition of non-self pathogen molecules and damage-associated molecular patterns of plant structures. Pathogen molecules recognized by the plant system can be conserved or hyper-variable. Conserved molecules are known as pathogen-associated molecular patterns (PAMPs), which are generally structural components of pathogens, such as the flagellin protein from bacterial flagella, and chitin and glucan polysaccharides from fungal cell walls. The plant's innate immunity consists of two layers (Jones and Dangl, 2006). The first layer is known as PAMP-triggered immunity (PTI), which uses cell membrane receptors (receptor-like kinase and receptor-like proteins) to recognize PAMPs and trigger defense responses, such as callose deposition at the cell periphery. PTI is the front line of defense that is effective against non-adapted pathogens and therefore also contributes to non-host resistance and basal resistance. Introgression of PTI components from wild germplasm to cultivars can confer resistance against invasive lineages of pathogens. This is the idea behind lineage-exclusion breeding, where resistance genes are pyramided into cultivars with specificity to the prevalent pathogen lineages (Bhadauria and Banniza, 2014). The second layer of the plant's innate immunity is known as effector triggered immunity (ETI), an amplified version of PTI, which uses intracellular receptors encoded by dominant or semi-dominant resistance (*R*) genes to recognize effectors (avirulence proteins) and triggers defense responses, such as hypersensitive response cell death.

Resistance proteins possess a tripartite domain organization: N-terminal coiled-coil or Toll/Interleukin-1 receptor (TIR), centrally localized conserved nucleotide binding (NB)-ARC (APAF-1 [apoptotic protease-activating factor-1], R proteins and a CED-4 [*Caenorhabditis elegans* death-4 protein]) domain and a highly polymorphic C-terminal leucine rich repeat (LRR) domain. Therefore, these proteins are also known as NB-ARC-LRR proteins (NBLRs) (Van der Biezen and Jones, 1998) (Figure 1A). NBLRs are structurally and functionally similar to mammalian NOD (nucleotide-binding oligomerization domain)-LRR proteins commonly known as CATERPILLER proteins (Caspase activation and recruitment domain [CARD], transcription enhancer, R (purine)-binding, pyrin, lots of leucine repeats) that trigger inflammatory and immune responses (Inohara et al., 2005), suggesting a convergent evolution of plant and animal immune systems.

Table 1. Legume NBLRome.

Species (common name)	Genome architecture	Genome size (Mb)	Gene models	*NBLR*
Cicer arietinum (Chickpea)	2n = 16	738	28,269	187 (0.66%)
Medicago truncatula (Barrel medic)	2n = 16	470	47,529	764 (1.61%)
Glycine max (Soybean)	2n = 40	1,115	46,430	506 (1.09%)
Cajanus cajan (Pigeonpea)	2n = 22	833	48,680	406 (0.83%)
Lotus japonicus	2n = 12	472	22,650	158 (0.69%)
Lens culinaris (Lentil)	2n = 14	~4000*	-	~400 (1%)*

*estimated number

*NBLR*s constitute nearly 1% of the total gene models in legumes (Table 1).

NBLRs of plants that recognize pathogen effectors have an analogous function to antibodies that recognize antigens in vertebrates, but whereas the vertebrate system is an adaptive immune system, plants rely on the multilayer innate immune system. The molecular interaction between NBLRs and pathogen effectors leads to an antagonistic co-evolution between plants and their pathogens as diversifying selection is operating to increase the recognition specificity of NBLRs and the detection evasion capacity of pathogens potentially by point mutation in effectors (direct interaction) or jettisoning of recognizable effectors (indirect interaction; Silverstein et al., 2005; Stergiopoulos and deWit, 2008). This evolutionary process is commonly known as molecular arms race between plants and their pathogens (Figure 1A).

Lentil (*Lens culinaris* Medikus) is a cool-season grain legume crop and an important source of dietary proteins and complex carbohydrates in the Mediterranean and Indian subcontinent. Lentil was domesticated around 9000 years ago along with wheat, barley and peas in the Fertile Crescent, a historical term used to describe a region spreading in-between the Persian Gulf and the Mediterranean Sea. Canada, India and Turkey are now the leading lentil producers (FAOSTAT, 2013). In Canada, fungal diseases (Anthracnose, Ascochyta blight and Stemphylium blight) are the main constraints to lentil production and account for significant yield and quality losses.

Lentil has a narrow genetic base presumably due to a bottleneck during domestication and as a result, any resistance in the cultivated gene pool is gradually overcome by pathogens, probably because of a compromised pathogen recognition capacity of the lentil immune system encoded by the *NBLR* class of *R* genes. Typical NBLR proteins contain 9 to 40 leucine rich repeats, each of which forms a beta strand as exemplified by an *Colletotrichum lentis*-responsive NBLR protein LcNBLR21 (GenBank accession JG293836) from the partially resistant lentil cultivar CDC Robin (Figure 1B; Bhadauria et al., 2013). This protein belongs to the TIR-NBLR class of R proteins and has 11 beta strands, which are arranged into a horseshoe shaped structure in the R protein. The concave surface of this structure recognizes pathogen avirulence protein(s) through physical interaction as originally envisaged under the gene-for-gene hypothesis (Flor 1942; Flor 1955), or as thought of in indirect interactions through the guard (Van der Biezen and Jones, 1998; Dangl and Jones, 2001) and the decoy hypotheses (van der Hoorn and Kamoun, 2008). Within the LRR motif, two leucine (L) residues are highly conserved whereas 5 amino acid residues (X) are highly polymorphic and define the recognition specificity of a NBLR protein for specific avirulence recognition. To counter this detection, pathogens mutate their recognizable avirulence proteins. As a result, new recognition specificities of the *NBLR* class of *R* genes are required to sustain lentil production. New sources of resistance for fungal pathogens have been identified in *L. ervoides* (Brign.) Grande, a member of the tertiary gene pool of lentil (Tullu et al., 2006). Next-generation genomics approaches, such as genotyping-by-sequencing, which has been successfully demonstrated in *Lens* spp. (Wong et al., 2015) are being used to generate a single nucleotide polymorphism (SNP)- based genetic linkage map for *L. ervoides*, and to identify loci on the map harbouring resistance to *C. lentis* races 0 and 1, *Ascochyta*

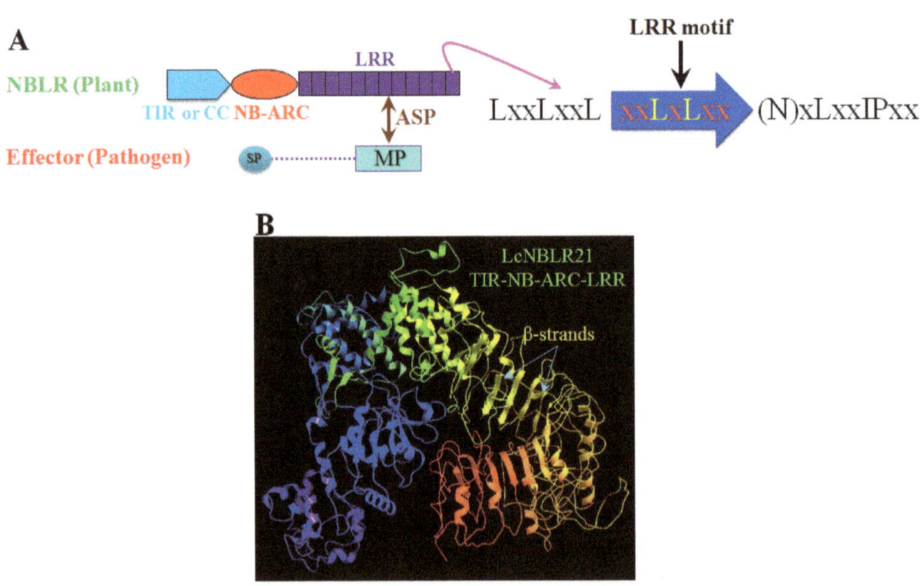

Figure 1. NBLRs and effector. (A) Antagonistic co-evolutionary arms race between NBLR (Plant arm) and effector (Pathogen arm). (B) A NBLR protein LcNBLR21 (GenBank accession JG293836) from lentil cultivar CDC Robin that shows up-regulation during anthracnose infection. TIR, Toll/Interleukin Receptor; CC, Coiled-Coil; NB-ARC, Nucleotide Binding-(APAF-1 [apoptotic protease-activating factor-1], R proteins and CED-4 [*Caenorhabditis elegans* death-4 protein]) domain; LRR, Leucine Rich Repeat Domain; SP, Signal Peptide; MP, Mature Protein; L, Leucine; X, any amino acid residue; I, Isoleucine; P, Proline; N, Asparagine; and ASP, Antagonistic selection pressure.

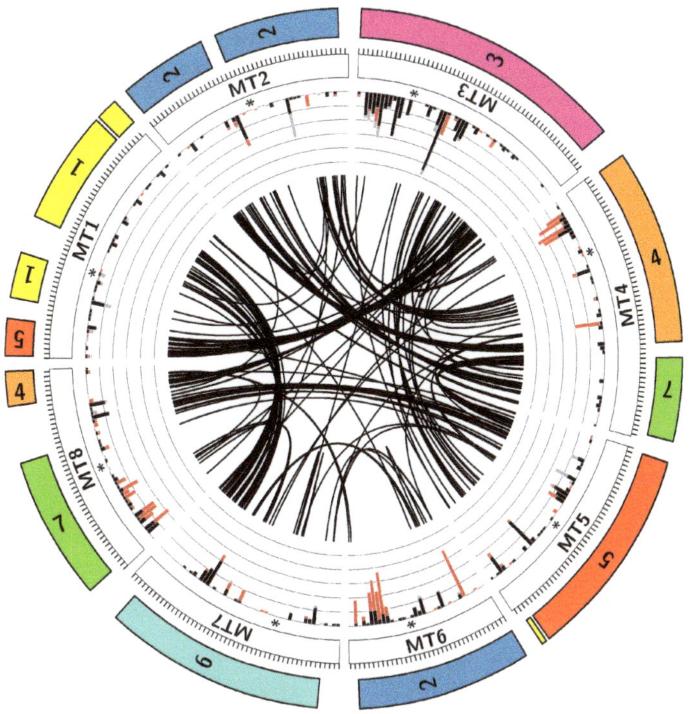

Figure 2. Circos plot showing the distribution of *NBLR* genes in *Medicago truncatula* Mt4.0 genome and their synteny regions in lentil. The ideogram represents the *M. truncatula* Mt4.0 chromosomes while the outer tiles represent syntenic regions of seven *lens culinaris* linkage groups. The *lens culinaris* linkage groups show high synteny to *M. truncatula* chromosomes except a small translocation between LG 1 and LG 5 and an expected translocation in LG 4 and LG 7 due to a translocation in *M. truncatula* genome. The distribution of genes encoding Toll/Interleukin-1 receptor, coiled-coil NBLR and regular NBLR, which was inferred from IMGAG annotations based on window size of 1 MB is shown red, grey and black bars, respectively. The innermost links represent the syntenic blocks within *M. truncatula* genome and the asterisk indicates centromeric region.

lentis and *Stemphylium botryosum*. SNP markers flanking these resistance loci can then be used in marker-assisted genome introgression to generate *L. culinaris* cultivars with superior genetic resistance against these fungal diseases. In addition, comparative genomics can be particularly useful for both cultivated and wild lentil species, such as *L. ervoides* whose linkage map and syntenic information relative to model legumes are available (Sharpe et al., 2013, Gujaria-Verma et al., 2014). For example, comparison of the *L. culinaris* genetic map consisting of seven linkage groups (Sharpe et al., 2013) with the eight chromosomes of the model legume *Medicago truncatula* (*Mt*; Young et al., 2011, Tang et al., 2014) revealed that the *NBLR* mega clusters on *Mt* chromosome 3 and 6 (Ameline-Torregrosa et al., 2008) are syntenic to regions on linkage group 3 and 2 of *L. culinaris*, respectively (Figure 2). Such synteny-based *in silico* comparative approaches can facilitate the identification of homologous genes and positional cloning of disease resistance genes in cultivated and wild lentil. This method has been successfully employed in the pasture legume alfalfa (*M. sativa*) for the transfer of *RCT1*, which confers resistance to *C. trifolii* (Yang et al., 2008). Similarly, comparison of genetic maps revealed a high degree of genomic collinearity or synteny between *L. culinaris* and *L. ervoides* with the exception of one major translocation between the linkage groups 1 and 5 (Gujaria-Verma et al., 2014). This information will facilitate efficient use of wild germplasm by tracking the disease resistance-responsive *L. ervoides* alleles during introgression into cultivated lentil. Rapid utilization of resistance genes of wild germplasm is expected once the full lentil genome sequence becomes publicly available.

Acknowledgements

We acknowledge the financial support from the Saskatchewan Agriculture Development Fund, the Saskatchewan Pulse Growers and Western Grain Research Foundation.

References

Ameline-Torregrosa, C., Wang, B., O'Bleness, M.S., Deshpande, S., Zhu, H., Roe, B., Young, N.D., and Cannon., S.B. (2008). Identification and Characterization of Nucleotide-Binding Site-Leucine-Rich Repeat Genes in the Model Plant Medicago truncatula. Plant Physiol. *146*(1): 5-21

Bhadauria, V., and Banniza, S. (2014). What lies ahead in post-genomics era: A perspective on genetic improvement of crops for fungal disease resistance. Plant Signal. Behav. *9*, e28503.

Bhadauria, V., Bett, K.E., Zhou, T., Vandenberg, A., Wei, Y., and Banniza, S. (2013). Identification of *Lens culinaris* defense genes responsive to the anthracnose pathogen *Colletotrichum truncatum*. BMC Genet. *14*, 31.

Dangl, J.L., and Jones, J.D.G. (2001). Plant pathogens and integrated defense responses to infection. Nature *411*, 826-833.

FAOSTAT. (2013). Food Agriculture Organization of The United Nations. Available at http://faostat.fao.org (Accessed July 13, 2015).

Flor, H. H. (1942). Inheritance of pathogenicity in a cross between physiologic races 22 and 24 of *Melampsora lini*. Phytopathology *32*, 5.

Flor, H.H. (1955). Host-parasite interaction in flax rust-its genetics and other implications. Phytopathology *45*, 680-85.

Gujaria-Verma, N., Vail, S. L., Carrasquilla-Garcia, N., Penmetsa, R. V., Cook, D. R., Farmer, A. D., Vandenberg, A., Bett, K. E. (2014). Genetic mapping of legume orthologs reveals high conservation of synteny between lentil species and the sequenced genomes of *Medicago* and chickpea. Frontiers in Plant Science, *5*, 676.

Inohara, N., Chamaillard, M., McDonald, C., and Nunez, G. (2005). NOD-LRR proteins: role in host-microbial interactions and inflammatory disease. Annu. Rev. Biochem. *74*, 355-383.

10. Jones, J.D.G., and Dangl J.L. (2006). The plant immune system. Nature *444*, 323-329.

Sharpe, A.G., Ramsay, L., Sanderson, L.A., Fedoruk, M.J., Clarke, W.E., Li, R., Kagale, S., Vijayan, P., Vandenberg, A., and Bett, K.E. (2013). Ancient orphan crop joins modern era: gene-based SNP discovery and mapping in lentil. BMC Genomics *14*, 92.

Silverstein, K.A.T., Graham, M.A., Paape, T.D., and VandenBosch, K.A. (2005) Genome organization of more than 300 defensin-like genes in Arabidopsis. Plant Physiol. *138(2)*, 600-610.

Tang, H., Krishnakumar, V., Bidwell, S., Rosen, B., Chan, A., Zhou, S., Gentzbittel, L., Childs, K.L., Yandell, M., Gundlach, H., Mayer, K.F., Schwartz, D.C., Town, C.D. (2014). An improved genome release (version Mt4.0) for the model legume *Medicago truncatula*. BMC Genomics. *15*, 312.

Tullu, A., Buchwaldt, L., Lulsdorf, M., Banniza, S., Barlow, B., Slinkard, A.E., et al. (2006) Sources of resistance to anthracnose (*Colletotrichum truncatum*) in wild *Lens* species. Genet. Resour. Crop Ev. *53*, 111-119.

Van der Biezen, E.A., and Jones, J.D.G. (1998) The NB-ARC domain: A novel signalling motif shared by plant resistance gene products and regulators of cell death in animals. Current Biol. *8*, R226-228.

Van der Biezen, E.A., and Jones, J.D.G. (1998). Plant disease-resistance proteins and the gene-for-gene concept. Trends Biochem. Sci. *23(12)*, 454-456.

Van der Hoorn R.A.L., and Kamoun, S. (2008) From Guard to decoy: A new model for perception of plant pathogen effectors. Plant Cell *20(8)*, 2009-2017.

Wong, M.M., Gujaria-Verma, N., Ramsay, L., Yuan, H.Y., Caron, C., Diapari, M., Vandenberg, A., and Bett, K.E. (2015). Classification and characterization of species within the genus lens using genotyping-by-sequencing (GBS). PLoS One *10(3)*, e0122025

Yang, S., Gao, M., Xu, C., Gao, J., Deshpande, S., Lin, S., Roe, B.A., and Zhu, H. (2008). Alfalfa benefits from *Medicago truncatula*: The *RCT1* gene from M. truncatula confers broad-spectrum resistance to anthracnose in alfalfa. PNAS *105(34)*, 12164-12169.

Young, N.D., Debellé, F., Oldroyd, G.E., Geurts, R., Cannon, S.B., Udvardi, M.K., Benedito, V.A., Mayer, K.F., Gouzy, J., Schoof, H., Van de Peer, Y., Proost, S., Cook, D.R., Meyers, B.C., Spannagl, M., Cheung, F., De Mita, S, Krishnakumar, V., Gundlach, H., Zhou, S., Mudge, J., Bharti, A.K., Murray, J.D., Naoumkina, M.A., Rosen, B., Silverstein, K.A., Tang, H., Rombauts, S., Zhao, P.X., Zhou, P., Barbe, V., Bardou, P., Bechner, M., Bellec, A., Berger, A., Bergès, H., Bidwell, S., Bisseling, T., Choisne, N., Couloux, A., Denny, R., Deshpande, S., Dai, X., Doyle, J.J., Dudez, A.M., Farmer, A.D., Fouteau, S., Franken, C., Gibelin, C., Gish, J., Goldstein, S., González, A.J., Green, P.J., Hallab, A., Hartog, M., Hua, A., Humphray, S.J., Jeong, D.H., Jing, Y., Jöcker, A., Kenton, S.M., Kim, D.J., Klee, K., Lai, H., Lang, C., Lin, S., Macmil, S.L., Magdelenat, G., Matthews, L., McCorrison, J., Monaghan, E.L., Mun, J.H., Najar, F.Z., Nicholson, C., Noirot, C., O'Bleness, M., Paule, C.R., Poulain, J., Prion, F., Qin, B., Qu, C., Retzel, E.F., Riddle, C., Sallet, E., Samain, S., Samson, N., Sanders, I., Saurat, O., Scarpelli, C., Schiex, T., Segurens, B., Severin, A.J., Sherrier, D.J., Shi, R., Sims, S., Singer, S.R., Sinharoy, S., Sterck, L., Viollet, A., Wang, B.B., Wang, K., Wang, M., Wang, X., Warfsmann, J., Weissenbach, J., White, D.D., White, J.D., Wiley, G.B., Wincker, P., Xing, Y., Yang, L., Yao, Z., Ying, F., Zhai, J., Zhou, L., Zuber, A., Dénarié, J., Dixon, R.A., May, G.D., Schwartz, D.C., Rogers, J., Quétier, F., Town, C.D., and Roe, B.A. (2011). The Medicago genome provides insight into the evolution of rhizobial symbioses. Nature *480(7378)*, 520-524.

Other Publications of Interest

Xu, J.-P. (2014). Next-generation Sequencing: Current Technologies and Applications. Caister Academic Press, U.K. ISBN: 978-1-908230-33-1

He, Z. (2014). Microarrays: Current Technology, Innovations and Applications. Caister Academic Press, U.K. ISBN: 978-1-908230-49-2

Poptsova, M.S. (2014). Genome Analysis: Current Procedures and Applications. Caister Academic Press, U.K. ISBN: 978-1-908230-29-4

Fuentes, M. and LaBaer, J. (2014). Proteomics: Targeted Technology, Innovations and Applications. Caister Academic Press, U.K. ISBN: 978-1-908230-46-1

Murillo, J., Vinatzer, B.A., Jackson, R.W. and Arnold, D.L. (2015). Bacteria-Plant Interactions: Advanced Research and Future Trends. Caister Academic Press, U.K. ISBN: 978-1-908230-58-4

Caranta, C., Aranda, M.A., Tepfer, M. and Lopez-Moya, J.J. (2011). Recent Advances in Plant Virology. Caister Academic Press, U.K. ISBN: 978-1-904455-75-2

Nannipieri, P., Pietramellara, G. and Renella, G. (2014). Omics in Soil Science. Caister Academic Press, U.K. ISBN: 978-1-908230-32-4

Current Status of Proteomic Studies on Defense Responses in Rice

Xifeng Chen[1], Vijai Bhadauria[2] and Bojun Ma[1*]

[1]College of Chemistry and Life Sciences, Zhejiang Normal University, Jinhua, 321004, China
[2]Crop Development Centre/Dept. of Plant Sciences, University of Saskatchewan, Saskatoon, Saskatchewan S7N 5A8 Canada
*Corresponding author: E-mail: mbj@zjnu.cn;
Tel.: 86-579-82282949; Fax: 86-579-82282269

Abstract
Biotic stresses are constraints to plant growth and development negatively impacting crop production. To counter such stresses, plants have developed stress-specific adaptations as well as simultaneous responses. The efficacy and magnitude of inducible adaptive responses are dependent on activation of signaling pathways and intracellular networks by modulating expression, or abundance, and/or post-translational modification of proteins associated with defense mechanisms. Proteomics plays an important role in elucidating plant defense mechanisms by mining the differential regulation of proteins to various biotic stresses. Rice, one of the most widely cultivated food crops in world, is constantly challenged by a variety of biotic stresses, and high-throughput proteomics approaches have been employed to unravel the molecular mechanism of the biotic stresses-response in rice. In this review, we summarize the latest advances of proteomic studies on defense responses and discuss the potential relevance of the proteins identified by proteomic means in rice defense mechanism. Furthermore, we provide perspective for proteomics in unraveling the molecular mechanism of rice immunity.

Introduction

Plants often suffer from biotic stresses caused by various pathogens, such as fungi, bacteria and viruses, which lead to significant damage, and even death. In evolution, plants have developed effective defense mechanisms to cope with various pathogenic microorganisms encountered at almost every stage of their development. In the process of defense response, plants regulate the expression of related intracellular resistance proteins to adjust their physiological state or morphology to resistant and inhibit the invasive pathogens. Many proteins are subjected to modifications at their C- or N-termini following synthesis, and various types of post-translational modification (PTM) play an important role in the plant defense signaling network. Therefore, the search for resistance-related proteins may significantly improve our understanding of plant defense mechanisms, which, in turn, will help to design rational strategies to improve plant disease resistance.

High-throughput proteomics studies can provide information in quantitative protein profiling, dynamic PTM, protein subcellular localization/compartmentalization, dissection of signaling pathways and protein-protein interactions (Agrawal et al., 2011; Agrawal et al., 2013; Hossain et al., 2012; Matros et al., 2011). In current high-throughput proteomics, plant proteins are classically separated on the two-dimensional electrophoresis (2-DE) gel, and followed by mass spectrometry (MS) or tandem MS (MS/MS) or liquid chromatography (LC)-MS/MS (Rakwal and Agrawal, 2003). Many plant species have been used to study quantitative changes in protein abundance in response to different abiotic and biotic stresses (Fang et al., 2015; Hu et al., 2015). Proteomic data have been used to determine the quantitative and qualitative changes in related proteins in plants and to understand the relationship between these proteins and biological processes. Hence, plant proteomics not only can identify the proteins involved in the direct defense responses but also can reveal the relationships between defense responses and biological processes, in which the identified proteins participating.

Rice is a seminal model plant of monocotyledons species and is the one of the most widely cultivated food crops throughout the world. Rice development and production is constrained by the diseases caused by various fungal, bacterial, and viral pathogens (Gnanamanickam et al., 1999). Proteomics studies in rice has made considerable progress in providing functional information of proteins under various physiological conditions in the various developmental stages, tissues, cells, and abiotic and biotic stress environments (Komatsu and Tanaka, 2005). In this review, we summarize the recent advances in rice proteomic studies on defense responses and discuss the potential relevance of the proteins identified by proteomic means in rice defense mechanism.

Proteins identified by proteomic approaches that are responsive to various pathogens

1. Fungi

Defense response of rice to the fungal pathogens like *Magnaporthe oryzae*, *Rhizoctonia solani* and *Cochliobolus miyabeanus*, have been investigated by proteomic studies. Rice blast, caused by *M. oryzae*, is one of the most destructive diseases of rice. Using the combined approaches of 2-DE and MS, Kim *et al.* (2004) identified 8 proteins including receptor-like protein kinases (RLK), β-1,3-glucanase, peroxidase, probenazole-inducible protein (PBZ1), and pathogenesis-related protein (PR5, PR10), induced by *M. oryzae* in rice inoculated leaves, all of which are defense-related. Li *et al.* (2015) conducted a time-course phosphoproteomic analysis on the leaves of both resistant and susceptible rice cultivars infected with *M. oryzae*, and 56 phosphoproteins belonging to 12 functional categories have been identified by MS/MS. The defense-related proteins, defense signaling-related proteins, and microtubule-associated proteins are differentially phosphorylated between the compatible and incompatible interactions, and such differences apparently contributed to the difference in resistance against fungal infection in the two cultivars (Li et al., 2015). However, the proteins involved in photosynthesis, antioxidation, and protein folding were not differentially accumulated between the resistant and the susceptible cultivars. These proteins could substantially enhance reactive oxygen species (ROS) production and contribute to the basal resistance of the plant against the pathogens (Li et al., 2015).

Sheath blight, caused by the fungus *R. solani*, is another major fungal disease of rice world-wide. Lee et al. (2006) used 2-DE and ESI Q-TOF MS to identify proteins upregulated in resistant and susceptible rice in response to *R. solani*. Six proteins identified with presumed anti-fungal (β-1,3-glucanase), photosynthetic (RuBisCO) and proteolytic (proteasome) activities were increased in both susceptible and resistant plants, suggesting a common baseline defense response for different host-pathogen interactions; An additional 8 proteins with anti-fungal activity (3-β-hydroxysteriod dehydrogenase, 3-β HSD; chitinase), glycolysis (glyceraldehyde 3-phosphate dehydrogenase, GAPDH), and anti-oxidation (ascorbate peroxidase, APX) were increased only in the resistant line (Lee et al., 2006). The induction of 3-β HSD was detected for the first time in resistant rice plants in response to pathogen challenge, suggesting a defensive role of this enzyme in rice against *R. solani*. (Lee et al., 2006).

Fungus *C. miyabeanus* causes brown spot disease in rice leaves resulting in yield loss. To understand the interaction between rice and *C. miyabeanus*, Kim et al. (2014) employed proteomic approaches of 2DE and nESI (nano-electrospray ionization)-LC-MS/MS in the *C. miyabeanus*-infected rice leaves, and successfully identified the sequences of 63 protein spots (39 increased and 24 decreased). In response to *C. miyabeanus*, the levels of enzymes involved in Calvin cycle (fructose bisphosphate aldolase; sedoheptulose-1, 7-bisphosphatase; RuBisCO; transketolase, TKL) and glycolysis (phosphoglycerate kinase; enolase; GAPDH) decreased in rice leaves, whereas the levels of enzymes in the TCA cycle (oxaloacetate; aspartate aminotransferase), ethylene biosynthesis (homocysteine S-methyltransferase), antioxidant process (APX; superoxide dismutase, SOD; peroxiredoxin, PRDX) and PR proteins (β-1,3-glucanase; PR5; PR10; salt induced protein, SalT) increased in rice leaves (Kim et al., 2014).

2. Bacteria

Current proteomic studies of rice defense responses against bacterial pathogens mainly focus on *Xanthomonas oryzae pv. oryzae* (*Xoo*), which is the second most important rice pathogen causing bacterial blight. Using 2-DE and MS, Yu *et al.* (2008) identified 64 differentially regulated proteins proteins from rice leaves infected with *Xoo*, and these proteins were those with various functions in photosynthesis (RuBisCO; oxygen-evolving enhancer protein, OEE), protein turnover (translational elongation factor; proteasome), glycolysis (GAPDH) and antioxidant defense (APX; glutathione S-transferase, GST), *etc.* Interestingly, 51 of these 64 proteins were involved in photosynthesis. Notably, the reduction in the level of larger fragment of RuBisCO large subunit (RcbL) was concomitant with increasing in the level of smaller fragments of RcbL, indicating that intact RcbL are degraded following *Xoo* attack (Yu et al., 2008).

Mahmood et al. (2006) analyzed the proteome of plasma membrane (PM) and cytosol from rice leaves inoculated with compatible and incompatible races of *Xoo*, respectively, and identified 20 differentially expressed proteins in response to *Xoo*, in which 9 were induced and 10 were depressed by compatible race, while 7 were induced and 3 were depressed by incompatible race. These proteins were related to photosynthesis (RuBisCO; OEE; ATP synthase, chloroplast), metabolism (TKL; GADPH) and defense (PR5; PBZ1; SOD; PRDX). Interestingly, RcbL was also found to be fragmented into two smaller proteins upon infections by two different *Xoo* races (Mahmood et al., 2006).

By using an improved method of PM protein purification, Chen *et al.* (2007) conducted PM proteomics of the rice suspension cells expressing a disease resistance gene *Xa21*, to identify PM components involved in the early defense response to bacterial blight. Eleven protein spots with predicted functions in plant defense were identified by MS/MS, including H^+-ATPase, phosphatase, hypersensitive-induced response protein, prohibitin, zinc finger protein, universal stress protein, and heat shock protein (HSP), demonstrating that direct proteomics analysis of rice PM can be applied to identify potential PM components involved in the rice defense response to microbes (Chen et al., 2007).

3. Virus

Despite large number of studies on viral diseases to date, the knowledge of the proteins that control viral diseases is limited. The rice yellow mottle virus (RYMV) is a member of the *Sobemovirus* genera, and destructive for rice production (Pinel et al., 2000). The RYMV is considered as a model virus to study the molecular mechanisms of viral resistance (Sire and Brugidou, 2002). Using 2-DE and LC-MS/MS, Ventelon-Debout *et al.* (2004) identified 32 differentially regulated proteins involved in metabolism, defense, and translation in the rice cells infected by RYMV, and these proteins belonged to three functional categories: defense/stress-related proteins, metabolism, and translation and protein turnover. A number of proteins regulated by abiotic stress response pathway (such as PR10, HSP, SOD, SalT, Dehydrin, Ethylene-inducible protein) were activated by RYMV in rice leaves, and proteins involved in glycolysis pathway such as phosphoglycerate mutase, phosphoglycerate dehydrogenase, aldolase C1 and GADPH were also detected during RYMV infection, suggesting that the glycolysis pathway plays an important role in defense/stress response.

Unlike fungi or bacteria, viruses generally encode a significantly smaller set of proteins, *e.g.* RYMV with only four open reading frames (Siré, 2002). It is likely that the virus recruits host proteins to finish their life cycle and cope with host defenses. To prove these multiple interactions, Brizard et al. (2006) extracted the protein complexes formed in the infected rice leaves by RYMV, and identified a large number of proteins from the complexes by MS. The proteins identified in the complexes are those involved in plant defense, metabolism, translation, protein synthesis and intracellular transport, presumably important for viral replication and movement. Surprisingly, no viral proteins were identified in the complex, indicating that viruses are able to recruit various host proteins with different functions to fulfill their life cycle (Brizard et al., 2006).

4. Elicitor

Fungal pathogens often release oligosaccharides, known as the elicitors, that can be specifically recognized by receptors on the host plasma membrane and this specific recognitions can trigger intracellular immune responses (Basse et al., 1993). Kim *et al.* (2003) discovered elicitor-induced 14 proteins in rice cells cultured in a suspension containing crude extracts of the rice blast fungus, and most were PR proteins including PBZ1, PR10, Isoflavone reductase (IRL) and SalT. Agrawal et al. (2002) revealed that proteins APX, PR5 and PR10 were accumulated in rice leaves treated with chitosan, a major component of many fungal cell walls, and demonstrated that chitosan triggers defense-signaling pathways in leaves of rice. Liao et al. (2009) used CSB I, an elicitor purified from fungus *M. grisea*, to treat rice leaves, and detected the differentially expressed proteins by 2-DE and LTQ (linear ion trap)-MS/MS. The identified proteins were functionally classified into defense proteins (PR10a, PR5 and SalT), ROS metabolites enzymes (SOD, GST and CAT), cell death related (translationally controlled tumor protein), signal transduction, molecule biosynthesis and metabolism categories (Liao et al., 2009). Commonly, these proteins were induced faster and stronger in incompatible interactions than compatible ones (Kim et al., 2003; Kim et al., 2009; Liao et al., 2009).

5. Lesion mimic mutants

Lesion mimic mutant (LMM) plants spontaneously display necrotic lesion on leaves without any pathogenic infection (Lorrain et al., 2003). The phenotype of LMM is similar to hypersensitive response (HR), which is a form of programmed cell death (PCD) associated with resistance to pathogens through rapid induction of local cell death around the infection site (Morel and Dangl, 1997). Most LMMs constitutively activate immune responses, including deposition of callose, induction of *PR* genes, production of ROS and accumulation of phytoalexins (Staskawicz et al., 1995), and display enhanced resistance to pathogens, such as rice blast and/or bacterial blight (Mizobuchi et al., 2002; Wu et al., 2008). Therefore, LMMs constitutes a powerful tool to unravel the pathways of PCD and disease response in plants.

Recently, the comparative proteomics between wild type (WT) rice and LMMs of *cdr1* (cell death and resistance; Takahashi et al., 2003), *cdr2* (Tsunezuka et al., 2005), *blm* (blast lesion mimic; Jung et al., 2006), *spl1* (spotted leaf; Kim et al., 2008) and *spl6* (Kang et al., 2007) have been reported. We also analyzed the proteomics of a rice LMM of *spl5* (Chen et al., 2013). These proteomic results showed that proteins involved in ROS scavenging, defense responses or cell death, such as SOD (*spl1*, *blm*), CAT (*spl6*, *blm*), Peroxidase (*spl1*), APX (*spl5*, *blm*), GST (*cdr2*, *spl5*), were induced in LMMs, whereas proteins involved in photosynthesis and glycolysis, such as RuBisCO (*spl1*, *spl5*, *spl6*, *blm*, *crd2*) and GAPDH (*spl1*, *spl5*, *cdr2*), were often depressed in LMMs. As expected, defense-related protein like PR5 (*spl1*), PR10 (*blm*), PBZ1 (*spl1*, *cdr2*, *blm*) and Chia2a (*spl5*) were significantly over-accumulated in LMMs. These results suggest that the decreased photosynthesis and energy metabolism were coupled with the increased oxidative stress and defense responses in LMMs, which might be a common feature for PCD or defense response in plant (Jung et al., 2006).

Proteomics-based identification of the factors in rice immunity

Plant makes a transition from growth and reproductive posture to a defensive posture, must undergo a global up- and down-expression of genes in defense response (Bilgin et al., 2010). Most proteins identified in rice by the proteomic approaches in response to different pathogens are those abundance of PR proteins (fungus, Kim et al., 2004; bacterial, Mahmood et al., 2006; virus, Ventelon-Debout et al., 2004) and ROS-related proteins (fungus, Kim et al., 2014; bacterial, Yu et al., 2008; virus, Ventelon-Debout et al., 2004), reflecting the general roles rather than a specific role in response to a specific pathogen. These results just confirmed the previously known conclusions drawn from using traditional approaches and are therefore not particularly useful in identifying new factors or unraveling new pathways in rice immunity. Identification of the low abundance proteins with critical roles in rice immunity is usually hampered by both the presence of these high abundance "noise" proteins and the limited sensitivities of the available proteomic methodologies (Wagner et al., 2002). Despite these limitations, some biologically meaningful information has been extracted from these proteomic studies in rice. Firstly, it has been shown for the first time that 3-β HSD with anti-fungal activity was induced only in resistant rice plants in response to *R. solani*, suggesting a defensive role of this enzyme in rice against *R. solani*. (Lee et al., 2006); secondly, the protein complex formed in response to RYMV infection in rice cells contains no viral proteins but only the host proteins presumably required for the viral life cycle (Brizard et al., 2006); thirdly, the rice RcbL was found to be fragmented upon *Xoo* infections seemingly in a race-independent manner (Mahmood et al., 2006; Yu et al., 2008). This may be possibly explained by the observation that ROS induces fragmentation of the RuBisCO protein *in vitro* (Ishida et al., 1999; Ishida et al., 1997). However, the functional role of RcbL fragmentation in rice immunity remained to be determined; lastly, the photosynthesis and energy metabolism are attenuated in favor of the enhanced oxidative and defense responses in different LMM mutants (Chen et al., 2013; Jung et al., 2006; Kim et al., 2008; Tsunezuka et al., 2005), which might be a common feature for PCD or plant immunity.

At present, besides the aforementioned fragmented information, little systematical new information has been obtained in rice immunity using high through-put proteomic approaches. This largely due to the limited proteomic data generated so far. With the saturation of the rice proteomic data of rice-pathogen interactions (both compatible and incompatible interactions) at different stages of the entire infection processes, new protein factors or enzymes involved in different stages of rice immunity including those in pathogen recognition, signal transduction initiated or exerted through protein-protein interactions/protein modifications/degradations, transcriptome reprogramming and production/biosynthesis of metabolites with anti-pathogen activities or cell wall strengthening will be eventually identified. Proteomic studies in rice, especially in defense response or immunity are still at its rudimentary stage. Adequate data are required to draw a clear picture of rice immunity. However, given the complex nature of the proteomic data, challenges remain even with the saturated dataset available. The ability to extract the biologically meaningful information from the hierarchy massive dataset and the ability to distinguish the primary casual events from the interference of the secondary consequent events are critical to dissect rice immunity pathways or networks. The application of combinatory approaches of functional genomics, proteomics, and metabolomics will certainly be helpful in speeding up this identification process.

Concluding remarks and prospects

High through-put proteomic studies are playing increasingly important roles in deciphering virtually every aspect of cellular functions in plant stress responses, as well as in establishing possible relationships between protein abundance/modification and plant stress tolerance. However, many challenges remain in proteomic studies. One of the constant challenges for proteomics is inadequate protein identification because of the interference of high abundance proteins (Wagner et al., 2002). For example, RuBisCO consists of 30 to 50% of total plant protein from green tissues, and causes less analytical sensitivity and dynamic range in protein identification of plant proteomics (Krishnan and Natarajan, 2009; Rose et al., 2004; Widjaja et al., 2009). This challenge can be resolved with the removal of the abundant proteins from proteomic samples and increasing the sensitivities of proteomic detections for the low abundant proteins. The application of the newly developed advanced methodologies such as gel-free protein separation approaches, multidimensional protein identification technology, quantitative proteomic approaches including isotope-coded affinity tags, targeted mass tags, and isobaric tags for relative and absolute quantitation will certainly help to resolve this problem. Another challenge is how to translate the massive data of proteomics from different researches into knowledge, which can interpret the pathway and/or network mechanism in plants and be readily applied into crop improvement programs. The solution might largely hinge on developing interdisciplinary approaches, creating sufficient genetic resources, and establishing robust bioinformatics tools with novel algorithmic solutions. In that case, with the saturation of proteomic data, the potential biomarkers in the linear pathways or intertwined networks common or unique to each biotic stress will be revealed, and these biomarkers will be useful for generating the next-generation crop plants using molecular breeding and/or coupled with genetic engineering with enhanced crop yield and seed quality.

Acknowledgements
We thank Dr. Jian-Zhong Liu for critical reading of this manuscript. This work was supported by the National Key Programs for Transgenic Crops (2014ZX08009003-001); the National Natural Science Foundation of China (31171519, 31371593); the Natural Science Foundation of Zhejiang Province (LZ14C130001, LY13C130008, Y14C130004); the Open Fund of State Key Laboratory of Rice Biology (120103).

References

Agrawal, G.K., Bourguignon, J., Rolland, N., Ephritikhine, G., Ferro, M., Jaquinod, M., Alexiou, K.G., Chardot, T., Chakraborty, N., Jolivet, P., *et al.* (2011). Plant organelle proteomics: collaborating for optimal cell function. Mass Spectrom Rev *30*, 772-853.

Agrawal, G.K., Rakwal, R., Tamogami, S., Yonekura, M., Kubo, A., and Saji, H. (2002). Chitosan activates defense/stress response(s) in the leaves of *Oryza sativa* seedlings. Plant Physiol Biochem *40*, 1061-1069.

Agrawal, G.K., Sarkar, A., Righetti, P.G., Pedreschi, R., Carpentier, S., Wang, T., Barkla, B.J., Kohli, A., Ndimba, B.K., Bykova, N.V., *et al.* (2013). A decade of plant proteomics and mass spectrometry: translation of technical advancements to food security and safety issues. Mass Spectrom Rev *32*, 335-365.

Basse, C.W., Fath, A., and Boller, T. (1993). High affinity binding of a glycopeptide elicitor to tomato cells and microsomal membranes and displacement by specific glycan suppressors. J Biol Chem *268*, 14724-14731.

Bilgin, D.D., Zavala, J.A., Zhu, J., Clough, S.J., Ort, D.R., and DeLucia, E.H. (2010). Biotic stress globally downregulates photosynthesis genes. Plant Cell Environ *33*, 1597-1613.

Brizard, J.P., Carapito, C., Delalande, F., Van Dorsselaer, A., and Brugidou, C. (2006). Proteome analysis of plant-virus interactome: comprehensive data for virus multiplication inside their hosts. Mol Cell Proteomics *5*, 2279-2297.

Chen, F., Yuan, Y., Li, Q., and He, Z. (2007). Proteomic analysis of rice plasma membrane reveals proteins involved in early defense response to bacterial blight. Proteomics *7*, 1529-1539.

Chen, X., Fu, S., Zhang, P., Gu, Z., Liu, J., Qian, Q., and Ma, B. (2013). Proteomic analysis of a disease-resistance-enhanced lesion mimic mutant *spotted leaf 5* in rice. Rice *6*, 1.

Fang, X., Chen, J., Dai, L., Ma, H., Zhang, H., Yang, J., Wang, F., and Yan, C. (2015). Proteomic dissection of

plant responses to various pathogens. Proteomics *15*, 1525-1543.
Gnanamanickam, S.S., Priyadarisini, V.B., Narayanan, N.N., Vasudevan, P., and Kavitha, S. (1999). An overview of bacterial blight disease of rice and strategies for its management. Curr Sci *77*, 1435-1443.
Hossain, Z., Nouri, M.Z., and Komatsu, S. (2012). Plant cell organelle proteomics in response to abiotic stress. J Proteome Res *11*, 37-48.
Hu, J., Rampitsch, C., and Bykova, N.V. (2015). Advances in plant proteomics toward improvement of crop productivity and stress resistance. Front Plant Sci *6*, 209.
Ishida, H., Makino, A., and Mae, T. (1999). Fragmentation of the large subunit of ribulose-1,5-bisphosphate carboxylase by reactive oxygen species occurs near Gly-329. J Biol Chem *274*, 5222-5226.
Ishida, H., Nishimori, Y., Sugisawa, M., Makino, A., and Mae, T. (1997). The large subunit of ribulose-1,5-bisphosphate carboxylase/oxygenase is fragmented into 37-kDa and 16-kDa polypeptides by active oxygen in the lysates of chloroplasts from primary leaves of wheat. Plant Cell physiol *38*, 471-479.
Jung, Y.H., Rakwal, R., Agrawal, G.K., Shibato, J., Kim, J.A., Lee, M.O., Choi, P.K., Jung, S.H., Kim, S.H., Koh, H.J., *et al.* (2006). Differential expression of defense/stress-related marker proteins in leaves of a unique rice blast lesion mimic mutant (*blm*). J Proteome Res *5*, 2586-2598.
Kang, S.G., Matin, M.N., Bae, H., and Natarajan, S. (2007). Proteome analysis and characterization of phenotypes of lesion mimic mutant *spotted leaf 6* in rice. Proteomics *7*, 2447-2458.
Kim, J.Y., Wu, J., Kwon, S.J., Oh, H., Lee, S.E., Kim, S.G., Wang, Y., Agrawal, G.K., Rakwal, R., Kang, K.Y., *et al.* (2014). Proteomics of rice and *Cochliobolus miyabeanus* fungal interaction: insight into proteins at intracellular and extracellular spaces. Proteomics *14*, 2307-2318.
Kim, S.T., Cho, K.S., Yu, S., Kim, S.G., Hong, J.C., Han, C.D., Bae, D.W., Nam, M.H., and Kang, K.Y. (2003). Proteomic analysis of differentially expressed proteins induced by rice blast fungus and elicitor in suspension-cultured rice cells. Proteomics *3*, 2368-2378.
Kim, S.T., Kang, Y.H., Wang, Y., Wu, J., Park, Z.Y., Rakwal, R., Agrawal, G.K., Lee, S.Y., and Kang, K.Y. (2009). Secretome analysis of differentially induced proteins in rice suspension-cultured cells triggered by rice blast fungus and elicitor. Proteomics *9*, 1302-1313.
Kim, S.T., Kim, S.G., Hwang, D.H., Kang, S.Y., Kim, H.J., Lee, B.H., Lee, J.J., and Kang, K.Y. (2004). Proteomic analysis of pathogen-responsive proteins from rice leaves induced by rice blast fungus, *Magnaporthe grisea*. Proteomics *4*, 3569-3578.
Kim, S.T., Kim, S.G., Kang, Y.H., Wang, Y., Kim, J.Y., Yi, N., Kim, J.K., Rakwal, R., Koh, H.J., and Kang, K.Y. (2008). Proteomics analysis of rice lesion mimic mutant (*spl1*) reveals tightly localized probenazole-induced protein (PBZ1) in cells undergoing programmed cell death. J Proteome Res *7*, 1750-1760.
Komatsu, S., and Tanaka, N. (2005). Rice proteome analysis: a step toward functional analysis of the rice genome. Proteomics *5*, 938-949.
Krishnan, H.B., and Natarajan, S.S. (2009). A rapid method for depletion of Rubisco from soybean (*Glycine max*) leaf for proteomic analysis of lower abundance proteins. Phytochemistry *70*, 1958-1964.
Lee, J., Bricker, T.M., Lefevre, M., Pinson, S.R., and Oard, J.H. (2006). Proteomic and genetic approaches to identifying defence-related proteins in rice challenged with the fungal pathogen *Rhizoctonia solani*. Mol Plant Pathol *7*, 405-416.
Li, Y., Ye, Z., Nie, Y., Zhang, J., Wang, G.L., and Wang, Z. (2015). Comparative phosphoproteome analysis of *Magnaporthe oryzae*-responsive proteins in susceptible and resistant rice cultivars. J Proteomics *115*, 66-80.
Liao, M., Li, Y., and Wang, Z. (2009). Identification of elicitor-responsive proteins in rice leaves by a proteomic approach. Proteomics *9*, 2809-2819.
Lorrain, S., Vailleau, F., Balague, C., and Roby, D. (2003). Lesion mimic mutants: keys for deciphering cell death and defense pathways in plants? Trends Plant Sci *8*, 263-271.
Mahmood, T., Jan, A., Kakishima, M., and Komatsu, S. (2006). Proteomic analysis of bacterial-blight defense-responsive proteins in rice leaf blades. Proteomics *6*, 6053-6065.
Matros, A., Kaspar, S., Witzel, K., and Mock, H.P. (2011). Recent progress in liquid chromatography-based separation and label-free quantitative plant proteomics. Phytochemistry *72*, 963-974.
Mizobuchi, R., Hirabayashi, H., and Kaji, R. (2002). Isolation and characterization of rice lesion-mimic mutants with enhanced resistance to rice blast and bacterial blight. Plant Sci *163*, 345-353.
Morel, J.B., and Dangl, J.L. (1997). The hypersensitive response and the induction of cell death in plants. Cell Death Differ *4*, 671-683.
Pinel, A., N'Guessan, P., Bousalem, M., and Fargette, D. (2000). Molecular variability of geographically distinct isolates of rice yellow mottle virus in Africa. Arch Virol *145*, 1621-1638.
Rakwal, R., and Agrawal, G.K. (2003). Rice proteomics: Current status and future perspectives. Electrophoresis *24*, 3378-3389.
Rose, J.K., Bashir, S., Giovannoni, J.J., Jahn, M.M., and Saravanan, R.S. (2004). Tackling the plant proteome: practical approaches, hurdles and experimental tools. Plant J *39*, 715-733.
Sire, C., and Brugidou, C. (2002). Rice viruses: Current knowledge, agronomic impact and control methods. Virologie *6*, 431-443.
Staskawicz, B.J., Ausubel, F.M., Baker, B.J., Ellis, J.G., and Jones, J.D. (1995). Molecular genetics of plant disease resistance. Science *268*, 661-667.
Takahashi, A., Kawasaki, T., Wong, H.L., Suharsono, U., Hirano, H., and Shimamoto, K. (2003). Hyperphosphorylation of a mitochondrial protein, prohibitin, is induced by calyculin A in a rice lesion-mimic mutant *cdr1*. Plant Physiol *132*, 1861-1869.
Tsunezuka, H., Fujiwara, M., Kawasaki, T., and Shimamoto, K. (2005). Proteome analysis of programmed cell death and defense signaling using the rice lesion mimic mutant *cdr2*. Mol Plant Microbe In *18*, 52-59.

Ventelon-Debout, M., Delalande, F., Brizard, J.P., Diemer, H., Van Dorsselaer, A., and Brugidou, C. (2004). Proteome analysis of cultivar-specific deregulations of *Oryza sativa indica* and *O. sativa japonica* cellular suspensions undergoing rice yellow mottle virus infection. Proteomics *4*, 216-225.

Wagner, K., Miliotis, T., Marko-Varga, G., Bischoff, R., and Unger, K.K. (2002). An automated on-line multidimensional HPLC system for protein and peptide mapping with integrated sample preparation. Anal Chem *74*, 809-820.

Widjaja, I., Naumann, K., Roth, U., Wolf, N., Mackey, D., Dangl, J.L., Scheel, D., and Lee, J. (2009). Combining subproteome enrichment and Rubisco depletion enables identification of low abundance proteins differentially regulated during plant defense. Proteomics *9*, 138-147.

Wu, C., Bordeos, A., Madamba, M.R., Baraoidan, M., Ramos, M., Wang, G.L., Leach, J.E., and Leung, H. (2008). Rice lesion mimic mutants with enhanced resistance to diseases. Mol Genet Genomics *279*, 605-619.

Yu, C.L., Yan, S.P., Wang, C.C., Hu, H.T., Sun, W.N., Yan, C.Q., Chen, J.P., and Yang, L. (2008). Pathogenesis-related proteins in somatic hybrid rice induced by bacterial blight. Phytochemistry *69*, 1989-1996.

Other Publications of Interest

Xu, J.-P. (2014). Next-generation Sequencing: Current Technologies and Applications. Caister Academic Press, U.K. ISBN: 978-1-908230-33-1

He, Z. (2014). Microarrays: Current Technology, Innovations and Applications. Caister Academic Press, U.K. ISBN: 978-1-908230-49-2

Poptsova, M.S. (2014). Genome Analysis: Current Procedures and Applications. Caister Academic Press, U.K. ISBN: 978-1-908230-29-4

Fuentes, M. and LaBaer, J. (2014). Proteomics: Targeted Technology, Innovations and Applications. Caister Academic Press, U.K. ISBN: 978-1-908230-46-1

Murillo, J., Vinatzer, B.A., Jackson, R.W. and Arnold, D.L. (2015). Bacteria-Plant Interactions: Advanced Research and Future Trends. Caister Academic Press, U.K. ISBN: 978-1-908230-58-4

Caranta, C., Aranda, M.A., Tepfer, M. and Lopez-Moya, J.J. (2011). Recent Advances in Plant Virology. Caister Academic Press, U.K. ISBN: 978-1-904455-75-2

Nannipieri, P., Pietramellara, G. and Renella, G. (2014). Omics in Soil Science. Caister Academic Press, U.K. ISBN: 978-1-908230-32-4

Herold, K.E. and Rasooly, A. (2009). Lab-on-a-Chip Technology: Fabrication and Microfluidics. Caister Academic Press, U.K. ISBN: 978-1-904455-46-2

Metabolomics of Disease Resistance in Crops

Vicent Arbona and Aurelio Gómez-Cadenas*

Ecofisiologia i Biotecnologia, Dept. Ciències Agràries i del Medi Natural, Universitat Jaume I, Castelló de la Plana, Spain
*Corresponding author. Email: aurelio.gomez@uji.es
Tel: +34 964 729402

Abstract
Plants are continuously exposed to the attack of invasive microorganisms, such as fungi or bacteria, and also viruses. To fight these attackers, plants develop different metabolic and genetic responses whose final outcome is the production of either toxic compounds that kill the pathogen or deter its growth, and/or semiotic molecules that alert other individuals from the same plant species. These molecules are derived from the secondary metabolism and their production is induced upon detection of a pathogen-associated molecular pattern (PAMP). These PAMPs are different molecules that are perceived by the host cell triggering defense responses. PAMP-elicited compounds are highly diverse and specific of every plant species and can be divided into preformed metabolites or phytoanticipins that are converted into toxic molecules upon pathogen perception, and toxic metabolites or phytoalexins that are produced only upon pathogen attack. Moreover, plant volatile emissions are also modified in response to pathogen attack to alert neighboring individuals or to make plants less attractive to pathogen vector arthropods. Plant metabolite profiling techniques have allowed the identification of novel antimicrobial molecules that are induced upon elicitation. However, more studies are required to assess the specific function of metabolites or metabolite blends on plant-microbe interactions.

Introduction
Opposite to animals, vascular plants do not possess an immune system which involves the presence of cells specialized in detection, prosecution and elimination of invasive microorganisms. However, plants are still able to respond to the attack of microorganisms and predatory arthropods relying on different mechanisms, first an ***innate immunity*** comprising a sensitive system to detect Pathogen-Associated Molecular Patterns (PAMPs) based on membrane-bound specific pattern recognition receptors (PRR) (PAMP-Triggered Immunity or PTI), and a second layer named Effector-Triggered Immunity or ETI, involving intracellular proteins that respond to pathogen effectors or virulence factors (Schwessinger and Zipfel, 2008).

PAMP-Triggered Immunity (PTI)
PAMPs are characteristic molecular structures shared by large groups of pathogens that act as plant defense elicitors. Among PAMPs perceived by plants, fungal structures such as cell wall β-glucan; conserved epitopes from cell wall transglutaminases and elicitins (secreted lipotransfer proteins); or bacteria-derived molecules such as lipopolysaccharides, flagellin, bacterial cold-shock proteins, and EF-Tu are found (Zipfel et al., 2006). Perception of PAMPs is not necessarily a specific event as some of them can elicit defense responses in a wide range of plant species (e.g. flagellin or the isolated epitope flg22). Nevertheless, most specific PRRs responsible of this pattern recognition remain elusive and only a few have been so far characterized. Most plant PRRs are membrane-bound receptor-like kinases comprising a ligand-binding domain on the cell surface. These molecules can also contain a cytoplasmic kinase domain (Receptor-Like Kinases or RLKs) or not (Receptor-Like Protein types or RLPs) that initiates PRR-triggered immunity upon PAMP perception (Macho and Zipfel, 2014). For a proper immune response induction, these receptors dimerize upon perception of the PAMP. To this respect, several models showing homo-, hetero-dimerization and heteromulti-merization exist. Chitin perception in Arabidopsis is mediated by LysM-RLK CERK1/RLK1/LYK1. This CERK1 receptor contains three extracellular domains that bind long fungal chitin oligomers (comprising seven to eight N-acetyl-glucosamine residues) leading to its homodimerization and formation of an active receptor complex (shorter chitin oligomers do not induce receptor homodimerization and do not trigger immune response). Upon homodimerization, CERK1 cytoplasmic kinase domains are brought together enabling intermolecular transphosphorylation, following a similar mechanism to that of animal tyrosine kinases (Macho and Zipfel, 2014). Interestingly, chitin perception in rice is mediated by a mechanism known as hetero-multimerization in which a chitin binding protein CEBiP (a predicted GPI-anchored protein) with no kinase cytoplasmic domain (as an RLP) is involved. Therefore, transduction signal after chitin perception requires the cooperation with additional proteins. CEBiP homodimerizes in the presence of active chitin and also forms a hetero-oligomeric complex with OsCERK1 (an ortholog of Arabidopsis CERK1 but its only extracellular domain does not bind chitin) (Shimizu et al., 2010). Flagellin detection and signal transduction carried out by PRRs is well studied in the model plant *Arabidopsis thaliana.* In this plant species, LRR-RLK-FLS2 is the PRR responsible for binding flagellin from flagellated phytopathogenic procariotes (such as *Pseudomonas syringae* or *Xanthomonas spp.*) or the 22-aminoacid epitope flg22 (Dardick et al., 2012). This is a Leucine Repeat-Rich RLK

that forms heterodimers with LRR RLK-BAK1/SERK3 regulatory complex upon epitope binding. Although, dimerization of FLS2 and BAK1 is not required for a proper flg22 binding both molecules act as co-receptors for epitope binding and signal activation. A similar activation mechanism has been described for the brassinosteroid receptor BRI1 (another LRR-RLK) that binds to co-receptors BAK1 or SERK1. This interaction between LRR-RLKs (and LRR-RLP) and BAK1 or related SERK proteins seems to be a common trend as supported by increasing evidence (Macho and Zipfel, 2014). As for rice CERK1, some mechanistic differences are also found in this case as well: OsSERK2, an *Oryza sativa* ortholog of BAK1 binds constitutively to the LRR-RLK XA21 conferring resistance to *Xanthomonas oryzae* pv. *oryzae*, but no ligand has been confirmed yet therefore it cannot be precluded how PAMP interaction would modify OsSERK2-XA21 interaction (Bahar et al., 2014; Sun et al., 2013). A more complex interaction has been observed in tomato, where the BAK1 ortholog interacts with an LRR-RLP named Eix1 and negatively regulates signaling mediated by Eix2 (a related LRR-RLP). Upon ligand perception, receptor complexes interact with different cytoplasmic partners to trigger intracellular signaling leading to a rapid ROS burst and expression of genes involved in defense (Bar et al., 2010). One of these partners is the kinase BIK1 that has several protein targets in Arabidopsis such as AtRBOHD NADPH oxidase involved in apoplastic ROS production in response to PAMP perception. Another protein target in rice that responds to chitin-triggered signaling through OsCERK1 is the Rac GDP/GTP exchange factor (OsRacGEF1) that activates OsRac1 leading to resistance to fungal pathogens (Akamatsu et al., 2013). Moreover, the activation of defense responses on PAMP perception needs to be properly shut down after the pathogenic threat is over. To this respect, as signaling activation is driven by phosphorylation negative regulation modulated by protein phosphatases is a plausible explanation. Hence, several protein phosphatases 2C have been shown to interact with members of different PRR complexes: KAPP interacts with FLS2 (flg22-triggered responses) as well as with other plant RLKs (Gomez-Gomez et al., 2001), in rice XB15 phosphatase and XB24 ATPase associate with XA21 thus regulating the phosphorylation status of PRR (Park et al., 2008). Additionally, PAMP-bound activated RLKs are also degraded to enable the restoration of free receptors at the plasma membrane. Degradation of receptors is carried out through polyubiquitination catalyzed by the E3-ubiquitin ligases PUB12 and PUB13 that constitutively interact with BAK1 and are recruited into the FLS2 receptor complex after perception of flg22 (Lu et al., 2011).

Effector-Triggered Immunity (ETI)

However, these defense responses can be inhibited by virulence factors that successful pathogens have evolved. In turn, adapted plants have also developed resistance proteins that detect these virulence factors and inhibit their action on cell targets. This response is mediated by resistance proteins (R proteins, most of them encoding highly polymorphic NB-LRR proteins) that recognize these virulence factors leading to the induction of a hypersensitive response (HR) including programmed death of infected cells, inhibiting pathogen spread. This is considered an enhanced PTI response that leads to disease resistance. It is worth noting that this ETI branch is effective on obligate biotrophs (microorganisms that can only feed on living cells) or hemibiotrophs but not on pathogens that kill host cells during colonization or necrotrophs (Jones and Dangl, 2006). It is hypothesized that these NB-LRR proteins act as guards of intracellular targets of pathogen effectors (this recognition can be carried out directly or indirectly by detecting products of the action on host target molecules). As mentioned above, this layer of defense leads to HR and induction of programmed cell death contributing to arrest pathogen expansion. This programmed cell death has been associated to the intracellular production of ROS by mitochondria and chloroplasts and the activation of NB-LRR and metacaspases (Coll et al., 2011).

Metabolite profiling techniques in plant-pathogen interactions

In recent years, phytopathologists have gained access to modern metabolite profiling platforms that allow monitoring hundreds of metabolites in a single analysis. Moreover, these techniques are exhaustive enough to allow researchers to identify novel metabolites with potential antimicrobial activity (Allwood et al., 2008). Hence, comparison of resistant versus susceptible varieties, genotypes or mutants is a fundamental tool to decipher resistance mechanisms and identify defense compounds. In this context, instead of monitoring a few (known toxic compounds) the array of low molecular weight metabolites (metabolome) are analyzed and quantitated constituting the whole metabolic phenotype of a plant. To this respect, it is then possible to identify pre- and post-invasion resistant and susceptible metabolic phenotypes and infer relationships among different pathways.

Most non-volatile secondary metabolites are semipolar chemical compounds that can be easily separated on a C18 column (or specifically modified versions of this packing) using reversed phase liquid chromatography coupled to a high resolution mass spectrometer (usually a hybrid quadrupole/time-of-flight or QTOF-MS) through an atmospheric pressure ionization source (either electrospray ESI or chemical ionization, APCI). This instrumental coupling is known as LC/MS. High resolution QTOF-MS instruments provide accurate mass measurements (less than 0.02 arbitrary mass units of difference) and real isotopic pattern (related to the particular elemental composition of each compound). In addition, these instruments also allow fragmentation of selected compounds, thanks to the presence of a quadrupole segment and a collision cell before the flight tube, to gain structural information. This configuration is usually sufficient to analyze different types of secondary metabolites: flavonoids and triterpenoids (Arbona et al., 2015; Böttcher et al., 2008), phenolics (Böttcher et al., 2009a), glucosinolates and related compounds (Böttcher et al., 2009b; Zandalinas et al., 2012), lipids (Zoeller et al., 2012), etc... Due to the high specificity of secondary metabolites and the little information available on them (mass spectra, fragmentation pattern, elution time, etc...), together with the poor cross-platform exchangeability of LC/MS data, make very often necessary their *de novo* identification using in-house built databases or matching the annotation (MS spectra and retention time) with pure standards. For these reasons, ensuring high quality spectral information is of key importance to facilitate mass spectra curation and interpretation (Figure 1).

Complementarily, volatiles and polar compounds are analyzed by means of gas chromatography coupled to mass spectrometry (GC/MS). Polar compounds such as carbohydrates, free aminoacids and carboxylic acids of the Krebs cycle can be analyzed by GC/MS as methoxime/trimethylsilyl derivatives (Roessner et al., 2001) whereas volatiles do not need any further processing (Beck et al., 2014). Identification of polar metabolites and volatiles in this case is carried out in a targeted fashion using publicly-available databases such as the Gölm Metabolite Database (http://gmd.mpimp-golm.mpg.de/), NIST or the Fiehn metabolite database (http://fiehnlab.ucdavis.edu/db). This is possible for several reasons: 1) In GC analyses, retention index markers, such as fatty acid methyl esters or alkanes, are added to standardize retention times of all eluted compounds to retention indices; this makes GC a more robust chromatographic technique than LC, where retention index markers are not employed; 2) the ionization in GC/MS is achieved through electron impact (EI), essentially an electron beam collides with each eluted metabolite and generates a population of ions. Retention indices (RI) as well as fragment ions are then used to match metabolite identities in metabolite databases.

Analysis of metabolomics data (both LC/MS and GC/MS) requires the aid of bioinformatic tools that allow extraction of chromatographic and spectral data in a manageable file format, such as MS Excel spreadsheets. This step can be achieved using different software tools developed so far: xcms (Smith et al., 2006), metAlign (Lommen, 2009), MzMine (Katajamaa et al., 2006; Pluskal et al., 2010) or, more recently, GridMass (Treviño et al., 2015), etc...perform peak picking and alignment of mass chromatographic peaks (Arbona et al., 2009). There are also tools for integrated GC/MS analysis that allow performing peak picking and alignment, RI calculation and matching of MS spectra in databases: TargetSearch (Cuadros-Inostroza et al., 2009), TagFinder (Luedemann et al., 2008) or FiehnLib (Strehmel et al., 2013). In addition, it is possible to annotate unknown derivatized metabolites by

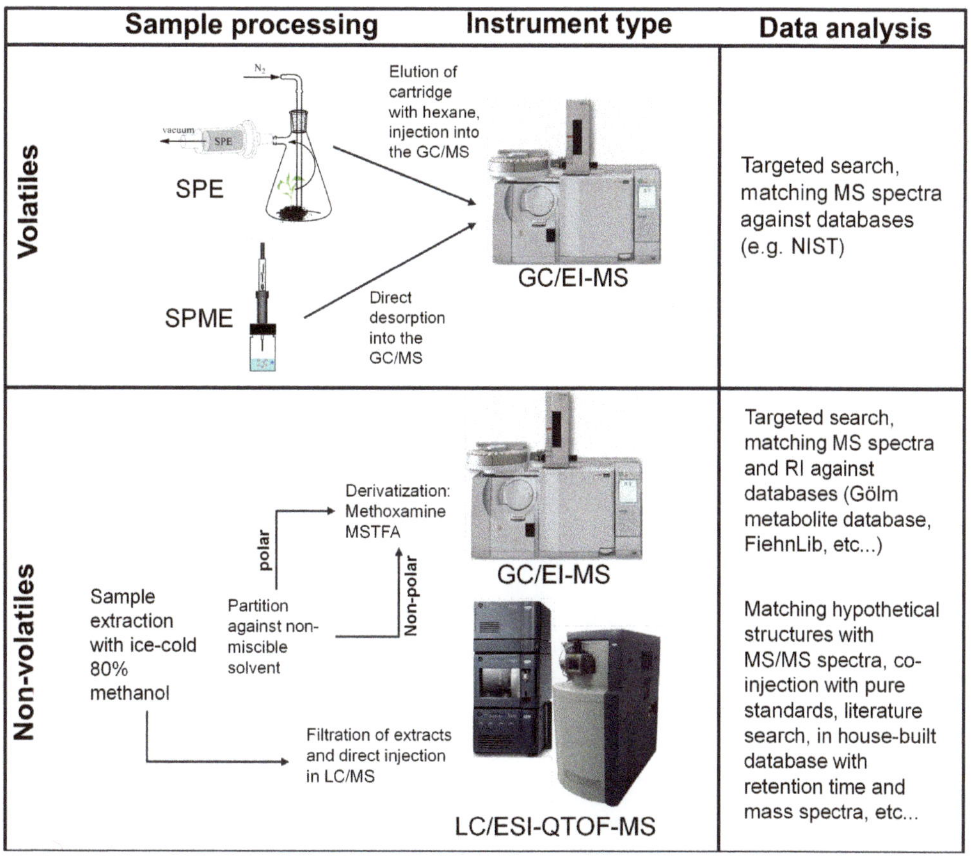

Figure 1. Different mass spectrometry-based metabolite profiling strategies according to sample type and metabolite class. Abbreviations: GC/EI-MS, gas chromatography/electron impact-mass spectrometry; LC/ESI-QTOF-MS, liquid chromatography/electrospray-quadrupole time-of-flight-mass spectrometry; MSTFA, methyl silyl trifluoroacetamide; SPE, solid phase extraction; SPME, solid phase microextraction.

using GC/APCI(+)-QTOF-MS. The APCI ion source generates predominantly protonated ions and very few fragments or none. Annotation of unknowns requires the evaluation of adduct ion formation and calculation of elemental composition. In addition, an estimation of the number of methoxime and trimethylsilyl units per derivatized molecule is needed followed by a deep examination of the corresponding collision-induced dissociation mass spectrum (Strehmel et al., 2013).

Analysis of volatiles does not require previous derivatization and due to the nature of the matrix detection limits can be very low. The main problem is that these compounds need to be collected *in vivo* from plant samples because they are released to the atmosphere as they are synthesized. There are two main collection strategies: 1) sampling the headspace of a hermetically closed vial or flask in which the plant material is contained and 2) force an inert gas in this hermetically closed flask to carry the metabolites that are collected in a special adsorbent material (a solid phase extraction technique, SPE, applied to gas matrices). The first option implies that the surrounding headspace gas of the vial is not renewed. Then, the equilibrium between the plant and the atmosphere is quickly reached and the migration of volatiles from plant tissues to the headspace gas is therefore reduced. In addition, although the matrix (air) is quite inert to MS detectors, emission of volatiles is often very slow and, therefore, long incubation periods are required (Figure 1). The second option is more advantageous in terms of extraction efficiency and the possibility of including a pre-concentration step. Following this strategy, gas surrounding the plant is continuously forced through a SPE cartridge and emitted volatiles retained. Cartridges are subsequently eluted with *n*-hexane and injected in the GC/MS. An improved version is the SPME, the M standing for micro, in which cartridge elution volumes are significantly reduced (Beck et al., 2014; Tikunov et al., 2005).

Metabolic responses to pathogen infection

Plants produce a vast array of low molecular weight compounds (an estimated of 100,000 different metabolites in the plant kingdom). These molecules, collectively known as secondary metabolites, are not essential for cell survival but do have a specific role in the adaptation to a changing environment. Most of them can be primarily grouped in four branches: isoprenoids, phenylpropanoids, alkaloids and fatty acid/polyketides (Dixon, 2001). Importantly, most secondary metabolites are specific of a plant family/genus or even species, making their functional characterization in plant defense by conventional molecular and genetic techniques a daunting task. Although most antimicrobial metabolites synthesized by plants have a broad spectrum activity, their relative activity relies on the presence of detoxification enzymes in a pathogen strain. The synthesis and accumulation of these compounds is induced after perception of pathogen attack (perception of PAMPs or pathogen effector molecules). Defense compounds can be divided into ***phytoalexins***, that are produced upon pathogen perception, and ***phytoanticipins*** that are synthesized in an inactive form and pathogen recognition triggers its biochemical modification into a toxic derivative (Piasecka et al., 2015). It is important to recall that plants challenged with pathogenic fungal or bacterial strains modify their metabolome in the attacked cells (local response) and this response can be extended to the rest of the plant (systemic response). However, not all secondary molecules (synthesized *de novo* or modified during pathogen attack) have a toxic role, and not only one molecule is responsible for the toxic effect. Therefore, to ascertain the role of each elicited metabolite, it should be isolated and tested for toxic activity under control conditions, which is not always possible.

In this review, we described some compounds whose *in vivo* function in plant immunity against bacterial and fungal pathogens has been studied in detail (see Figure 2 for the most important groups of plant metabolites involved in defense).

Phytoanticipins

Saponins are glycosides from lipophilic polycyclic structures derived from the isoprenoid pathway, occurring as triterpenoids or steroids (sapogenin). These molecules are widely distributed in the plant kingdom (mainly in dicots but also in some monocots) (Huhman et al., 2005; Piasecka et al., 2015). This group includes triterpenoids from *Avena spp.* named avenacins (A1, A2, B1 and B2) only accumulated in roots whereas avenacosides, with steroidal sapogenins, are accumulated in leaves (Piasecka et al., 2015). In addition, steroidal glycoalkaloids found in solanaceous plant species constitute important representatives of this group (e.g. α-tomatine of *Solanum lycopersicum* L. Mill). This compound is highly abundant in green tomato fruits showing values of 500 mg kg^{-1} fresh fruit weight and it is degraded as the fruit ripens (Friedman, 2002). This compound has been shown to have an important activity against fungal pathogens. Hence, expression of β_2-tomatinase gene from *Septoria lycopersici* in *Nectria haematococca*, a fungus that can colonize red tomato fruits but not green ripe ones, enabled it to detoxify α-tomatine and its ability to parasitize green tomato fruits (Sandrock and VanEtten, 2001). More recently, it has been shown that mutant *Cladosporium fulvum* lines with suppressed CfTom1 glycosyl hydrolase activity (GH10 tomatinase) exhibited significantly lower virulence in tomato (Ӧkmen et al., 2013). This is due to the accumulation of tomatidine upon cleavage of α-tomatine by CfTom1 hydrolase that suppresses induced defense responses in tomato (Ito et al., 2004). Moreover, the role of avenacins in immunity in oat has been studied by means of a genetic screening of a mutant population of *Avena strigosa* leading to the identification of *saponin deficient* (*sad*) mutants. These mutants are deficient in enzymes that are specifically involved in avenacin biosynthesis and not avenacosides. Therefore, roots of homozygous sad lines exhibited hyper-susceptibility to adapted and non-adapted fungal strains. (Mugford et al., 2013).

Glucosinolates are the most studied secondary metabolites. Globally, these molecules are β-D-thioglucosides-*N*-hydroxysulfates with different aliphatic, aromatic or indolic substitutions and are primarily present

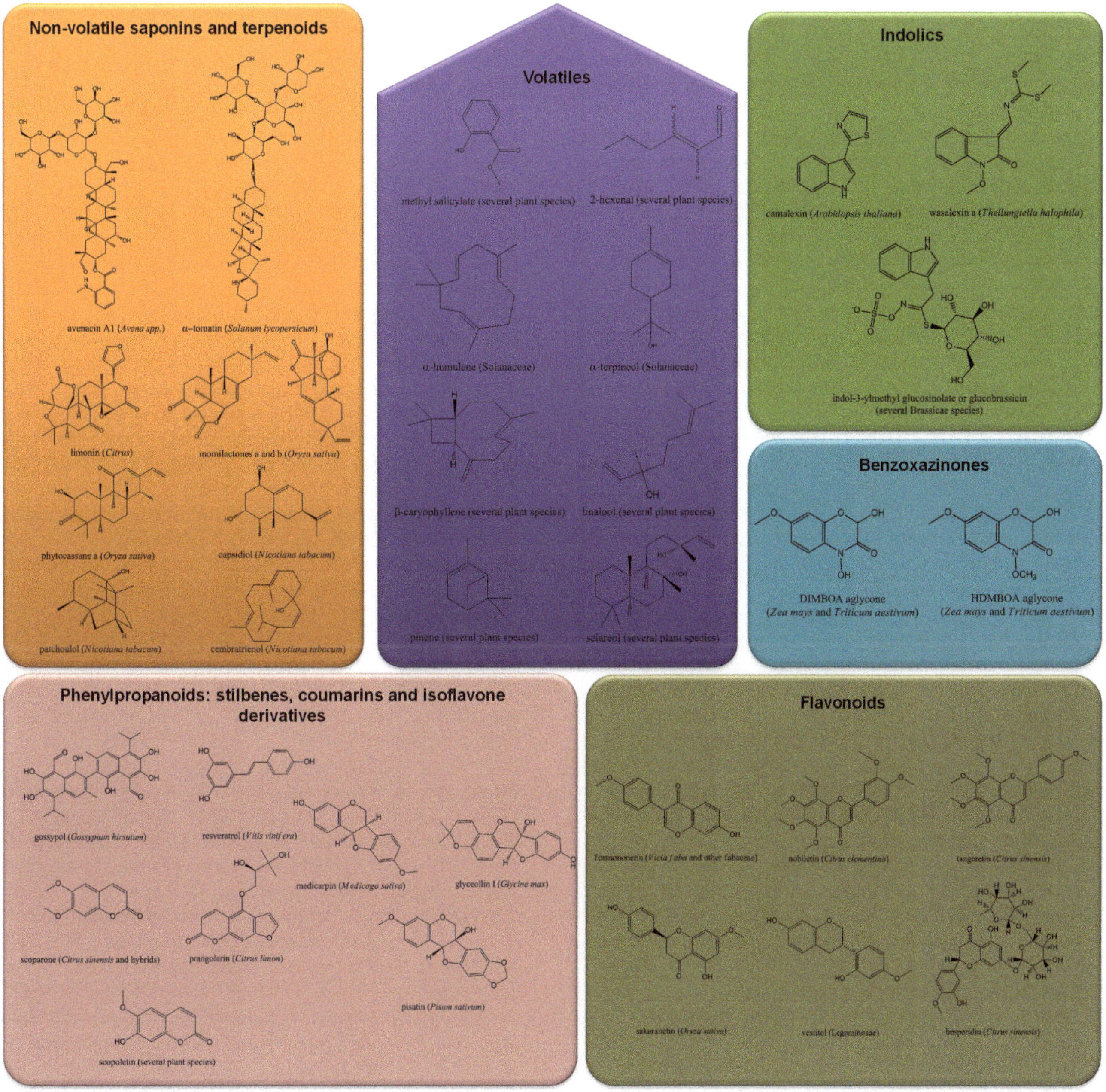

Figure 2. Compound classes and structures of metabolites involved in plant defense.

in Brassicales and some species of the Euphorbiaceae family. Glucosinolates are derived from different aminoacids such as alanine, valine/leucine, isoleucine, methionine, phenylalanine/tyrosine and tryptophan. Nevertheless, not all possible structures are present in all brassicaceae species but a particular subset. For instance, in the model species *Arabidopsis thaliana*, aliphatic glucosinolates derived from methionine and indolic glucosinolates derived from tryptophan are abundant (Piasecka et al., 2015); in leaves of cultivated rapeseed, levels of some aliphatic glucosinolates such as progoitrin, gluconapin and glucobrassicanapin and indolics such as neoglucobrassicin are predominant whereas in kale, cabbage (Velasco et al., 2011) and wild radish, indolic glucosinolates were less abundant compared to aliphatics (Malik et al., 2010). There is a wealth of information on the interaction of glucosinolates with the environment. In their glycosylated form, glucosinolates are quite stable and biologically inactive but the activation of β-thioglucoside hydrolases or myrosinases upon pathogen attack initiates glucosinolate hydrolysis and release of toxic breakdown products (Grubb and Abel, 2006). Breakdown products are unstable and decompose in different types of molecules including isothiocyanates (ITCs), nitriles and epithioalkanes that have the toxic effect on microbes. Particularly, ITCs derived from aliphatic glucosinolates have a deterrent activity against predating arthropods such as spider mites (Piasecka et al., 2015; Zhurov et al., 2014); those derived from indolic glucosinolates have antifeedant effects on aphids (Kim et al., 2008). In *Arabidopsis thaliana*, tryptophan-derived metabolites conferred non-host resistance to *Plectosphaerella cucumerina*. For instance, *cyp79b2/cyp79b3* double mutant, that cannot convert tryptophan to the indol-3-acetaldoxime the first committed step in indol glucosinolate biosynthesis, exhibited an increased *P. cucumerina* entry rate compared to wt or phytoalexin deficient 3 (*pad3*) mutants (Sanchez-Vallet et al., 2010). In addition, challenging Arabidopsis or other brassicaceae with pathogens induces expression of PEN2 that encodes a lysosome-associated atypical myrosinase that metabolizes indolic glucosinolates generating particular breakdown products such as raphanusamic acid and indol-3-amine and also induces the expression of CYP81F2, a P450 cytochrome monooxygenase that catalyzes the hydroxylation of indol-3-ylmethyl glucosinolate into hydroxyindol-3-ylmethyl glucosinolate. Both loss-of-function *pen2* and *cyp81f2* Arabidopsis mutants exhibit similar penetration phenotypes in non-adapted fungal pathogens such as *Blumeria graminis* f. sp. *hordei* (Bednarek et al., 2009).

Cyanogenic glucosides are hypothesized to be the precursors of glucosinolates and are widespread in the plant kingdom including ferns and gimnosperms. Chemically, these molecules are β-D-glucosides of α-hydroxynitriles derived from tyrosine, phenylalanine, valine, leucine and isoleucine. As glucosinolates, these metabolites are quite stable and biologically inactive as synthesized, but after hydrolysis the α-hydroxynitriles released decompose spontaneously to generate hydrogen cyanide. These metabolites are known to have an influence on plant-arthropod interaction but their role in defense against microbial pathogens is not yet clear. To this respect, highly cyanogenic *Phaseolus lunatus* accessions showed an elevated susceptibility to the hemibiotroph *Colletotrichum gloeosporiodes* compared to genotypes with a lower cyanogenic potential. These accessions also showed a significantly lower polyphenol oxidase activity; indeed, exogenous HCN application also depressed the activity of this enzyme. Likewise, highly cyanogenic varieties of *Hevea brasiliensis* (rubber tree) exhibited low scopoletin (a coumarin phytoalexin) accumulation and, hence, a strong susceptibility to the blight fungus *Microcyclus ulei* (Lieberei et al., 1989). This generalized depression of plant immunity upon hydrolysis of cyanogenic glucosides has been associated to the toxicity of HCN gas on neighboring plant cells that inhibits enzyme activity. On the contrary, in barley, cyanogenesis has proved to be beneficial against pathogenic microbes. However, accumulation of leucine-derived epìhetero-dendrin does not render cyanogenic barley plants since no specific β-glucosidase is expressed. Transient expression of the sorghum cyanogenic β-glucosidase *dhurrinase2* in barley reduced notably its colonization by *Blumeria graminis* f. sp. *hordei*, indicating that cyanogenesis is effective against this biotrophic fungus (Lipka et al., 2005; Nielsen et al., 2006).

Benzoxazinone glucosides are another group of phytoanticipins found primarily in crops of the Poaceae family including corn (*Zea mays*), wheat (*Triticum spp.*), rye (*Secale cereale*) and some barley wild relatives. In corn and wheat, the main accumulating benzoxazinone glucoside is DIMBOA (2,4-dihydroxy-7-methoxy-1,4-benzoxazin-3-one) glucoside. Their role in plant immunity has been assigned based on a positive correlation between their accumulation in tissues and pathogen resistance. As in cyanogenic glucosides or glucosinolates, their function is attributable to aglycones whereas glycosylated forms are stable and harmless. As aglycones, these molecules have shown a remarkable antifungal activity in vitro against a wide range of plant pathogens. In vivo, reverse genetics experiments also support the role of these metabolites in plant immunity. That is the case of the *bx1* corn mutant, unable to synthesize DIMBOA glucoside, which showed an extreme susceptibility to *Septosphaeria turcica*, responsible for corn leaf blight. Moreover, challenging corn plants with different pathogens not only induces accumulation and hydrolysis of DIMBOA glucosides but also its metabolism to other forms, such as the monohydroxylated and dimethoxylated HDMBOA, that can also be synthesized *de novo*, as demonstrated by its accumulation in *bx1* plants (Ahmad et al., 2011; Huffaker et al., 2011).

Phytoalexins

As mentioned above, phytoalexins are antimicrobial compounds that are toxic *per se* and produced upon pathogen attack.

Camalexin is the typical phytoalexin from *Arabidopsis thaliana* which is synthesized from tryptophan after its metabolism to indol-3-yl acetaldoxime catalyzed by CYP79B2 and CYP79B3, being the only step shared with indole glucosinolate biosynthesis (Böttcher et al., 2009b; Glawischnig, 2007). Afterwards, indol-3-yl acetaldoxime is converted to indol-3-yl acetonitrile (IAN) by CYP71A13, constituting another limiting step. This metabolite is then conjugated to cysteine and then converted spontaneously or in a reaction catalyzed by CYP71B15 to dihydrocama-lexic acid, the direct precursor of camalexin and substrate of PAD3. Camalexin accumulates upon elicitation with pathogens (Mert-Türk et al., 2003; Schlaeppi et al., 2010) but also after subjecting Arabidopsis to Cd^{2+} or Ag^{+} stress (Mert-Türk et al., 2003). Although not all cruciferous plants accumulate camalexin on pathogen attack, most phytoalexins of this species are derived from tryptophan and, therefore, harbor an indole moiety. To this respect, cruciferous phytoalexins can be divided into six groups depending on the initial core structure: group I phytoalexins comprise indole-substituted compounds with thiocarbamate or dithiocarbamate and dithioimidate moieties, group II are based on cyclobrassinin sulfoxide, core structure of group III phytoalexins is indol-3-ylacetonitrile whereas group IV contain indol-3-ylacetaldehyde; phytoalexins from group V are known as 'spiro' and contain a methyl thiodi-hydrothiazole group attached to the indole moiety. Finally, group VI phytoalexins, in which camalexin is included, is constituted of relatively stable compounds with different structures and MS/MS fragmentation patterns from those in the rest of groups (Pedras et al., 2006). As an example, rapeseed seedlings infected with *Plasmodio-phora brassicae* produced at least six different indole phytoalexins: spirobrassinin (V), brassilexin (VI), brassicanate (IV), rutalexin (II), cyclobrassinin (II) and 4-methoxybrassinin (I) (Pedras et al., 2008). The model plant *Thellungiella halophila* accumulated wasalexins (I) and methoxybrassenin (has a dimethyl acetylcarbo-nimido-dithioate substituent on the methoxyindole ring and not a methyl ethylcarbamodithioate as methoxybrassinin) upon elicitation with $CuCl_2$ whereas camalexin could not be detected (Pedras and Adio, 2008).

In other plant species, compounds derived from distinct pathways act as phytoalexins and are produced upon pathogen elicitation: phenols, acetophenones, biphenyls and dibenzofurans, stilbenes, coumarins, flavonoids, terpenes, etc. (Gottstein and Gross, 1992). In *Vitis vinifera*, the stilbenoid phytoalexin resveratrol can be present alone or forming different oligomers. Moreover, this metabolite has gained scientific relevance due to its attributed health properties (Vogt, 2010). Resveratrol synthesis is catalyzed by stilbene synthase (STS) from coenzyme A esters of cinnamic acid derivatives (particularly, resveratrol derives from *p*-coumaroyl-CoA derivatives). This STS enzyme is encoded by a multigene family in grapevine and is phylogenetically related to chalcone synthase that competes with STS by the phenolic acid-CoA derivatives. The expression of this gene is induced by biotic or abiotic elicitors, increasing phytoalexin levels (Jeandet, 2015; Jeandet et al., 2002). Mulberry (*Morus alba*) also produce stilbene phytoalexins (oxyresveratrol) upon elicitation with fungal pathogens such as *Fusarium solani* f. sp. *mori.* (Gottstein and Gross, 1992).

Another group of phytoalexins widely distributed in angiosperms are coumarins and the most important representative is scopoletin. This phytoalexin was first identified in rubber tree after infection with pathogenic fungi (Gottstein and Gross, 1992). Its synthesis arises from feruloyl-CoA in a single hydroxylation step catalyzed by a dioxygenase, subsequent isomerization, rotation and lactonization. Moreover, scopoletin can be further modified by UDP-glucosyl transferases rendering scopolin (Vogt, 2010). In *Citrus*, irradiation and infection with pathogens also induces accumulation of several coumarin-type phytoalexins. Scoparone was isolated from the bark tissue and fruit rind of sweet orange after infection with *Phytophthora citrophthora* (Afek and Sztejnberg, 1988). Its accumulation was associated to resistance to *P. citrophothora* and antifungal treatments such as fosetyl-Al also increased scoparone levels in *Citrus* tissues (Gottstein and Gross, 1992). Nevertheless, some coumarins might have an opposite effect on pathogen growth. Inoculation of wounded *C. limon* fruits with *Penicillium digitatum* rendered a 2% of fruits exhibiting green mold symptoms. However, when epicarp oil from lemons was isolated and applied to washed fruit wounds, green mold development increased up to 92%. Fractionation and spectrometric assays allowed identification of the coumarin prangolarin that could act as a facilitator of green mold penetration and invasion (Arimoto et al., 1995).

Flavonoids also constitute a widely distributed group of plant secondary metabolites. They are highly structurally diverse and are involved in many essential processes including defense against pathogens and protection against stress-induced damage. To this respect, there are 'preformed' flavonoids, synthesized during the normal development of a plant, and could be involved in pathogen defense, but other are synthesized on pathogen attack (Treutter, 2006). This group of compounds comprises anthocyanins (active against bacterial blight agent *Xanthomonas spp.* in cotton leaves), flavones, flavonols (that occur in the lesion margins after infection in *Eucalyptus globulus*), flavanones, and dihydroflavonols (aglycones and glycosylated forms), chalcones, dihydrochalcones and aurones (Williams and Grayer, 2004). Flavonoids, as stilbenes and coumarins, are synthesized from *p*-coumaroyl-CoA. This compound is metabolized by chalcone synthase (CHS) into chalcone, the first precursor molecule in flavonoid biosynthesis (Vogt, 2010). The involvement of flavonoids in plant defense has been extensively studied: isoflavonoids glyceollins in soybean (Ebel et al., 1976), pisatin in pea (Perrin and Bottomley, 1961), medicarpin in alfalfa (He, 2000), etc. They are effective not only against fungal pathogens but also against bacteria (Piasecka et al., 2015). In grass species, several examples of flavonois after pathogen inoculation are found: flavanone sakuranetin in rice (Hasegawa et al., 2014) or 3-deoxyanthocyanidins in sorghum (e.g. apigenindin and luteolinidin) (Snyder and Nicholson, 1990). Indeed, impairment in the synthesis of any of these metabolites was associated to enhanced susceptibility (Piasecka et al., 2015). In legumes, the most important phytoalexins are of the isoflavonoid type including medicarpin, glyceollin II, kievitone and vestitol (Harborne, 1999). In mulberry, a group of flavanoids named kuwanons including approximately 12 members were identified as antifungal metabolites. Moreover, infection of mulberry with *Fusarium solani* or *Stigmina mori* allowed the identification of morusin, a diphenylpropane derivative whose accumulation inhibits fungal and bacterial growth (Gottstein and Gross, 1992). Infection of sweet orange citrus fruits with *P. citrophthora* induced the accumulation polymethoxylated flavones (nobiletin, sinensetin and tangeretin) and the hydrolysis of flavanone glycosides (hesperidin and isonaringin) to increase aglycone levels. These compounds exhibited a significant antimicrobial activity *in vitro* and its induction appeared as a plausible defense mechanism in citrus (del Río et al., 2004). In addition, modulation of polymethoxylated flavone levels in citrus fruits by hormonal treatments increased resistance to *P. citrophthora* (Ortuño et al., 2002). More recently, in harvested citrus fruits, it has been shown the induction of several flavonoid glycosides and aglycones after a heat treatment, resulting in reduced germination of *Penicillium italicum* in the fruit rind tissue. Among the induced flavonoids, quercetin glycosides, hesperetin, naringenin, diosmin and rutin were found (Yun et al., 2013) which further supports their role in disease resistance. As mentioned above, flavonoids are broad spectrum antimicrobial metabolites being active not only against fungal pathogens but also bacteria. The disease known as Huanglongbing (HLB) (or citrus greening) is caused by different strains of the *Candidatus liberibacter* (Hijaz et al., 2013). A preliminary study carried out on symptomatic and control leaves of sweet orange plants indicated important changes in secondary metabolites (Manthey, 2008). Subsequent metabolomics studies revealed a primary accumulation of several hydroxycinnamates and certain selected flavonoids such as hesperidin, naringenin and quercetin (Cevallos-Cevallos et al., 2009; Hijaz et al., 2013). Although the specific role of these metabolites on bacterial growth and expansion has not been assessed, hesperidin accumulation in sweet orange has also been associated to defense mechanism against the bacterial pathogen *Xylella fastidiosa* (Soares et al., 2015).

Terpenoid phytoalexins often coexist with other of different origin; such is the case of the phenylpropanoid sakuranetin that is accumulated upon pathogen attack along with momilactones, oryzalexins and phytocassanes in rice, and zealexin and kauralexin in corn, all of them diterpenes derived from geranyl geranyl pyrophosphate (Horie et al., 2015; Piasecka et al., 2015). About 20 genes involved in the biosynthesis of these compounds have been identified and investigated in rice, including several diterpene synthases (copalyl diphosphate synthases or CPS, kaurene synthases-like or KSL and P450 monooxygenases). Among all, CPS4 is involved in the biosynthesis of momilactones and oryzalexin S, as extracted from studies with a rice T-DNA insertion mutant *cps4-tos*, devoid in both groups of terpenoid phytoalexins, that displays significant sensitivity to rice blast fungus *Magnaporthe oryzae* and *Fusarium fujikuroi* responsible of the 'bakanae' or 'foolish seedling' disease (Piasecka et al., 2015). Nevertheless, although these plants showed significantly reduced levels of momilactones and oryzalexin S, they are not defective in the production of other terpenoid phytoalexins such as phytocassanes A to E and oryzalexins A to F; therefore, the putative role of these metabolites in rice immunity needs a further validation. The synthesis of zealexins and kauralexins is less understood, although it is thought that mediates interactions with microbial pathogens and insects in aerial tissues. However, recent studies have shown that *anther ear 2* (*an2)* corn mutants deficient in kauralexin production are more prone to drought-induced damage (Vaughan et al., 2014). Under water deprivation, phytoalexin production is root-specific and does not have any effect on phytoalexin levels in shoots but could have an influence on the ability to induce their production in aboveground tissues. Moreover, treatment with the plant hormone ABA caused the accumulation of terpenoid phytoalexins in maize roots (Vaughan et al., 2015). Interestingly, in *Nicotiana plumbaginifolia*, the production of the sesquiterpenoid capsidiol was inhibited by ABA treatment (Mialoundama et al., 2009). Other phytoalexins are induced by jasmonates (Naoumkina et al., 2007) or SA (Durango et al., 2013). Citrus produce a large number of triterpenoids that have been associated to fruit and juice quality as well as responses of vegetative tissues to the environment. These compounds are synthesized from squalene by formation of a polycyclic molecule containing a furanolactone core structure; the main representative is limonin and its respective glucosylated form (Arbona et al., 2015). These compounds have been found to accumulate in vegetative *Citrus* tissues in response to HLB and could be part of the metabolic defense response (Manthey, 2008; Slisz et al., 2012).

ROS production and lipid peroxidation are major hallmarks of the HR in plants. The formation of lipid peroxides is catalyzed by lipoxygenases (LOX) and these molecules are involved in the execution of programmed cell death and also in signal transduction processes. In the cytoplasm, LOXs oxidize fatty acids rendering oxylipins that can diverge into the jasmonate or the hydroperoxide lyase biosynthesis pathways (Zoeller et al., 2012). Elicitation by pathogen attack or wounding induces cell membrane lipases that release linoleic acid from phospholipids into the cytoplasm (Aliferis and Jabaji, 2012; Dat et al., 2005; Zoeller et al., 2012). Subsequently, LOX enzymes catalyze the formation of lipid hydroperoxides. A central molecule is 13-hydroperoxy octadecatrienoic acid (13-HPOT) that can be further metabolized into 13-keto octadecatrienoic acid, cleaved into *n*-hexenal, or diverted into jasmonate biosynthetic pathway (Schaller, 2001). Besides jasmonates that do have a role in stress responses (De Ollas et al., 2015), fatty acid hydroperoxides have been shown to have a role in the induction of HR after pathogen attack (Dat et al., 2005; Zoeller et al., 2012) although cell death is not always associated with lipid peroxidation (Dat et al., 2005). In animal systems, lipid signaling plays an important role in the development of inflammatory response. As examples, the induction of leukotrienes and thromboxanes from arachidonic acid occurs in acute and chronic inflammatory processes (Hammond and O'Donnell, 2012). These responses are also downregulated after application of non-steroidal anti-inflammatory drugs such as acetyl salicylic acid or ibuprofen (Savchenko et al., 2010). Unlike animal systems, higher plants do not accumulate arachidonic acid; however, certain moss species such as the model bryophyte *Physcomitrella patens* are reported to contain significant amounts of this fatty acid; in addition, they also contain linoleic and linolenic acids (Beike et al., 2014) and accumulate cyclopentenone oxylipins (such as 12-oxo-phytodienoic acid, OPDA) but not jasmonic acid (Beike et al., 2014; Stumpe et al., 2010). Nevertheless, this fact does not seem to affect their ability to cope with biotic threats (Ponce de León et al., 2007), what means that other oxylipins might have taken over the role that JA and JA-Ile have in higher plants, reinforcing the hypothesis of lipid signaling as a trans-kingdom key feature in the response to infections.

In plant-pathogen interactions, much attention is paid to soluble semipolar metabolites as most phytoalexins and phytoanticipins fall within this category. However, successful pathogens subvert plant metabolism in order to enable the efficient uptake, sequestration and utilization of nutrients derived from the host photosynthetic activity. Under these circumstances, a reduction in net CO_2 assimilation as well as a reprogramming of host carbon partitioning is often observed within the first stages of infection (Parker et al., 2009). It is expected that levels of polar metabolites from the primary metabolism: carbohydrates, aminoacids, tricarboxylic acids, etc., are affected by pathogen attack. In *Arabidopsis thaliana*, inoculation with *Pseudomonas syringae* caused metabolic changes that affected glucosinolates, phenolic compounds but also aminoacids (phenylalanine and tyrosine) and disaccharides (sucrose) (Ward et al., 2010). Similarly, in symptomatic leaves of HLB affected citrus plants, phenylalanine accumulated whereas sucrose levels decreased (Chin et al., 2014), which indicates a different mode of action to that of *P. syringae*. Indeed, citrus varieties with enhanced susceptibility to HLB showed higher basal levels of L-proline, L-serine, and L-aspartic acid, galactose as well as butanedioic and tetradecanoic acids whereas resistant varieties showed higher levels of L-glycine and mannose. In addition, a stronger metabolic response was observed in sensitive varieties respect to resistant ones (Cevallos-Cevallos et al., 2012). Changes in primary metabolism indeed might influence the occurrence of secondary metabolites as their direct precursors. Therefore, microbes reprogram plant primary metabolism to provide themselves with nutrients; additionally, adapted plants might use this reprogramming to feed the secondary metabolism with precursors to synthesize lethal phytoalexins (Sanchez-Vallet et al., 2010). Moreover, it is also likely that plants respond to pathogen infection by reprogramming their own primary metabolism to deplete infected cells of the nutrients necessary to support microbial growth.

In response to pathogen elicitation, plants emit a number of volatile organic compounds (VOCs) that have several roles such as semiochemicals in plant-plant, plant-insect or plant-microbe interactions, responses to herbivore, biological control of entomological pests or invasive plants and as sensory and flavor attributes (Beck et al., 2014). Herbivory and fungal infections in different plant species induce the emission of several terpenes and fatty acid-derived volatiles that could be involved, directly or indirectly, in plant defense. For instance, when corn plants are attacked by herbivores synthesize β-caryophyllene from farnesyl diphosphate that is released to the atmosphere and attracts natural enemies of the herbivores (Beck et al., 2014). The induced VOC blend seems to exhibit attacker specificity and different strains of the same pathogen could induce different combinations of volatiles. Although their specific role is not yet clear, certain pathogen inducible volatile compounds such as 3-hexenol, 2-hexenal, methyl salicylate or linalool have been shown to inhibit pathogen growth (Ponzio et al., 2013). In corn, infection with different species of *Fusarium* induced the production of diterpene VOCs (β- and α-selinenes, β-macrocarpene, β-bisabolene and trichodiene) and their production correlated with the induction of zealexins (Becker et al., 2014). The specific role of these volatiles in infection response is not known although C6 green leaf volatiles have been associated to priming neighboring plants against pest attack (Beck et al., 2014). To this respect, it has been recently reported that the green leaf volatile 3-hexenyl acetate enhances defense against *Fusarium graminearum* in wheat (Ameye et al., 2015). In citrus, downregulation of D-limonene synthase altered monoterpene levels and also induced resistance to *Penicillium digitatum* infection (Rodriguez et al., 2013). Conversely, overproduction of plant volatiles by transgenic Arabidopsis plants overexpressing a terpene synthase involved in monoterpene production (35S:TPS23/27) were not resistant to *Verticillium longisporum*. Released monoterpenes had a positive effect on conidial germination and hyphal growth, subsequently allowing a faster colonization (Roos et al., 2015). As mentioned above, the role of VOCs as semiochemicals transferring information between individuals of the same plant species, different plant species and also different kingdoms is of ecological relevance (Morrell and Kessler, 2014). For instance, *Candidatus* Liberibacter alters the volatile emission spectra in the host tomato plant. A number of VOCs were more abundant in the headspace of infected plants: α-farnesene, α-pinene, terpinen-4-ol, β-terpineol, α-terpineol, and γ-terpineol along with aldehydes nonanal and octanal. This altered volatile emission influenced settlement behaviour of the bacterial vector (psyllid), as non-infected psyllids preferentially settle on an infected plant and non-infected plants attract infected psyllids (Mas et al., 2014). Other tri-trophic interactions have been described involving VOCs as mediators of this interaction, corn inoculation with the endophytic bacterium *Enterobacter aerogenes* increased resistance to blight fungus *Setosphaeria turcica* due to the increased production of 2,3-butanediol. However, these plants were more susceptible to the caterpillar *Spodoptera litoralis* (D'Alessandro et al., 2014)¸ reinforcing the specificity of the VOCs blend production. Further research will focus in deciphering the specific roles of different VOCs blends and their implications in ecological interactions.

Metabolic response of plants to viruses

Viruses are invasive nucleic acids that use the host cell enzyme machinery to propagate throughout the entire organism. Infection of plants by viruses has different effects depending on virus-plant host compatibility, overall health status of plants, etc. Some viruses improve host plant quality and attractiveness to vectors (e.g. aphids), but other (e.g. Cucumber Mosaic Virus or CMV) significantly reduce plant 'appealing' through dramatic changes in phloem carbohydrate and aminoacid redistribution, resulting in massive dispersal of vectors. Interestingly, in CMV-infected squash there is an increase in volatile emission (but not in the particular VOCs blend) that exerts an attractive effect on vector aphids (Mauck et al., 2014). Therefore, it seems clear that also viruses subvert plant metabolism in order to facilitate their own dispersal. Another example found in the literature is the interaction of tobacco with *Bemisia tabaci*, the tobacco whitefly. Tobacco is a relatively poor host for *B. tabaci* but infection with *tomato yellow leaf curl virus* (TYLCV) increased host suitability associated to changes in volatile terpenoid synthesis (Luan et al., 2013). A similar response was observed in tomato, whiteflies preferred settling on TYLCV-infected plants that on healthy ones that have higher emissions of the volatile terpenes β-myrcene, thymene, β-phellandrene, caryophyllene, (+)-4-carene, and α-humulene (Fang et al., 2013). In addition, there are also differences between resistant and susceptible cultivars. Tomato cultivars resistant to TYLCV exhibited a more coordinated response in the primary metabolism than susceptible ones, leading to the production of defensive secondary metabolites aimed to cope with the infection

Table 1. Transgenic strategies to enhance secondary metabolite production and plant defense against pathogens (adapted from Dubey et al., 2014; Großkinsky et al., 2012).

Metabolite class	Plant species	Target secondary metabolite	Transgenic strategy/genetic modification	Observed result
Alkaloids				
	Arabidopsis thaliana	Camalexin	Knock out of phytoalexin deficient 3 (*PAD3*), responsible of the last step in camalexin biosynthesis (Böttcher, Westphal, et al., 2009)	Decreased camalexin, biosynthesis, no effect on *Pseudomonas syringae* infection
			EtOH-inducible oomycete elicitor (PaNie)	Time-dependent induction of tryptophan and camalexin biosynthesis genes by PaNie
			Overexpression of *miR393*	Re-direction of secondary metabolism from camalexin to glucosinolates, increased resistance against *Pseudomonas syringae* and *Hyaloperonospora parasitica*, increased susceptibility against *Alternaria brassicicola*
			Dexamethasone inducible overexpression of *MEK2*; *mpk3*/*mpk6* knock out	Camalexin accumulation after *MEK2* induction, reduced/delayed camalexin biosynthesis in *mpk3* and *mpk6* mutant lines, increased susceptibility against *Botrytis cinerea* of *mpk3* mutant line
			Knock out of *WRKY33*	Reduced camalexin levels in *wrky33* mutant lines
	Catharanthus roseus	Strictosidine	Overexpression of tryptophan decarboxylase and strictosidine synthase	Increased alkaloid production.
Terpenoids				
	Oryza sativa	Momilactones	Overexpression of transcription factor (e.g. *OsTGAP1*) and biosynthesis genes Knock down of biosynthesis genes	Increased momilactone levels in overexpression lines, constitutive momilactone accumulation in *OsTGAP1* overexpression lines, reduced momilactone levels in knock down lines
			CIPK14/15 RNAi and overexpression of CIPK15	Decreased momilatone accumulation in *CIPK14/15* RNAi-line, increased momilactone accumulation in *CIPK15* overexpression line
		Momilactone A	Overexpression of Accelerated Cell Death and Resistance 1 (*ACDR1)* a putative Raf-like MAPKKK.	Increased momilactone A accumulation and increased resistance against *Magnaporthe grisea* in *ACDR1* overexpression lines
			Spotted leaf 18 *Spl18* mutant	Increased momilactone A accumulation and increased resistance against *Magnaporthe grisea* in *spl18* mutant
			Overexpression of Rac GTPAse (*RAC1)*	Increased momilactone A accumulation and increased resistance against *Xanthomonas oryzae* in *RAC1* overexpression lines
			Overexpression of *SBP*, a Se binding protein involved in plant defense.	Increased momilactone A accumulation an increased resistance against *Magnaporthe grisea* and *Xanthomonas oryzae* in *SBP* overexpression lines
		β-caryophyllene	Overexpression of *OsTPS3* encoding a beta-caryophyllene synthase	Plants with increased emission rates of (E)-β-caryophyllene.
		Phytocassanes	*CIPK14/15* RNAi and overexpression of *CIPK15*	Decreased phytocassane accumulation in *CIPK14/15* RNAi-line, increased phytocassane accumulation in *CIPK15* overexpression line
	Nicotiana tabacum	Capsidiol	Overexpression of isopentenyl transferase *(ipt)*	Induction of capsidiol biosynthesis and increased resistance against *Pseudomonas syringae*, increased susceptibility to *Botrytis cinerea*
		Trichodiene	Over expression of trichodiene synthase gene, that converts farnesyl pyrophosphate to trichodiene - Inserted under constitutive action of CaMV35S	Expression of active enzyme, increases the production of the sesquiterpenoid product in leaves
		Monoterpenes	Expression of *Perilla frutescens* limonene synthase	Increased synthesis of limonene.
		Patchoulol and sesquiterpenes	Overexpression of patchoulol synthase.	Catalyses the conversion of 2E,6E-Farnesyl diphosphate to patchoulol and diphosphate. Synthesis of patchoulol (volatile) and 13 additional sesquiterpene products.
		Cembratrienol	Antisense co-suppression of Cytochrome P450 hydroxylase gene specific to the trichome gland Involved in metabolism of sesquiterpenoids	Exudates of transgenic plants contain high concentrations of cembratrieneol and showed lower aphid predation.
	Nicotiana plumbaginifolia		ABA deficient mutants *Npaba1* and 2	Increased capsidiol accumulation in *Npaba1* and *Npaba2* mutant lines
	Mentha x piperita L.	Several terpenoids	Co-suppression of 1-deoxy-d-xylulose-5-phosphate reducto-isomerase (DXR). En enzyme that catalyzes the first committed step of methyl erythritiol phosphate (MEP) pathway	Transgenic plants showed a lower yield loss due to pests and pathogens.
	Petunia hybrida, *Solanum lycopersicum*, *Dianthus caryophyllus*	Linalool	Constitutive over expression in flowers of S-linalool synthase that catalyzes the conversion of geranyl diphosphate to 3S-linalool	Linalool and its derivatives produced by the transgenic plants act as repellents of *Myzus persicae* aphids and associated pathogens.

	Nicotiana spp.	Volatile terpenoids, sclareol	*NpPDR1* overexpression, an ABC transporter.	Increased volatile diterpene emission and sclareol secretion upon elicitation with pathogens.
	Artmisia annua	sesquiterpenoids	Overexpression of farnesyl diphosphate synthase	Increased metabolic flux in the sesquiterpenoid biosynthetic pathway
	Lotus japonicus Nicotiana tabacum	(E,E)-geranyl linalool and 4,8,12-trimethyltrideca-1,3,7,11-tetraene (TMTT)	Terpene synthase gene – PITPS2 from *Phaseolus lunatus*	Transgenic plants produce (E,E)-geranyl linalool and 4,8,12-trimethyltrideca-1,3,7,11-tetraene (TMTT). Attracts predatory mites to infested plants.
	Gossypium hirsutum	Gossypol	Overexpression of *NPR1*, that controls systemic acquired resistance (SAR).	Increased gossypol accumulation and increased resistance against *Rhizoctonia solani* and *Alternaria alternata* in *NPR1* overexpression lines
Phenylpropanoid pathway				
Cinnamates/ Coumarins	*Nicotiana tabacum*	Scopoletin	Overexpression of isopentenyl transferase (*ipt*) gene.	Induction of capsidiol biosynthesis and increased resistance against *Pseudomonas syringae*, increased susceptibility to *Botrytis cinerea*
			Antisense expression of Tobacco O-glucosyl Transferase (*TOGT*) gene.	Reduced scopoletin accumulation and increased TMV susceptibility in *TOGT* antisense lines
	Solanum lycopersicum	Increased oxidation of phenolic substrates	cDNA transfer of polyphenol oxidase gene from *Solanum tuberosum* under constitutive action of CMV35S	Transgenic plants showed increased resistance to *Pseudomonas syringae* and strong inhibition of bacterial growth.
Flavones/ Isoflavones	*Arabidopsis thaliana, Solanum lycopersicum,*	Medicarpin	Overexpression of chalcone isomerase (*CHI) gene.*	Increased flavonol levels in *CHI* overexpression lines
	Medicago sativa		Overexpression of isoflavone *O*-methyl transferase (*IOMT*).	Increased medicarpin accumulation and increased resistance against *Phoma medicaginis* in *IOMT* overexpression lines
	Several plant species	e.g. Coumestrol, Glyceollins, Medicarpin	Various modulations of biosynthesis genes and regulatory elements	Diverse
	Lotus corniculatus	Vestitol	Antisense expression of chalcone synthase (*CHS*)	Reduced vestitol accumulation in *CHS* antisense lines
	Medicago truncatula	Formononetin, Medicarpin	Overexpression of isoflavone synthase (*IFS*)	Increased formononetin and medicarpin accumulation in *IFS* overexpression lines
	Pisum sativum	Pisatin	RNAi lines for (+)6a-hydroxymaackiain 3-O-methyltransferase (*HMM*), isoflavone reductase (*IFR*) and sophorol reductase (*SOR*).	Reduced pisatin accumulation in *HMM*, *IFR* and *SOR* RNAi lines
			HMM antisense, overexpression of pisetin demethylating activity *(PDA)*.	Reduced pisatin accumulation in *HMM* antisense and *PDA* overexpression lines
	Oryza sativa	Sakuranetin	Overexpression of *ACDR1* gene.	Increased sakuranetin accumulation and increased resistance against *Magnaporthe grisea* in *ACDR1* overexpression lines
			Spotted leaf 18 (Spl18) mutant	Increased sakuranetin accumulation and increased resistance against *Magnaporthe grisea* in *spl18* mutant
Stilbenes	Several plant species	e.g. Resveratrol	Overexpression of *STS gene*	Increased stilbene accumulation and increased resistance to various pathogens in *STS* overexpression lines
	Lactuca sativa	Resveratrol	Overexpression of *STS*	Transfer of resveratrol accumulation
	Oryza sativa, Solanum lycopersicum		Overexpression of *STS*	Increased resistance against *Magnaporthe grisea* and *Phytophthora infestans*

(Sade et al., 2014). There are also differences in locally-infected tissues or in organs that have developed a systemic response to the infection. For instance, in response to *tomato mosaic virus* (ToMV), tomato leaves responded to local infection by decreasing aminoacid, sucrose and phenolic acid levels and increasing tricarboxylic acid contents, systemic leaves showed similar profiles but also exhibited increased levels of tryptophan, sucrose and caffeoyl esters of glucaric acid. This was associated to a requirement of carbohydrates in locally-infected leaves to support the biosynthesis of defensive compounds such as rutin (López-Gresa et al., 2012). However, the specific role of changes in secondary metabolites on virus infection is not clear, as viruses are not real 'living' organisms and therefore their live cycle should not be affected by induced toxic compounds. A putative role of these metabolites could be directed to reduce virus spread by acting against the entomological vector. To ascertain the specific role, more *in vitro* and *in vivo* assays of potential defense compounds need to be performed both on virus replication and vector fitness.

Conclusions and future prospects

The high-throughput metabolite profiling techniques (LC/MS and GC/MS) have allowed the identification of a great number of novel molecules with putative antimicrobial, pathogen-growth deterring or semiotic functions. However, the great challenge is to undoubtedly assess their function *in vivo*. Modulation of toxic plant products to respond to pathogens could constitute a unique tool to enhance resistance, improving crop productivity and reducing pesticide application. Hence, metabolomics in combination with other -omics techniques have significantly contributed to the identification of genes and pathways responsible for the biosynthesis and accumulation of a plethora of secondary metabolites. As a result, a number of candidate genes for transgenic approaches are now available. However, it is important to properly assess any potential side effects of increasing phytoalexin production as these metabolites could be significantly toxic for human consumption. In addition, the host-pathogen interaction plasticity has also to be taken into account, as phytoalexin-overproducing transgenic crops could in turn drive adaptation of pathogens (Großkinsky et al., 2012). For these reasons, although substantial progress has been made towards the identification of genes and metabolites with potential antimicrobial activities and their implementation in crop improvement strategies (Table 1, see also Dubey et al., 2014), there is still a long way until researchers could fully understand the complex plant-pathogen interactions.

Acknowledgments
Funding from the Spanish Ministerio de Economia y Competitividad (MINECO) and Universitat Jaume I through grants AGL2013-42038-R and P1 1B2013-23, respectively, to Aurelio Gómez-Cadenas is greatly acknowledged.

References

Afek, U., and Sztejnberg, A. (1988). Accumulation of scoparone, a phytoalexin associated with resistance of citrus to *Phytophthora citrophthora*. Phytophathol *78*, 1678-1682.

Aharoni, A., Giri, A.P., Deuerlein, S., Griepink, F., de Kogel, W.-J., Verstappen, F.W.A., Verhoeven, H.A., Jongsma, M.A., Schwab, W., and Bouwmeester, H.J. (2003). Terpenoid metabolism in wild-type and transgenic Arabidopsis plants. Plant Cell *15*, 2866-2884.

Ahmad, S., Veyrat, N., Gordon-Weeks, R., Zhang, Y., Martin, J., Smart, L., Glauser, G., Erb, M., Flors, V., Frey, M., et al. (2011). Benzoxazinoid metabolites regulate innate immunity against aphids and fungi in maize. Plant Physiol *157*, 317-327.

Ahuja, I., Kissen, R., and Bones, A.M. (2012). Phytoalexins in defense against pathogens. Trends Plant Sci *17*, 73-90.

Akamatsu, A., Wong, H.L., Fujiwara, M., Okuda, J., Nishide, K., Uno, K., Imai, K., Umemura, K., Kawasaki, T., Kawano, Y., et al. (2013). An *OsCEBiP/OsCERK1-OsRacGEF1-OsRac1* module is an essential early component of chitin-induced rice immunity. Cell Host Microbe *13*, 465-476.

Aliferis, K. A., and Jabaji, S. (2012). FT-ICR/MS and GC-EI/MS metabolomics networking unravels global potato sprout's responses to *Rhizoctonia solani* infection. PLoS One *7*, 1-13.

Allwood, J.W., Ellis, D.I., and Goodacre, R. (2008). Metabolomic technologies and their application to the study of plants and plant-host interactions. Physiol Plant *132*, 117-135.

Ameye, M., Audenaert, K., De Zutter, N., Steppe, K., Van Meulebroek, L., Vanhaecke, L., De Vleesschauwer, D., Haesaert, G., and Smagghe, G. (2015). Priming of wheat with the green leaf volatile Z-3-hexenyl acetate enhances defense against *Fusarium graminearum* but boosts deoxynivalenol production. Plant Physiol *167*, 1671-1684.

Arbona, V., Iglesias, D.J., Talón, M., and Gómez-Cadenas, A. (2009). Plant phenotype demarcation using nontargeted LC-MS and GC-MS metabolite profiling. J Agric Food Chem *57*, 7338-7347.

Arbona, V., Iglesias, D.J., and Gómez-Cadenas, A. (2015). Non-targeted metabolite profiling of citrus juices as a tool for variety discrimination and metabolite flow analysis. BMC Plant Biol *15*, 1-16.

Bahar, O., Pruitt, R., Luu, D.D., Schwessinger, B., Daudi, A., Liu, F., Ruan, R., Fontaine-Bodin, L., Koebnik, R., and Ronald, P. (2014). The *Xanthomonas* Ax21 protein is processed by the general secretory system and is secreted in association with outer membrane vesicles. PeerJ *2*, e242.

Bar, M., Sharfman, M., Ron, M., and Avni, A. (2010). BAK1 is required for the attenuation of ethylene-inducing xylanase (Eix)-induced defense responses by the decoy receptor LeEix1. Plant J *63*, 791-800.

Beck, J.J., Smith, L., and Baig, N. (2014). An overview of plant volatile metabolomics, sample treatment and reporting considerations with emphasis on mechanical damage and biological control of weeds. Phytochem Anal *25*, 331-341.

Becker, E.-M., Herrfurth, C., Irmisch, S., Köllner, T.G., Feussner, I., Karlovsky, P., and Splivallo, R. (2014). Infection of corn ears by *Fusarium spp.* induces the emission of volatile sesquiterpenes. J Agric Food Chem *62*, 5226-5236.

Bednarek, P., Pislewska-Bednarek, M., Svatos, A., Schneider, B., Doubsky, J., Mansurova, M., Humphry, M., Consonni, C., Panstruga, R., Sanchez-Vallet, A., et al. (2009). A glucosinolate metabolism pathway in living plant cells mediates broad-spectrum antifungal defense. Science *323*, 101-106.

Beike, A.K., Jaeger, C., Zink, F., Decker, E.L., and Reski, R. (2014). High contents of very long-chain polyunsaturated fatty acids in different moss species. Plant Cell Rep *33*, 245-254.

Böttcher, C., von Roepenack-Lahaye, E., Schmidt, J., Schmotz, C., Neumann, S., Scheel, D., and Clemens, S. (2008). Metabolome analysis of biosynthetic mutants reveals a diversity of metabolic changes and allows identification of a large number of new compounds in Arabidopsis. Plant Physiol *147*, 2107-2120.

Böttcher, C., von Roepenack-Lahaye, E., Schmidt, J., Clemens, S., and Scheel, D. (2009a). Analysis of phenolic choline esters from seeds of *Arabidopsis thaliana* and Brassica napus by capillary liquid chromatography/electrospray- tandem mass spectrometry. J Mass Spectrom *44*, 466-476.

Böttcher, C., Westphal, L., Schmotz, C., Prade, E., Scheel, D., and Glawischnig, E. (2009b). The multifunctional enzyme CYP71B15 (PHYTOALEXIN DEFICIENT3) converts cysteine-indole-3-acetonitrile to camalexin in the indole-3-acetonitrile metabolic network of *Arabidopsis thaliana*. Plant Cell *21*, 1830-1845.

Brillada, C., Nishihara, M., Shimoda, T., Garms, S., Boland, W., Maffei, M.E., and Arimura, G. (2013). Metabolic engineering of the C16 homoterpene TMTT in *Lotus japonicus* through overexpression of (E,E)-geranyllinalool synthase attracts generalist and specialist predators in different manners. New Phytol *200*, 1200-1211.

Canel, C., Lopes-Cardoso, M.I., Whitmer, S., van der Fits, L., Pasquali, G., van der Heijden, R., Hoge, J.H.C., and Verpoorte, R. (1998). Effects of over-expression of strictosidine synthase and tryptophan decarboxylase on alkaloid production by cell cultures of *Catharanthus roseus*. Planta *205*, 414-419.

Cevallos-Cevallos, J.M., Rouseff, R., and Reyes-De-Corcuera, J.I. (2009). Untargeted metabolite analysis of healthy and Huanglongbing-infected orange leaves by CE-DAD. Electrophoresis *30*, 1240-1247.

Cevallos-Cevallos, J.M., Futch, D.B., Shilts, T., Folimonova, S.Y., and Reyes-De-Corcuera, J.I. (2012). GC-MS metabolomic differentiation of selected citrus varieties with different sensitivity to citrus huanglongbing. Plant Physiol Biochem *53*, 69-76.

Chin, E.L., Mishchuk, D.O., Breksa, A.P., and Slupsky, C.M. (2014). Metabolite signature of *Candidatus* Liberibacter asiaticus infection in two citrus varieties. J Agric Food Chem *62*, 6585-6591.

Chong, J. (2002). Downregulation of a pathogen-responsive tobacco UDP-Glc:phenylpropanoid glucosyltransferase reduces scopoletin glucoside accumulation, enhances oxidative stress, and weakens virus resistance. Plant Cell *14*, 1093-1107.

Coll, N.S., Epple, P., and Dangl, J.L. (2011). Programmed cell death in the plant immune system. Cell Death Differ *18*, 1247-1256.

Colliver, S.P., Morris, P., and Robbins, M.P. (1997). Differential modification of flavonoid and isoflavonoid biosynthesis with an antisense chalcone synthase construct in transgenic *Lotus corniculatus*. Plant Mol Biol *35*, 509-522.

Cuadros-Inostroza, A., Caldana, C., Redestig, H., Kusano, M., Lisec, J., Peña-Cortés, H., Willmitzer, L., and Hannah, M. (2009). TargetSearch--a Bioconductor package for the efficient preprocessing of GC-MS metabolite profiling data. BMC Bioinformatics *10*, 428.

D'Alessandro, M., Erb, M., Ton, J., Brandenburg, A., Karlen, D., Zopfi, J., and Turlings, T.C.J. (2014). Volatiles produced by soil-borne endophytic bacteria increase plant pathogen resistance and affect tritrophic interactions. Plant Cell Environ *37*, 813-826.

Dardick, C., Schwessinger, B., and Ronald, P. (2012). Non-arginine-aspartate (non-RD) kinases are associated with innate immune receptors that recognize conserved microbial signatures. Curr Opin Plant Biol *15*, 358-366.

Dat, J., Cotte, B. Van De, Montillet, J., Chamnongpol, S., Ruste, C., Breusegem, F. Van, Triantaphylide, C., Agnel, J., Battesti, C., and Inze, D. (2005). Fatty acid hydroperoxides and H_2O_2 in the execution of hypersensitive cell death in tobacco leaves. Plant Physiol. *138*, 1516-1526.

Deavours, B.E., and Dixon, R.A. (2005). Metabolic engineering of isoflavonoid biosynthesis in alfalfa. Plant Physiol *138*, 2245-2259.

Delaunois, B., Cordelier, S., Conreux, A., Clément, C., and Jeandet, P. (2009). Molecular engineering of resveratrol in plants. Plant Biotechnol J *7*, 2-12.

Dixon, R.A. (2001). Natural products and plant disease resistance. Nature *411*, 843-847.

Dixon, R.A., Achnine, L., Kota, P., Liu, C.-J., Reddy, M.S.S., and Wang, L. (2002). The phenylpropanoid pathway and plant defence-a genomics perspective. Mol Plant Pathol *3*, 371-390.

Doares, S.H., Narvâez-Vâsquez, J., Conconi, A., and Ryan, C.A. (1995). Salicylic acid inhibits synthesis of proteinase inhibitors in tomato leaves induced by systemin and jasmonic acid. Plant Physiol *108*, 1741-1746.

Du, H., Huang, Y., and Tang, Y. (2010). Genetic and metabolic engineering of isoflavonoid biosynthesis. Appl Microbiol Biotechnol *86*, 1293-1312.

Dubey, S., Manduri, A., Arora, N., and Deepa Sankar, P. (2014). Transgenics in phytochemical defence. Res J Pharm Biol Chem Sci *5*, 763-774.

Ebel, J., Ayers, A.R., and Albersheim, P. (1976). Host-pathogen interactions: XII. response of suspension-cultured soybean cells to the elicitor isolated from *Phytophthora megasperma* var. *sojae*, a fungal pathogen of soybeans. Plant Physiol. *57*, 775-779.

Fang, Y., Jiao, X., Xie, W., Wang, S., Wu, Q., Shi, X., Chen, G., Su, Q., Yang, X., Pan, H., et al. (2013). Tomato yellow leaf curl virus alters the host preferences of its vector *Bemisia tabaci*. Sci Rep *3*, 2876.

Friedman, M. (2002). Tomato glycoalkaloids: role in the plant and in the diet. J Agric Food Chem *50*, 5751-5780.

Glawischnig, E. (2007). Camalexin. Phytochemistry *68*, 401-406.

Glazebrook, J., and Ausubel, F.M. (1994). Isolation of phytoalexin-deficient mutants of *Arabidopsis thaliana* and characterization of their interactions with bacterial pathogens. Proc Natl Acad Sci U S A *91*, 8955-8959.

Gomez-Gomez, L. Bauer Z, and Boller T. (2001). Both the extracellular leucine-rich repeat domain and the kinase activity of FLS2 are required for flagellin binding and signaling in Arabidopsis. Plant Cell *13*, 1155-1163.

Gottstein, D., and Gross, D. (1992). Phytoalexins of woody plants. Trees *6*, 55-68.

Großkinsky, D.K., Naseem, M., Abdelmohsen, U.R., Plickert, N., Engelke, T., Griebel, T., Zeier, J., Novák, O., Strnad, M., Pfeifhofer, H., et al. (2011). Cytokinins mediate resistance against *Pseudomonas syringae* in tobacco through increased antimicrobial phytoalexin synthesis independent of salicylic acid signaling. Plant Physiol *157*, 815-830.

Großkinsky, D.K., van der Graaff, E., and Roitsch, T. (2012). Phytoalexin transgenics in crop protection-Fairy tale with a happy end? Plant Sci *195*, 54-70.

Grubb, C.D., and Abel, S. (2006). Glucosinolate metabolism and its control. Trends Plant Sci *11*, 89-100.

Hammerschmidt, R. (1999). PHYTOALEXINS: What have we learned after 60 years? Annu Rev Phytopathol *37*, 285-306.

Hammond, V.J., and O'Donnell, V.B. (2012). Esterified eicosanoids: generation, characterization and function. Biochim Biophys Acta *1818*, 2403-2412.

Harborne, J.B. (1999). The comparative biochemistry of phytoalexin induction in plants. Biochem Syst Ecol *27*, 335-367.

Hasegawa, M., Mitsuhara, I., Seo, S., Okada, K., Yamane, H., Iwai, T., and Ohashi, Y. (2014). Analysis on blast fungus-responsive characters of a flavonoid phytoalexin sakuranetin; accumulation in infected rice leaves, antifungal activity and detoxification by fungus. Molecules *19*, 11404-11418.

He, X.-Z. (2000). Genetic manipulation of isoflavone 7-*O*-methyltransferase enhances biosynthesis of 4'-*O*-methylated isoflavonoid phytoalexins and disease resistance in alfalfa. Plant Cell *12*, 1689-1702.

Hijaz, F.M., Manthey, J. A., Folimonova, S.Y., Davis, C.L., Jones, S.E., and Reyes-De-Corcuera, J.I. (2013). An HPLC-MS characterization of the changes in sweet orange leaf metabolite profile following infection by the bacterial pathogen *Candidatus* liberibacter asiaticus. PLoS One *8*, 1-15.

Horie, K., Inoue, Y., Sakai, M., Yao, Q., Tanimoto, Y., Koga, J., Toshima, H., and Hasegawa, M. (2015). Identification of uv-induced diterpenes including a new diterpene phytoalexin, phytocassane F, from rice leaves by complementary GC/MS and LC/MS approaches. J Agric Food Chem *63*, 4050-4059.

Huffaker, A., Kaplan, F., Vaughan, M.M., Dafoe, N.J., Ni, X., Rocca, J.R., Alborn, H.T., Teal, P.E.A., and Schmelz, E.A. (2011). Novel acidic sesquiterpenoids constitute a dominant class of pathogen-induced phytoalexins in maize. Plant Physiol *156*, 2082-2097.

Huhman, D. V, Berhow, M. A, and Sumner, L.W. (2005). Quantification of saponins in aerial and subterranean tissues of *Medicago truncatula*. J Agric Food Chem *53*, 1914-1920.

Ito, S.I., Eto, T., Tanaka, S., Yamauchi, N., Takahara, H., and Ikeda, T. (2004). Tomatidine and lycotetraose, hydrolysis products of α-tomatine by *Fusarium oxysporum* tomatinase, suppress induced defense responses in tomato cells. FEBS Lett *571*, 31-34.

Jeandet, P. (2015). Phytoalexins: Current progress and future prospects. Molecules *20*, 2770-2774.

Jeandet, P., Douillet-Breuil, A.-C., Bessis, R., Debord, S., Sbaghi, M., and Adrian, M. (2002). Phytoalexins from the *Vitaceae*: biosynthesis, phytoalexin gene expression in transgenic plants, antifungal activity, and metabolism. J Agric Food Chem *50*, 2731-2741.

Jones, J.D.G., and Dangl, J.L. (2006). The plant immune system. Nature *444*, 323-329.

Kaimoyo, E., and VanEtten, H.D. (2008). Inactivation of pea genes by RNAi supports the involvement of two similar O-methyltransferases in the biosynthesis of (+)-pisatin and of chiral intermediates with a configuration opposite that found in (+)-pisatin. Phytochemistry *69*, 76-87.

Katajamaa, M., Miettinen, J., and Oresic, M. (2006). MZmine: toolbox for processing and visualization of mass spectrometry based molecular profile data. Bioinformatics *22*, 634-636.

Kim, J.-A., Cho, K., Singh, R., Jung, Y.-H., Jeong, S.-H., Kim, S.-H., Lee, J.-E., Cho, Y.-S., Agrawal, G.K., Rakwal, R., et al. (2009). Rice OsACDR1 (*Oryza sativa* accelerated cell death and resistance 1) is a potential positive regulator of fungal disease resistance. Mol Cells *28*, 431-439.

Kim, J.H., Lee, B.W., Schroeder, F.C., and Jander, G. (2008). Identification of indole glucosinolate breakdown products with antifeedant effects on *Myzus persicae* (green peach aphid). Plant J *54*, 1015-1026.

ÖKmen, B., Etalo, D.W., Joosten, M.H. J., Bouwmeester, H.J., de Vos, R.C.H., Collemare, J., and De Wit, P.J.G.M. (2013). Detoxification of α-tomatine by *Cladosporium fulvum* is required for full virulence on tomato. New Phytol *198*, 1203-1214.

Kurusu, T., Hamada, J., Hamada, H., Hanamata, S., and Kuchitsu, K. (2010). Roles of calcineurin B-like protein-interacting protein kinases in innate immunity in rice. Plant Signal Behav *5*, 1045-1047.

Lieberei, R., Biehl, B., Giesemann, A., and Junqueira, N.T.V. (1989). Cyanogenesis inhibits active defense reactions in plants. Plant Physiol. *90*, 33-36.

Lipka, V., Dittgen, J., Bednarek, P., Bhat, R., Wiermer, M., Stein, M., Landtag, J., Brandt, W., Rosahl, S., Scheel, D., et al. (2005). Pre- and postinvasion defenses both contribute to nonhost resistance in Arabidopsis. Science *310*, 1180-1183.

Liu, S., Hu, Y., Wang, X., Zhong, J., and Lin, Z. (2006). High content of resveratrol in lettuce transformed with a stilbene synthase gene of *Parthenocissus henryana*. J Agric Food Chem *54*, 8082-8085.

Lommen, A. (2009). MetAlign: interface-driven, versatile metabolomics tool for hyphenated full-scan mass spectrometry data preprocessing. Anal Chem *81*, 3079-3086.

López-Gresa, M.P., Lisón, P., Kim, H.K., Choi, Y.H., Verpoorte, R., Rodrigo, I., Conejero, V., and Bellés, J.M. (2012). Metabolic fingerprinting of Tomato Mosaic Virus infected *Solanum lycopersicum*. J Plant Physiol *169*, 1586-1596.

Lu, D., Lin, W., Gao, X., Wu, S., Cheng, C., Avila, J., Heese, A., Devarenne, T.P., He, P., and Shan, L. (2011). Direct ubiquitination of pattern recognition receptor FLS2 attenuates plant innate immunity. Science *332*, 1439-1442.

Luan, J.B., Yao, D.M., Zhang, T., Walling, L.L., Yang, M., Wang, Y.J., and Liu, S.S. (2013). Suppression of terpenoid synthesis in plants by a virus promotes its mutualism with vectors. Ecol Lett *16*, 390-398.

Lücker, J., Bouwmeester, H.J., Schwab, W., Blaas, J., Van Der Plas, L.H.W., and Verhoeven, H.A. (2001). Expression of *Clarkia* S-linalool synthase in transgenic petunia plants results in the accumulation of S-linalyl-β-d-glucopyranoside. Plant J *27*, 315-324.

Luedemann, A., Strassburg, K., Erban, A., and Kopka, J. (2008). TagFinder for the quantitative analysis of gas chromatography-mass spectrometry (GC-MS) based metabolite profiling experiments. Bioinformatics 1-7.

Macho, A.P., and Zipfel, C. (2014). Plant PRRs and the activation of innate immune signaling. Mol Cell *54*, 263-272.

Mahmoud, S.S., and Croteau, R.B. (2001). Metabolic engineering of essential oil yield and composition in mint by altering expression of deoxyxylulose phosphate reductoisomerase and menthofuran synthase. Proc Natl Acad Sci U S A *98*, 8915-8920.

Malik, M.S., Riley, M.B., Norsworthy, J.K., and Bridges, W. (2010). Variation of glucosinolates in wild radish (*Raphanus raphanistrum*) accessions. J Agric Food Chem *58*, 11626-11632.

Manthey, J.A. (2008). Differences in secondary metabolites in leaves from orange (*Citrus sinensis* L.) trees affected with greening disease (Huanglongbing) (HLB). Proc Florida State Hortic Soc *121*, 285-288.

Mas, F., Vereijssen, J., and Suckling, D.M. (2014). Influence of the pathogen *Candidatus* Liberibacter solanacearum on tomato host plant volatiles and psyllid vector settlement. J Chem Ecol *40*, 1197-1202.

Mauck, K.E., De Moraes, C.M., and Mescher, M.C. (2014). Biochemical and physiological mechanisms underlying effects of Cucumber mosaic virus on host-plant traits that mediate transmission by aphid vectors. Plant Cell Environ *37*, 1427-1439.

Mert-Türk, F., Bennett, M.H., Mansfield, J.W., and Holub, E.B. (2003). Camalexin accumulation in Arabidopsis thaliana following abiotic elicitation or inoculation with virulent or avirulent Hyaloperonospora parasitica. Physiol Mol Plant Pathol *62*, 137-145.

Mialoundama, A.S., Heintz, D., Debayle, D., Rahier, A., Camara, B., and Bouvier, F. (2009). Abscisic acid negatively regulates elicitor-induced synthesis of capsidiol in wild tobacco. Plant Physiol *150*, 1556-1566.

Mori, M., Tomita, C., Sugimoto, K., Hasegawa, M., Hayashi, N., Dubouzet, J.G., Ochiai, H., Sekimoto, H., Hirochika, H., and Kikuchi, S. (2007). Isolation and molecular characterization of a *Spotted leaf 18* mutant by modified activation-tagging in rice. Plant Mol Biol *63*, 847-860.

Morrell, K., and Kessler, A. (2014). Volatile-mediated information transfer and defence priming in plants: The scent of danger. Biochem *36*, 26-31.

Mugford, S.T., Louveau, T., Melton, R., Qi, X., Bakht, S., Hill, L., Tsurushima, T., Honkanen, S., Rosser, S.J., Lomonossoff, G.P., et al. (2013). Modularity of plant metabolic gene clusters: a trio of linked genes that are collectively required for acylation of triterpenes in oat. Plant Cell *25*, 1078-1092.

Naoumkina, M., Farag, M. a, Sumner, L.W., Tang, Y., Liu, C.-J., and Dixon, R. a (2007). Different mechanisms for phytoalexin induction by pathogen and wound signals in *Medicago truncatula*. Proc Natl Acad Sci USA *104*, 17909-17915.

Nielsen, K.A., Hrmova, M., Nielsen, J.N., Forslund, K., Ebert, S., Olsen, C.E., Fincher, G.B., and Møller, B.L. (2006). Reconstitution of cyanogenesis in barley (*Hordeum vulgare* L.) and its implications for resistance against the barley powdery mildew fungus. Planta *223*, 1010-1023.

Ohara, K., Ujihara, T., Endo, T., Sato, F., and Yazaki, K. (2003). Limonene production in tobacco with *Perilla* limonene synthase cDNA. J Exp Bot *54*, 2635-2642.

Okada, K. (2014). The Biosynthesis of isoprenoids and the mechanisms regulating it in plants. Biosci Biotechnol Biochem *75*, 1219-1225.

De Ollas, C., Arbona, V., and Gómez-Cadenas, A. (2015). Jasmonoyl isoleucine accumulation is needed for abscisic acid build-up in roots of Arabidopsis under water stress conditions. Plant Cell Environ DOI: 10.1111/pce.12536.

Ono, E., Wong, H.L., Kawasaki, T., Hasegawa, M., Kodama, O., and Shimamoto, K. (2001). Essential role of the small GTPase Rac in disease resistance of rice. Proc Natl Acad Sci U S A *98*, 759-764.

Ortuño, A., Arcas, M.C., Botía, J.M., Fuster, M.D., and del Río, J. A. (2002). Increasing resistance against *Phytophthora citrophthora* in tangelo Nova fruits by modulating polymethoxyflavones levels. J Agric Food Chem *50*, 2836-2839.

Park, C.-J., Peng, Y., Chen, X., Dardick, C., Ruan, D., Bart, R., Canlas, P.E., and Ronald, P.C. (2008). Rice XB15, a protein phosphatase 2C, negatively regulates cell death and XA21-mediated innate immunity. PLoS Biol *6*, e231.

Parker, D., Beckmann, M., Zubair, H., Enot, D.P., Caracuel-Rios, Z., Overy, D.P., Snowdon, S., Talbot, N.J., and Draper, J. (2009). Metabolomic analysis reveals a common pattern of metabolic re-programming during invasion of three host plant species by *Magnaporthe grisea*. Plant J *59*, 723-737.

Parkhi, V., Kumar, V., Campbell, L.M., Bell, A.A., Shah, J., and Rathore, K.S. (2010). Resistance against various fungal pathogens and reniform nematode in transgenic cotton plants expressing Arabidopsis NPR1. Transgenic Res *19*, 959-975.

Pedras, M.S.C., and Adio, A.M. (2008). Phytoalexins and phytoanticipins from the wild crucifers *Thellungiella halophila* and *Arabidopsis thaliana*: rapalexin A, wasalexins and camalexin. Phytochemistry *69*, 889-893.

Pedras, M.S.C., Adio, A.M., Suchy, M., Okinyo, D.P.O., Zheng, Q.-A., Jha, M., and Sarwar, M.G. (2006). Detection, characterization and identification of crucifer phytoalexins using high-performance liquid chromatography with diode array detection and electrospray ionization mass spectrometry. J Chromatogr A *1133*, 172-183.

Pedras, M.S.C., Zheng, Q.A., and Strelkov, S. (2008). Metabolic changes in roots of the oilseed canola infected with the biotroph *Plasmodiophora brassicae*: Phytoalexins and phytoanticipins. J Agric Food Chem *56*, 9949-9961.

Perrin, D.R., and Bottomley, W. (1961). Pisatin: an Antifungal Substance from Pisum sativum L. Nature *191*, 76-77.

Piasecka, A., Jedrzejczak-rey, N., and Bednarek, P. (2015). Secondary metabolites in plant innate immunity : conserved function of divergent chemicals. New Phytol *206*, 948-964.

Pluskal, T., Castillo, S., Villar-Briones, A., and Oresic, M. (2010). MZmine 2: modular framework for processing, visualizing, and analyzing mass spectrometry-based molecular profile data. BMC Bioinformatics *11*, 395.

Ponce de León, I., Oliver, J.P., Castro, A., Gaggero, C., Bentancor, M., and Vidal, S. (2007). *Erwinia carotovora* elicitors and *Botrytis cinerea* activate defense responses in *Physcomitrella patens*. BMC Plant Biol *7*, 52.

Ponzio, C., Gols, R., Pieterse, C.M.J., and Dicke, M. (2013). Ecological and phytohormonal aspects of plant volatile emission in response to single and dual infestations with herbivores and phytopathogens. Funct Ecol 27, 587-598.

Rasmann, S., Köllner, T.G., Degenhardt, J., Hiltpold, I., Toepfer, S., Kuhlmann, U., Gershenzon, J., and Turlings,

T.C.J. (2005). Recruitment of entomopathogenic nematodes by insect-damaged maize roots. Nature *434*, 732-737.

Rauhut, T., Luberacki, B., Seitz, H.U., and Glawischnig, E. (2009). Inducible expression of a Nep1-like protein serves as a model trigger system of camalexin biosynthesis. Phytochemistry *70*, 185-189.

Ren, D., Liu, Y., Yang, K.-Y., Han, L., Mao, G., Glazebrook, J., and Zhang, S. (2008). A fungal-responsive MAPK cascade regulates phytoalexin biosynthesis in Arabidopsis. Proc Natl Acad Sci U S A *105*, 5638-5643.

del Río, J., Gómez, P., Baidez, G., Arcas, M.C., Botía, J.M., and Ortuño A. (2004). Changes in the levels of polymethoxyflavones and flavanones as part of the defense mechanism of *Citrus sinensis* (cv. Valencia Late) fruits against *Phytophthora citrophthora*. J Agric Food Chem *52*, 1913-1917.

Robert-Seilaniantz, A., MacLean, D., Jikumaru, Y., Hill, L., Yamaguchi, S., Kamiya, Y., and Jones, J.D.G. (2011). The microRNA miR393 re-directs secondary metabolite biosynthesis away from camalexin and towards glucosinolates. Plant J *67*, 218-231.

Rodriguez, A., Shimada, T., Cervera, M., Alquezar, B., Gadea, J., Gomez-Cadenas, A., De Ollas, C.J., Rodrigo, M.J., Zacarias, L., and Pena, L. (2013). Terpene down-regulation triggers defense responses in transgenic orange leading to resistance against fungal pathogens. Plant Physiol *164*, 321-339.

Roessner, U., Luedemann, a., Brust, D., Fiehn, O., Linke, T., Willmitzer, L., and Fernie, A. (2001). Metabolic profiling allows comprehensive phenotyping of genetically or environmentally modified plant systems. Plant Cell *13*, 11-29.

Roos, J., Bejai, S., Mozūraitis, R., and Dixelius, C. (2015). Susceptibility to Verticillium longisporum is linked to monoterpene production by TPS23/27 in Arabidopsis. Plant J *81*, 572-585.

Sade, D., Shriki, O., Cuadros-Inostroza, A., Tohge, T., Semel, Y., Haviv, Y., Willmitzer, L., Fernie, A.R., Czosnek, H., and Brotman, Y. (2014). Comparative metabolomics and transcriptomics of plant response to Tomato yellow leaf curl virus infection in resistant and susceptible tomato cultivars. Metabolomics *11*, 81-97.

Sanchez-Vallet, A., Ramos, B., Bednarek, P., López, G., Piślewska-Bednarek, M., Schulze-Lefert, P., and Molina, A. (2010). Tryptophan-derived secondary metabolites in *Arabidopsis thaliana* confer non-host resistance to necrotrophic *Plectosphaerella cucumerina* fungi. Plant J *63*, 115-127.

Sandrock, R.W., and VanEtten, H.D. (2001). The relevance of tomatinase activity in pathogens of tomato: disruption of the β2-tomatinase gene in *Colletotrichum coccodes* and *Septoria lycopersici* and heterologous expression of the *Septoria lycopersici* β2-tomatinase in *Nectria haematococca*, a pathogen of tomato fruit. Physiol Mol Plant Pathol *58*, 159-171.

Savchenko, T., Walley, J.W., Chehab, E.W., Xiao, Y., Kaspi, R., Pye, M.F., Mohamed, M.E., Lazarus, C.M., Bostock, R.M., and Dehesh, K. (2010). Arachidonic acid: an evolutionarily conserved signaling molecule modulates plant stress signaling networks. Plant Cell *22*, 3193-3205.

Sawada, K., Hasegawa, M., Tokuda, L., Kameyama, J., Kodama, O., Kohchi, T., Yoshida, K., and Shinmyo, A. (2004). Enhanced resistance to blast fungus and bacterial blight in transgenic rice constitutively expressing *OsSBP*, a rice homologue of mammalian selenium-binding proteins. Biosci Biotechnol Biochem *68*, 873-880.

Schaller, F. (2001). Enzymes of the biosynthesis of octadecanoid-derived signalling molecules. J Exp Bot *52*, 11-23.

Schlaeppi, K., Abou-Mansour, E., Buchala, A., and Mauch, F. (2010). Disease resistance of Arabidopsis to *Phytophthora brassicae* is established by the sequential action of indole glucosinolates and camalexin. Plant J *62*, 840-851.

Schwessinger, B., and Zipfel, C. (2008) News from the frontline: recent insights into PAMP-triggered immunity in plants. Curr Opin Plant Biol *11*, 389-395.

Shimizu, T., Nakano, T., Takamizawa, D., Desaki, Y., Ishii-Minami, N., Nishizawa, Y., Minami, E., Okada, K., Yamane, H., Kaku, H., *et al.* (2010). Two LysM receptor molecules, CEBiP and OsCERK1, cooperatively regulate chitin elicitor signaling in rice. Plant J *64*, 204-214.

Slisz, A.M., Breksa, A.P., Mishchuk, D.O., McCollum, G., and Slupsky, C.M. (2012). Metabolomic analysis of citrus infection by "*Candidatus* Liberibacter" reveals insight into pathogenicity. J Proteome Res *11*, 4223-4230.

Smith, C. a, Want, E.J., O'Maille, G., Abagyan, R., and Siuzdak, G. (2006). XCMS: processing mass spectrometry data for metabolite profiling using nonlinear peak alignment, matching, and identification. Anal. Chem. *78*, 779-787.

Snyder, B.A., and Nicholson, R.L. (1990). Synthesis of phytoalexins in sorghum as a site-specific response to fungal ingress. Science *248*, 1637-1639.

Soares, M.S., da Silva, D.F., Forim, M.R., da Silva, M.F.D.G.F., Fernandes, J.B., Vieira, P.C., Silva, D.B., Lopes, N.P., de Carvalho, S.A., de Souza, A.A., et al. (2015). Quantification and localization of hesperidin and rutin in *Citrus sinensis* grafted on C. limonia after *Xylella fastidiosa* infection by HPLC-UV and MALDI imaging mass spectrometry. Phytochemistry.DOI: 10.1016/j.phytochem.2015.02.011.

Strehmel, N., Kopka, J., Scheel, D., and Böttcher, C. (2013). Annotating unknown components from GC/EI-MS-based metabolite profiling experiments using GC/APCI(+)-QTOFMS. Metabolomics 1-13.

Stukkens, Y., Bultreys, A., Grec, S., Trombik, T., Vanham, D., and Boutry, M. (2005). NpPDR1, a pleiotropic drug resistance-type ATP-binding cassette transporter from *Nicotiana plumbaginifolia*, plays a major role in plant pathogen defense. Plant Physiol *139*, 341-352.

Stumpe, M., Göbel, C., Faltin, B., Beike, A.K., Hause, B., Himmelsbach, K., Bode, J., Kramell, R., Wasternack, C., Frank, W., et al. (2010). The moss *Physcomitrella patens* contains cyclopentenones but no jasmonates: mutations in allene oxide cyclase lead to reduced fertility and altered sporophyte morphology. New Phytol *188*, 740-749.

Sun, Y., Han, Z., Tang, J., Hu, Z., Chai, C., Zhou, B., and Chai, J. (2013). Structure reveals that BAK1 as a co-receptor recognizes the BRI1-bound brassinolide. Cell Res *23*, 1326-1329.

Tijerino, A., Hermosa, R., Cardoza, R.E., Moraga, J., Malmierca, M.G., Aleu, J., Collado, I.G., Monte, E., and Gutierrez, S. (2011). Overexpression of the *Trichoderma brevicompactum tri5* gene: effect on the expression of the trichodermin biosynthetic genes and on tomato seedlings. Toxins *3*, 1220-1232.

Tikunov, Y., Lommen, A., Vos, C.H.R. De, Verhoeven, H.A., Bino, R.J., Hall, R.D., and Bovy, A.G. (2005). A novel approach for nontargeted data analysis for metabolomics. Large-scale profiling of tomato fruit volatiles. Plant Physiol *139*, 1125-1137.

Treutter, D. (2006). Significance of flavonoids in plant resistance: a review. Environ. Chem. Lett. *4*, 147-157.

Treviño, V., Yañez-Garza, I.-L., Rodriguez-López, C.E., Urrea-López, R., Garza-Rodriguez, M.-L., Barrera-Saldaña, H.-A., Tamez-Peña, J.G., Winkler, R., and Díaz de-la-Garza, R.-I. (2015). GridMass: a fast two-dimensional feature detection method for LC/MS. J Mass Spectrom *50*, 165-174.

Vaughan, M.M., Christensen, S., Schmelz, E.A., Huffaker, A., McAuslane, H.J., Alborn, H.T., Romero, M., Allen, L.H., and Teal, P.E.A. (2014). Accumulation of terpenoid phytoalexins in maize roots is associated with drought tolerance. Plant Cell Environ.DOI: 10.1111/pce.12482.

Velasco, P., Francisco, M., Moreno, D. A., Ferreres, F., García-Viguera, C., and Cartea, M.E. (2011). Phytochemical fingerprinting of vegetable *Brassica oleracea* and *Brassica napus* by simultaneous identification of glucosinolates and phenolics. Phytochem Anal *22*, 144-152.

Vogt, T. (2010). Phenylpropanoid biosynthesis. Mol Plant *3*, 2-20.

Wang, E., Wang, R., DeParasis, J., Loughrin, J.H., Gan, S., and Wagner, G.J. (2001). Suppression of a P450 hydroxylase gene in plant trichome glands enhances natural-product-based aphid resistance. Nat Biotechnol *19*, 371-374.

Ward, J.L., Forcat, S., Beckmann, M., Bennett, M., Miller, S.J., Baker, J.M., Hawkins, N.D., Vermeer, C.P., Lu, C., Lin, W., et al. (2010). The metabolic transition during disease following infection of *Arabidopsis thaliana* by *Pseudomonas syringae* pv. *tomato*. Plant J *63*, 443-457.

Williams, C. A, and Grayer, R.J. (2004). Anthocyanins and other flavonoids. Nat Prod Rep *21*, 539-573.

Wu, Q., and VanEtten, H.D. (2004). Introduction of plant and fungal genes into pea (*Pisum sativum* L.) hairy roots reduces their ability to produce pisatin and affects their response to a fungal pathogen. Mol Plant-Microbe Interact *17*, 798-804.

Wu, S., Schalk, M., Clark, A., Miles, R.B., Coates, R., and Chappell, J. (2006). Redirection of cytosolic or plastidic isoprenoid precursors elevates terpene production in plants. Nat Biotechnol *24*, 1441-1447.

Yun, Z., Gao, H., Liu, P., Liu, S., Luo, T., Jin, S., Xu, Q., Xu, J., Cheng, Y., and Deng, X. (2013). Comparative proteomic and metabolomic profiling of citrus fruit with enhancement of disease resistance by postharvest heat treatment. BMC Plant Biol *13*, 44.

Zandalinas, S.I., Vives-Peris, V., Gómez-Cadenas, A., and Arbona, V. (2012). A fast and precise method to identify indolic glucosinolates and camalexin in plants by combining mass spectrometric and biological information. J Agric Food Chem *60*, 8648-8658.

Zhurov, V., Navarro, M., Bruinsma, K., Arbona, V., Santamaria, M.E., Cazaux, M., Wybouw, N., Osborne, E.J., Ens, C., Rioja, C., et al. (2014). Reciprocal responses in the interaction between Arabidopsis and the cell-content-feeding chelicerate herbivore spider mite. Plant Physiol *164*, 384-399.

Zipfel, C., Kunze, G., Chinchilla, D., Caniard, A., Jones, J.D.G., Boller, T., Felix, G. (2006) Perception of the bacterial PAMP EF-Tu by the receptor EFR restricts Agrobacterium-mediated transformation. Cell *125*, 749-760.

Zoeller, M., Stingl, N., Krischke, M., Fekete, A., Waller, F., Berger, S., and Mueller, M.J. (2012). Lipid profiling of the Arabidopsis hypersensitive response reveals specific lipid peroxidation and fragmentation processes: biogenesis of pimelic and azelaic acid. Plant Physiol *160*, 365-378.

Other Publications of Interest

Xu, J.-P. (2014). Next-generation Sequencing: Current Technologies and Applications. Caister Academic Press, U.K. ISBN: 978-1-908230-33-1

He, Z. (2014). Microarrays: Current Technology, Innovations and Applications. Caister Academic Press, U.K. ISBN: 978-1-908230-49-2

Poptsova, M.S. (2014). Genome Analysis: Current Procedures and Applications. Caister Academic Press, U.K. ISBN: 978-1-908230-29-4

Fuentes, M. and LaBaer, J. (2014). Proteomics: Targeted Technology, Innovations and Applications. Caister Academic Press, U.K. ISBN: 978-1-908230-46-1

Murillo, J., Vinatzer, B.A., Jackson, R.W. and Arnold, D.L. (2015). Bacteria-Plant Interactions: Advanced Research and Future Trends. Caister Academic Press, U.K. ISBN: 978-1-908230-58-4

Caranta, C., Aranda, M.A., Tepfer, M. and Lopez-Moya, J.J. (2011). Recent Advances in Plant Virology. Caister Academic Press, U.K. ISBN: 978-1-904455-75-2

Nannipieri, P., Pietramellara, G. and Renella, G. (2014). Omics in Soil Science. Caister Academic Press, U.K. ISBN: 978-1-908230-32-4

Herold, K.E. and Rasooly, A. (2009). Lab-on-a-Chip Technology: Fabrication and Microfluidics. Caister Academic Press, U.K. ISBN: 978-1-904455-46-2

Omics Approach to Identify Factors Involved in *Brassica* Disease Resistance

Marta Francisco[1], Pilar Soengas[1], Pablo Velasco[1], Vijai Bhadauria[2], Maria E. Cartea[1] and Victor M. Rodríguez[1]*

[1]Group of Genetics, Breeding and Biochemistry of Brassicas. Misión Biológica de Galicia (MBG-CSIC), Apartado 28, 36080 Pontevedra, Spain
[2]Crop Development Centre / Department of Plant Sciences, University of Saskatchewan, Saskatoon, SK S7N 5A8 Canada
*Corresponding author. Email: vmrodriguez@mbg.csic.es
Tel: +34 986 85 48 00; Fax: +34 986 84 13 62.

Abstract
Understanding plant's defense mechanisms and their response to biotic stresses is of fundamental meaning for the development of resistant crop varieties and more productive agriculture. The *Brassica* genus involves a large variety of economically important species and cultivars used as vegetable source, oilseeds, forage and ornamental. Damage caused by pathogens attack affects negatively various aspects of plant growth, development, and crop productivity. Over the last few decades, advances in plant physiology, genetics, and molecular biology have greatly improved our understanding of plant responses to biotic stress conditions. In this regard, various 'omics' technologies enable qualitative and quantitative monitoring of the abundance of various biological molecules in a high-throughput manner, and thus allow determination of their variation between different biological states on a genomic scale. In this review, we have described advances in 'omic' tools (genomics, transcriptomics, proteomics and metabolomics) in the view of conventional and modern approaches being used to elucidate the molecular mechanisms that underlie *Brassica* disease resistance.

Introduction

The family *Brassicaceae* (syn. *Cruciferae*) is one of the crucial plant families for humans and animals and supplies several products from various plant parts. Since ancient times, *Brassica* crops have been used for many purposes, including vegetables, oilseeds, feed, condiments, fodder, green manure and even medical treatments. The genus *Brassica* includes a group of six interrelated species of worldwide economic importance. The principal vegetable species is *Brassica oleracea*, which includes vegetable and forage crops, such as kale, cabbage, broccoli, Brussels sprouts, cauliflower, and others; *Brassica rapa* includes vegetable forms, such as turnip, Chinese cabbage, and Pak Choi, along with forage and oilseed types; *Brassica napus* crops are mainly used as oilseed (rapeseed), although forage and vegetable types like leaf rape and "nabicol" are also included. Finally, the mustard group which is formed by three species, *Brassica carinata*, *Brassica nigra* and *Brassica juncea*, is mainly used as a condiment because of their seeds although leaves of *B. juncea* are also consumed as vegetables in Asian countries.

The productivity and quality of these crops are seriously affected by several diseases, which result in substantial economic losses every year worldwide. The increase in the economic importance of *Brassica* crops has led to an increase in research on the plant host responses to pathogens and the mechanisms underlying the resistance. Several diseases are known to occur on *Brassica* crops affecting yield and quality in the field and in storage (Table 1). Among them, black rot and clubroot could be considered the most important diseases. Black rot is caused by the bacterium *Xanthomonas campestris* pv. *campestris* (Pammel) Dowson and is considered to be one of the most devastating soilborne disease in *Brassica* crops worldwide (Lema et al., 2012; Vicente et al., 2001). The disease is favored by warm, humid conditions and can spread rapidly from rain dispersal and irrigation water. Typical disease symptoms include V-shaped yellow lesions starting from the leaf margins and blackening of the veins (Vicente et al., 2006; Vicente and Holub, 2013). Considerable research into the management of black rot in vegetable *Brassica* crops has been done, but the majority of commercial cultivars of *B. oleracea* and *B. rapa* are highly susceptible (Lema et al., 2012). Clubroot is caused by the protist *Plasmodiophora brassicae* which infects the majority of cruciferous and it is particularly important in horticultural *Brassica* production. *P. brassicae* causes galls on roots of infected plants which interfere with the plant uptake of water and nutrients. Under warm conditions, plants often wilt which is usually the first sign of clubroot infection (Rennie et al., 2015). Ideal conditions for clubroot infection are acid soils, high soil moisture, warm temperatures and the presence of a susceptible brassica host (Strhlow et al., 2014)

Besides Xanthomonas, other bacteria could cause diseases in *Brassica* spp. For instance, *Erwinia carotovora* and *Pseudomonas spp.* are common pathogens to most *Brassica* crops and cause a soft mushy breakdown (Table 1). *P. syringae* pv. *maculicola* (McCulloch) is specially virulent. The bacterium can be seed-borne and survive on plants and crop debris and infects early vegetative plant stages. Development of infection is favored by prolonged cool, wet conditions and can develop rapidly following plant stress.

Fungi could also cause important diseases in *Brassica* spp. *Alternaria* spp. cause leaf spot that affects the quality of the products and can become a major problem during storage (Lee and Hong, 2015). Most of the important economic

Table 1. Main diseases of brassica crops adapted from Tewari and Mithen (1999) and Rimmer *et al* (2007).

Bacterial diseases	**Species**	**Distribution**	**Host plants**
Black rot	*Xanthomonas campestris* pv. *campestris*	Worldwide	Cabbage, kale, cauliflower, broccoli, turnip greens, leaf rape, swede and oilseed rape
Bacterial leaf spot	*Xanthomonas campestris* pv. *armoraciae*	Worldwide	Cauliflower, Brussels sprouts, broccoli, collard, kale, and cabbage
Bacterial leaf spot	*Pseudomonas syringae* pv. *maculicola*	Worldwide	Cauliflower, broccoli, Brussels sprouts, kale, cabbage, turnip, mustard greens and oilseed rape
Bacterial soft rot	*Pseudomonas marginalis* pv. *marginalis* *Erwinia carotovora*	Worldwide	Chinese cabbage, pack choi, white cabbage, cauliflower, broccoli, kale, turnip, and mustard greens
Fungal diseases			
Leaf spot, black spot, gray leaf spot, target spot, pod spot	*Alternaria brassicae, A. brassicicola and A. japonica*	Europe, India, Canada	Kale, broccoli, Brussels sprouts, cabbage, cauliflower, Chinese cabbage, kohlrabi, kale, swede and oilseed rape
Black leg or phoma stem canker	*Leptosphaeria maculans*	Worldwide	Kale, cabbage, cauliflower, broccoli, cabbage, swede oilseed rape, oilseed turnip, rape and turnip
White mold	*Sclerotinia sclerotiorum*	Europe, Canada, India, China, and Australia	Cabbage, cauliflower, swede, kale, oilseed rape and oilseed turnip
White blister, white rust	*Albugo candida*	Worldwide	Kale, broccoli, Brussels sprouts, cabbage, Chinese cabbage, swede and mustards
Clubroot	*Plasmodiophora brassicae*	Worldwide	Cabbage, broccoli, Chinese cabbage, cauliflower, Brussels sprouts, swedes, turnip and oilseed rape
Ring spot	*Mycosphaerella brassicicola*	Worldwide	Cabbage, broccoli, cauliflower, Brussels sprouts and kale
White leaf spot	*Mycosphaerella capsellae*	Worldwide	Chinese cabbage, turnip rape, oilseed rape and mustards
Downy mildew	*Hyaloperonospora parasitica*	Worldwide	Cabbage, broccoli, cauliflower, Brussels sprouts, kale, kohlrabi, Chinese cabbage, turnip greens, turnip rape, oilseed rape, swede and mustards
White blister, white rust	*Albugo candida*	Worldwide	Kale, broccoli, Brussels sprouts, cabbage, Chinese cabbage, swede and mustards
Root rot and damping-off	*Rhizoctonia solani*	Worldwide	Cabbage, turnip rape, and oilseed rape
Yellows, Fusarium wilt	*Fusarium oxysporum, Fusarium conglutinans*	Europe, United States, Canada, Argentina, and Australia	Cabbage, broccoli, cauliflower, kale, kohlrabi, Chinese cabbage, turnip greens, turnip rape, oilseed rape and swede
Gray mold	*Botrytis cinerea*	Worldwide	Cabbage, cauliflower, kale, oilseed rape and oilseed turnip
Virus			
Cauliflower mosaic virus (CaMV)	*Caulimoviridae*	Europe, United States	Cauliflower, broccoli, cabbage, kale, Brussels sprouts and oilseed rape
Turnip mosaic virus (TuMV)	*Potyviridae*	Worldwide	Turnip, cabbage and swede
Cucumber mosaic virus (CMV)	*Bromoviridae*	Worldwide	Oilseed turnip and oilseed rape
Beet western yellows virus (BWYV)	*Luteoviridae*	Europe Australia	Broccoli, cabbage and oilseed rape
Radish mosaic virus (RaMV)	*Comoviridae*	America and Europe	Radish, swede, oilseed turnip and black mustard
Turnip crinckle virus (TCV)	*Tombusviridae*	Europe	Turnip and Chinese cabbage
Turnip yellow mosaic virus (TYMV)	*Tymoviridae*	Europe and Australia	Cabbage, cauliflower, broccoli, Chinese cabbage, oilseed rape, swede, kale, oilseed turnip and turnip greens

Brassica crops are hosts of *A. brassicae* (Table 1). This fungus produces small dark spots on older leaves in cool wet conditions, which are surrounded by a ring of lighter tissue. The disease can be carried over on crop residues and weeds. Other fungi such as *Leptosphaeria maculans* or *Sclerotinia sclerotiorum* (Lib.) de Bary attack a wide range of *Brassica* crops (Table 1). The former fungus causes dry rot/blackleg that commonly affects the stems and leaves of vegetable brassicas. Numerous vegetable and oilseed *Brassica* crops are hosts of this disease (Table 1). Stem damage can lead to cankering and severing of the plant at the base. *S. sclerotiorum* causes the so-called stem rot disease. The disease is mainly on aboveground parts, producing a cottony white mold. Finally, the mold is replaced by numerous hardened sclerotia that gradually change from white to black (Saharan and Mehta, 2008).

There are many viruses that affect *Brassica* crops. Symptoms vary depending on the plant host, age, variety, weather conditions and nutritional status. Viruses often have a number of alternative *Brassica* hosts and are usually spread from plant to plant by insects (e.g. aphids, thrips) or fungal vectors. The three main *Brassica* viruses are beet western yellows virus (BWYV), cauliflower mosaic virus (CuMV) and turnip mosaic virus (TuMV) (Table 1). Infected plants show patterns, distortion and mottling in the leaves and often die if infection occurs early (Hunter et al., 2002). The application of aphicides can prevent their introduction; however, correct timing is difficult to achieve. Seed treatments can give seedlings up to some weeks protection, often long enough to protect through the critical period when aphids are flying (McSorley and Frederick, 1995; Raybould et al., 1999).

'Omic' studies in *Brassica*-pathogen interactions

Plant infection by pathogenic microbes involves many dynamic changes in molecular communication and adaptation of host and pathogen physiology. Significant advances have been made in the past few decades in understanding the defense mechanisms associated with resistance in *Brassica* crops, and to date many genes governing resistance have been identified and used in crop improvement. However, due to the complexity of the genetic and molecular processes implicated in defense responses they need to be researched more extensively. In this regard, the rapid advances in 'omic' technologies provide an opportunity to generate new information either at the genomic, transcriptomic, proteomic or metabolomic levels. In the following sections, we provide an overview of integrated high-throughput genotyping technologies and functional 'omic' tools recently used for the identification of new genes, metabolic pathways and proteins involved in plant-pathogen resistance of *Brassica* crops.

Genomics

Genomics is a discipline in genetics that applies recombinant DNA, DNA sequencing methods, and bioinformatics to sequence, assemble, and analyze the function and structure of genomes. The field includes efforts to determine the entire DNA sequence of organisms and fine-scale genetic mapping. The first species of *Brassicaceae* being sequenced was *Arabidopsis thaliana* (Kaul et al., 2000). Genomes of the most important *Brassica* crops from the economical point of view have been recently sequenced and assembled combining Sanger and next-generation sequencing strategies (NGS). Genome sequences included those of Chinese cabbage (*B. rapa*), cabbage and kale *(B. oleracea)* and rapeseed (*B. napus*) (Chalhoub et al., 2014; Liu et al., 2014; Parkin et al., 2014; Wang et al., 2011). The genomes of *B. rapa* (A) and *B. oleracea* (C) have suffered a whole genome triplication (WGT), containing each three sub-genomes (Liu et al., 2014; Parkin et al., 2014; Wang et al., 2011), being this a characteristic of the *Brassicaceae* tribe (Li et al., 2014). While the genomes of *B. oleracea* and *B. rapa* are highly similar in terms of total gene clusters/sequences and the gene number in each cluster, there are also a large number of species-specific genes in the two species. *B. napus* is an amphidiploid between *B. rapa* and *B. oleracea* (AC genome). Most orthologous gene pairs in *B. rapa* and *B. oleracea* remain as homologous in *B. napus* (Chalhoub et al., 2014). Genomes sequencing projects for *Brassica* crops have produced vast amounts of sequence data that will provide useful information for genetic studies related to disease resistance. Sequences of the A, C and AC genomes, their assembly and predicted genes and associated annotations are available in several web pages which are summarized in: http://www.brassica.info/resource/genome.php. The availability of these reference genomes enhances our understanding of the genome architecture and the evolution of *Brassica* species, as well as facilitates identification of genes associated with important traits for breeding (Lee et al., 2015).

Besides, the genomes of other species from the *Brassicaceae* family closely related to *Brassica* crops, have been recently sequenced as those of *Eutrema salsugineum*, *Capsella rubella*, *Leavenworthia alabanica*, *Sisymbrium inio*, *Aethionema arabicum* or *Arabidopsis lyrata* (Hu et al., 2011; Slotte et al., 2013; Yang et al., 2013).

The availability of genome sequences of *Brassicaceae* species allows carrying on comparative genomics, to study basic biological similarities and differences as well as evolutionary relationships between organisms. The close phylogenetic relationship between the *Brassica* species and the model plant *A. thaliana* makes the transfer of knowledge from *A. thaliana* to *Brassica* crop improvement straightforward (Cho et al., 2015). Several studies have established chromosomal collinearity between *Brassica* and *A. thaliana*. The QTL of resistance to clubroot Crr1 is located on a region called A8 on *B. rapa* (Suwabe et al., 2012). After fine mapping, the region on A8 corresponded to a disease resistance gene cluster in the *A. thaliana* genome. The location and order on the markers showed good correspondence with those in *A. thaliana* with a few rearrangements such as inversions and insertions. This genomic region of *Brassica* is conserved and has a potential as a disease resistance gene cluster.

Moreover, the availability of whole genome sequences has led to the identification and validation of novel disease-resistance alleles (Hayward et al., 2012). Plant disease resistance (R) genes which specifically interact/recognize with corresponding pathogen avirulence (avr) genes are considered as plant genetic factors of a major layer. The interactions of this gene-for-gene manner activate the signal transduction cascades that turn on complex defense responses against pathogen attack and this is called incompatible interaction (Yu et al., 2014). Whole genome analysis of R gene families within cultivated and wild varieties of a host species, made available through genome sequencing, may not only enable the identification of novel genes, but extensive phylogenetic and evolutionary analyses may also infer the nature of R and Avr gene product interactions (Hayward et al., 2012). Most of the R genes in plant kingdom are members of NBS-LRR (nucleotide-binding site-leucine rich repeat) proteins. Recent whole genome sequence data enabled the genome wide identification, mapping and characterization of candidate 159 NBS-containing R genes in *A. thaliana* (Meyers et al., 2003). The genome sequence of *B. rapa* was used to identify 92 non-redundant NBS-encoding in approximately 100 Mbp of *B. rapa* genome sequence (Mun et al., 2009). More recently, 157 and 206 NBS-encoding genes in *B. oleracea* and *B. rapa* were identified, respectively (Yu et al., 2014). NBS-encoding genes exhibited differential expression pattern in different tissues and several genes were induced by wounding. Despite the fact that *B. rapa* has a significantly larger genome than *A. thaliana*, *B. rapa* contains relatively small number of NBS-encoding genes, presumably because of deletion of redundant genes related to genome diploidization (Mun et al., 2009).

Transcription factors (TF) play pivotal functions in signal transduction to activate or suppress defense response genes and regulate the interactions between different signaling pathways. The AP2/ERF superfamily is one of the largest groups of TFs in plants. Extensive research has confirmed that AP2/ERF TFs are involved in plant growth and development, hormone response, and biotic or abiotic stress responses (Liu et al., 2013). ERF genes could improve plant resistance because their overexpression enhanced resistance to various diseases and improved tolerance to drought, salt, and freezing in transgenic plants (Liu et al., 2013). The *Brassica* Database was surveyed to gain further information on the AP2/ERF superfamily and its subclade CRFs in Chinese cabbage. A total of 281 members were identified in this superfamily. Up to 35 tandem duplicated genes and 252 segmental duplicated genes were found among the 281 AP2/ERF TFs, while 30 tandem duplicated genes and only 75 segmental duplicated genes were found in *A. thaliana*, suggesting that the expansion of BrAP2/ERFTFs after speciation from *A. thaliana* is mainly attributed to segmental duplication events during the WGT.

With the availability of the *Brassica* A (Wang et al., 2011), C (Liu et al., 2014; Parkin et al., 2014) and AC genome (Chalhoub et al., 2014) sequences it is possible to identify high-density, genome-wide SNPs in *Brassica* spp. (Dalton-Morgan et al., 2014). High-density SNP arrays can be a cost-effective alternative for genome-wide polymorphism screens for genome-wide association (GWAS) studies of resistance to diseases, but also by breeders as a tool for comprehensive genome-wide screens of elite germplasm and breeding pools. Genome-wide high-density array in *B. napus* has been described recently (Dalton-Morgan et al., 2014) and it has been demonstrated its utility in genotyping R resistance genes to black leg on chromosome A07. A total of 21,311 SNP markers were developed by Park et al. (2010) by re-sequencing 1,398 STSs in eight genotypes of Chinese cabbage (*B. rapa*). 141 SNPs related to disease resistance and leaf traits were selected (Ahn et al., 2014). Among them, 20 SNP primers were polymorphic in a DH mapping population differing in resistance to clubroot, turnip mosaic virus and soft rot and they could be further used for the detection of QTLs and fine mapping studies of disease resistance of Chinese cabbage. From the same original set of SNPs markers designed by Park et al. (2010), 693 SNPs were tested for amplification in the *B. oleracea* genome, from this 425 were successfully applied into *B. oleracea*, suggesting that it is possible to apply SNPs markers developed base on the *B. rapa* genome to *B. oleracea* (Cho et al., 2015).

The application of GWAS to plants has greatly increased the resolution of QTL detection. Fopa Fomeju et al. (2014) by using a GWAS analysis identified genomic regions of *B. napus* linked to the resistance to stem canker, caused by the fungal pathogen *L. maculans*. A panel of 116 oilseed rape spring varieties was genotyped with 3228 SNPs. 321 markers were linked to the resistance, corresponding to 64 genomic regions. These genomic regions were relatively equally distributed on the A (53%) and C (47%) genomes of *B. napus*. 44% of these regions are duplicated homologous regions, which suggests structural and functional conservation of genetic factors involved in this trait in *B. napus*.

Efforts to understand crop genomes have been boosted in recent years by developments in NGS technologies. Dramatic cost reductions accompanying the rapid increase in scale of these systems today make DNA sequencing an accessible and powerful option for high-throughput analysis (Edwards et al., 2013). Re-sequencing is used to identify genetic variation between individuals, which can provide molecular genetic markers and insight into gene function (Tollenaere et al., 2012).

Using the sequencing-by-synthesis method, a total of 7.0 Gb Illumina pair-end reads were generated from each one of the parental lines of a mapping population of cabbage (Lv et al., 2013) with different levels of resistance to *Fusarium* wilt (caused by the fungus *Fusarium oxysporum*). InDel markers were designed according to the reference genome sequence of cabbage and the whole-genome resequencing of the two parents. A resistance gene (FOC1) was located on C06 flanked by two InDel markers. Afterwards Lv et al. (2014) designed additional InDel primers in that region to map the FOC1 gene more precisely. Two TIR-NBS-LRR type R genes were considered as candidate genes underlying the variation for *Fusarium* wilt resistance.

Black rot is one of the most devastating diseases to crucifers including *B. oleracea*. Up to 14 QTLs were described as linked to resistance in cabbage (Lee et al., 2015). Two parental lines of a mapping population of cabbage were re-sequenced up to 20x genome coverage and a genome-wide survey for SNPs was conducted. Approximately, 1.20 millions of SNPs were identified in the parents. Polymorphic SNPs between both parents were used to design dCAPS markers for those regions with low marker density. Finally, 103 markers were added to develop a higher density map. NBS-LRR genes within the QTLs regions were searched. Four QTL regions contained 21 candidate R genes. Two candidate disease resistance genes in BRQTL-C1 and seven genes in BRQTL-C3 were found as gene clusters (Lee et al., 2015).

Little information is available on the genetic control of quantitative resistance to blackleg. A QTL for blackleg resistance was mapped in a DH population of rapeseed (*B. napus*) and corresponds to the R gene Rlm4 placed on A07. This region was aligned with the reference genome sequence of *B. rapa* on which 18 candidate R genes were identified (Tollenaere et al., 2012). Parents of the DH population were re-sequenced (read coverage of 10x in both). Sequences were aligned with the reference genome of *B. rapa* and the 18 candidate genes were characterized in the parental lines following Illumina paired-end sequencing of the parents. Two candidate genes (BLR2 and BLR11) were selected based on the sequence identity of known resistance genes and sequence divergence in the NGS reads of both parents.

Transcriptomics

To effectively combat pathogen invasion, plants have developed sophisticated defense strategies that initiate with an extensive transcriptome reorganization that presages changes in plant metabolism. When the infection plays out, the plant metabolism often represents a shifting mixture of disease resistance responses and disease susceptibility responses. Over the last decades, studies on the transcriptome mechanisms that underlie the effects of plant pathogenic infection have received a great deal of attention. At a staggering pace the development of new technologies to perform genome-wide transcriptomic analyses (microarray based expression profiling methods, the availability of genomic and/or EST (expressed sequence tag) sequence data for some plant species and the recent development of the NGS) has allowed significant progress in the characterization of the transcriptomic plant-pathogen related response.

Speaking strictly the transcriptome is defined as the total set of transcripts (including mRNA, tRNA, rRNA and small non-coding RNAs) present in a given cell, tissue or organism. Techniques to analyze the transcriptome are mainly focused on the relative or absolute quantification of mRNA. These techniques could be divided in two major groups, those based on hybridization and those based on amplification. Microarrays are the paradigmatic example of the former group and were the first platform developed to study genome-wide expression. DNA microarrays provide a simple vehicle for exploring the transcriptome in a way that is both systematic and comprehensive (Brown and Botstein, 1999). This platform is based on the immobilization of discrete DNA sequences (probes) in an array which are used to interrogate the mRNA of a sample (target). Probe construction is a crucial step on microarray design. Regardless its chemical nature, probes have to hybridize very specific sequences and thus their design requires a basic knowledge of the genomic sequence of the species. For this reason the use of microarray technology has been limited to model organisms for years (Witzel et al., 2015). In the particular case of *Brassica* spp. early transcriptomic studies took advantage of the high degree of synteny in gene sequences between *A. thaliana* and *Brassica* spp. This highly genetic similitude has allowed the use of the *A. thaliana* microarrays to examine transcriptional changes associated with the response of *B. napus* cultivars to *S. sclerotiorum* infection at different time points following pathogen infection (Liu et al., 2005; Yang et al., 2007; Zhao et al., 2007) or to analyze the transcriptomic response to TuMV (turnip mosaic virus) infection in *B. rapa* (Li et al., 2014).

Nowadays, advances in the acquisition of *Brassica* spp. genome sequences have prompted to the development of specific microarrays. Two major groups of microarrays are available to perform genome-wide transcriptome studies in *Brassica* spp., spotted microarrays and oligonucleotide microarrays being both alternatives successfully used to characterize the *Brassica* response to different pathogens. Narusaka et al. (2006) used a spotted microarray with 1820 cDNA clones selected from 20166 non-redundant sequences of cDNA libraries of Chinese cabbage to investigate the transcriptome induced changes in tissues inoculated with the fungal pathogen *Colletotrichum higginsianum*. A custom-made cDNA microarray with 26090 oligonucleotides was also employed by Zhao et al. (2009) to monitor gene expression in *B. napus* in response to *S. sclerotiorum* insult. A different strategy was followed by Yin et al. (2006), who treated *B. napus* plants with oligochitosan to elicit the plant immune response and characterize this response using a microarray containing 8095 ESTs. A collection of *Brassica* specific oligonucleotide microarrays are nowadays commercially available (Love et al., 2010; Trick et al., 2009), but so far these have been seldom used in *Brassica*-pathogen interactions. Just recently, Jung et al. (2013) studied the transcriptional profiling of a transgenic Chinese cabbage using the *B. rapa* 24K oligo array challenged with *Pectobacterium carotovorum* subsp. *carotovorum* (*Pcc*).

The alternative to microarray platforms relies on technologies based on amplification. The recent acquisition of *Brassica* species genome sequences together with the advance in NGS, have allowed accurate measurement of expression levels of genes as well as detection of allele specific expression and alternative splicing variants. Two protocols have been developed to quantify level of expressions: tag-based sequencing approach (Digital Gene Expression (DGE)) and RNA sequencing (RNA-seq). DGE includes different techniques such as serial analysis of gene expression (SAGE) (Yamamoto et al., 2001), SuperSAGE (Matsumura et al., 2010) or cap analysis of gene expression (CAGE) (Kodzius et al., 2006). All these techniques are based on the generation of short cDNA tags from an mRNA population in several steps. More straightforward protocols are used for RNA-Seq. A typical RNA-Seq experiment starts with mRNA that is subsequently converted into cDNA to form an RNA-Seq library. By sequencing the millions of DNA fragments in the library (known as 'reads') with NGS, an accurate measure of the relative abundance of each transcript and splice variants can be obtained. In principle, any high-throughput sequencing technology could be used in both approaches.
Large-scale and genome-wide gene expression profiling methods have been applied to analyze *Brassica* crops infected by pathogens. Expression patterns of genetically transformed rapeseed plants with *hrf2,* a gene conferring rapeseed resistance to *S. sclerotiorum*, were compared with untransformed genotype, observing the breadth of molecular mechanism of hairpin-mediated plant responses related to defense, plant-hormone biosynthesis, catabolism, and signal transduction (Wang et al., 2015). RNA-Seq has been employed recently to elucidate resistance mechanisms involved in plant-pathogen interactions including fungi, bacteria and virus (Chu et al., 2014; Kalischuk et al., 2015; Kim et al., 2015; Lloyd et al., 2014; Shimizu et al., 2014).

Compared to other plants little information is available about the genome-wide expression response of *Brassica* spp. to pathogen insult. Special attention has been paid to the response of these species to fungal attack. Experiments on transcriptome profiling analysis of *Brassica* cultivars infected with fungal pathogens, led to the identification of more than 300 differentially expressed genes at least two-fold as compared to uninoculated controls and a large part of these genes exhibits temporal and quantitative differences between resistant genotypes and susceptible ones (Kim et al., 2015; Liu et al., 2005; Lowe et al., 2014; Narusaka et al., 2006; Shimizu et al., 2014; Yang et al., 2007; Zhao et al., 2007). Expressed genes included those encoding defense-associated proteins, transcription factors belonging to the zinc finger (WRKY, APETALA2 and MYB classes), phytohormone-responsive genes (jasmonic acid, ethylene and auxin synthesis enzymes) and cell wall structure genes. In addition, changes in the expression of genes encoding enzymes involved in carbohydrate and energy metabolism appeared to be directed towards shuttling carbon reserves to the tricarboxylic acid cycle and generating reactive oxygen species (Kim et al., 2015; Lowe et al., 2014; Shimizu et al., 2014; Zhao et al., 2009; Zhao et al., 2007). Transcripts from genes encoding enzymes involved in glucosinolate and phenylpropanoid biosynthesis were highly elevated after fungus infection, suggesting that secondary metabolites are also components of the defense response in *Brassica* spp. (Shimizu et al., 2014; Zhao et al., 2009). The integration of metabolomics with transcriptomic and genomic platforms has frequently been used as a strategy to identify candidate genes involved in the regulation of the levels of specific metabolites in plant systems. Thus, some of the candidate defense genes detected by transcriptomic approaches were also identified by integration with previously mapped QTLs (Zhao et al., 2007), and validated by qRT-PCR and Northerm blot analysis (Chu et al., 2014; Liu et al., 2005; Yang et al., 2007; Zhao et al., 2009).

Transcriptomic response to bacterial or virus pathogens has been extensively studied in *A. thaliana,* but not so in *Brassica* spp. Efforts to obtain *B. rapa* genotypes resistant to *P. carotovorum* prompted to the development of transgenic lines expressing a human cathelicidin antimicrobial peptide (Jung, 2013). The human peptide confers a moderate resistance to the pathogen by activating the expression of defensins, resistance-responsive proteins and LTP (lipid transfer proteins). Bacterial and virus pathogens have also been used to activate resistance gene expression of *Brassica* spp. (Kalischuk et al., 2015; Sarosh et al., 2009). Immunization of *B. napus* plants with *Bacillus amyloliquefaciens* prevents the attack of *Botrytis cinerea* (Sarosh et al., 2009). Interestingly, in spite of the root treatment of plants with *B. amyloliquefaciens* more transcripts were activated in leaves than in roots indicating a systemic effect in priming. Likewise, low dose of purified dsCaMV virus induce a complex transcriptome reactivation that allows a transgenerational resistance to pathogens. Though the mechanism involved in this response remain elusive, transcriptomic analysis point to epigenetic mechanisms (Kalischuk et al., 2015).

A special case of a transcriptomic study is the expression profile of small RNAs (including miRNAs) in response to pathogen challenge. Small RNAs are noncoding RNAs that play an important regulatory role in the growth and development of eukaryotes. In plants they are known to regulate the expression of a number of key developmental and stress-related genes and plant innate immune receptors (Eckardt, 2012). During the last years technological advances have allowed the study of the genome-wide expression profile of sRNAs. Shen et al. (2014) used this approach to analyzed the role of miRNAs in the oilseed rape-*Verticillium longisporum* interaction using two RNA libraries made from *V. longisporum* infected/ non-infected oilseed rape roots and employing *B. rapa* and *B. oleracea* genomes as references for miRNA prediction and characterization. A high throughput analysis identified two classes of sRNAs of *B. campestris* that modified their expression after infection with *E. carotovora*, host 28-nt sRNAs and sRNA homologous to the pathogen genome (Sun, 2014). Interaction with the protist *P. brassicae* also activates the expression of sRNAs in *B. napus* (Verma et al., 2014). This sRNAs are involved in the regulation of the expression of several transcription factors related to plant resistance.

Proteomics

Given that the language of plant-pathogen interactions is largely in proteins and is the final executors of most biological processes, proteomics is an important tool in translational genomics as it can translate plethora of genomic information into functional information. Proteomics is the large-scale functional analysis of proteins extracted from intact organisms, tissues, individual cells, or cell compartments, at defined time points during development or under specific conditions. Comparative proteome analysis, could lead to a more comprehensive understanding of biotic stresses in economical important *Brassica* crops as it can reveal factors involved in resistance against various diseases as proteins are directly related to functions. These factors can then be used in developing molecular markers to introgress resistance into cultivars through marker-assisted selection breeding or genome editing.

Many initial proteomic approaches were based on two-dimensional gel electrophoresis (2-DE), a good choice for rapidly identifying major proteome differences in healthy versus inoculated plants. However, it suffers from some ongoing concerns regarding quantitative reproducibility and limitations on the ability to study certain classes of proteins. Therefore, in recent years attention has been paid on the development of alternative approaches, such as promising gel-free proteomics. With the appearance of mass spectra (MS)-based proteomics, an entirely new toolbox has become available for quantitative analysis. Although these novel approaches were initially pitched as replacements for gel-based methods, they should probably be regarded as complements to rather than replacements of 2-DE. Protein spots selected from the gels are picked, trypsin digested, and subjected to ionization by ESI or matrix-assisted laser desorption/ionization (MALDI), both coupled to MS analyses. Subsequently, the peptide masses are used to query peptide mass databases in order to identify the proteins (Rabilloud et al., 2010). This proteomics approach has been successfully applied to some host-pathogen interaction studies in *Brassica* vegetables.

Canola (*B. napus*) is one of the important oil-seed crops worldwide, and clubroot and blackleg are the two main biological constraints to canola production. Clubroot caused by an obligate biotrophic protist *P. brassicae* is creating havoc in canola growing regions. Source of resistance has been identified in European fodder turnip *B. rapa* ssp. *rapifera* and successfully introgressed into oilseed rape (*B. napus*) (Bradshaw et al., 1997; Diederichsen et al., 2009; Hirai, 2006; Piao et al., 2009; Yoshikawa, 1981). This genetic resistance coupled with crop rotation can break the prolonged survival cycle of this pathogen in soil (up to 20 years). Kaido et al. (2007) carried out a comparative proteome analysis between a pair of resistant and susceptible turnip cultivars to the causal agent of culbroot disease. They achieved 2-DE to detect differences in protein profiles between resistant and susceptible cultured roots in the early responses against the given *P. brassicae* resting spores. These proteins, especially the up-regulated ones in resistant roots, must include specific agents involved in the resistance to clubroot infection.

Time-course proteomics revealed major changes in canola metabolism leading to susceptibility to *P. brassicae* (Cao et al., 2008). The authors used 2-DE coupled ESI-MS to identify 20 proteins in canola roots differentially regulated upon *P. brassicae* infection. Among these, there can be found reactive oxygen species-detoxification enzymes such as Cu/Zn sodium dismutase and cytochrome *c* oxidase, proteins implicated in metabolic pathway, enzymes such as S-adenosylmethionine synthetase, adenosine kinase, triose phosphate isomerase and Glycine-rich RNA-binding protein and the lignin biosynthesis enzyme caffeoyl-CoA O-methyltransferase. Blackleg is another major disease of oilseed rape worldwide. Twenty-four differentially expressed proteins in *B. napus* resistant cultivar Surpass 400 upon challenge with virulent and avirulent isolates of *L. maculans* were identified by the 2-DE coupled with matrix-assisted laser desorption ionization time-of-flight mass spectrometry (TOF-MS devices) (Marra et al., 2010). Among these there can be found stress- and defense associated proteins, such as heat shock protein, peptidyl-prolyl *cis-trans* isomerase, ribosome inactivating protein, ascorbate peroxidase and disulphide oxidoreductase. Functional validation of clubroot and blackleg resistance associated proteins can be achieved by generating transgenic plants using genome editing tools, such as Zn finger nuclease, TALENs and CRISPR/Cas9. Sun et al. (2014) studied proteome changes in the seedling leaves of the non-heading Chinese cabbage at different time points following inoculation with *H. parasitica* using 2-DE in combination with matrix-assisted laser desorption/ ionisation TOF-MS. They found 91 protein spots related to the resistance response. These proteins were assigned to different functional categories, such as amino acid and carbohydrate metabolism, photosynthesis and photorespiration, protein metabolism, signal transduction, redox homeostasis, and ethylene biosynthesis.

Recently developed shotgun proteomics approaches, such as iTRAQ (Isobaric tags for relative and absolute quantitation) have broader implication in plant disease resistance as these approaches allow multiplexing of samples and therefore, simultaneously compare and quantify differentially regulated proteins in susceptible and resistant genotypes. These technical advancements coupled to well-designed experiments will significantly reveal the protein function in plant stress- and defense and provide a wealth of information on plant proteome changes occurring in response to biotic stresses.

Metabolomics

For numerous organisms, complete genomes have been sequenced and transcriptome and proteome studies have been described. Only recently, metabolome analysis using MS platforms attracted attention (Hill and Roessner, 2013). Metabolomics is the analysis of the small molecules (metabolome) of an organism in a determined moment. This analysis provides the phenotypical response at the metabolic level under a particular environmental condition or stress. It is considered a very ambitious field due to the high number of chemically distinct molecules present in a typical plant sample, estimated to be at least 100,000 (Witzel et al., 2015). As stated by Sardans et al. (2011), metabolomics provides a better analysis of the different response capacities conferred by the phenotypic plasticity of each species, allowing ascertaining what metabolic pathways are involved in a phenotypic response. The metabolome consists of two types of compounds, the primary metabolites, compounds shared by different organisms, involved in the basic functions of the living cell, such as respiration, and the secondary metabolites. The later are species specific and play a role in the interaction with the environment, e.g. plant defense against pests and diseases (Verpoorte et al., 2007).

The analysis of metabolites is done by analytical platforms which combine several techniques like nuclear magnetic resonance spectroscopy (NMR), MS, gas chromatography (GC), liquid chromatography (LC), and capillary electrophoresis (CE). As explained by Witzel et al. (2015) changes in the primary and secondary metabolite pool have been characterized using a GC-MS approach, while CE-MS is more suitable for characterizing the products associated with the central metabolism. NMR is appropriate for the targeting of phenolic compounds, carbohydrates, organic acids and amino acids; ESI is used for the analysis of semi-polar metabolites. Together, these technologies can identify and quantify a wide range of primary and secondary metabolites. Besides, it is possible to add high mass resolution, MS/MS fragmentation patterns and UV spectra by the use of TOF-MS devices. Recent software developments have improved the capacity to recognize different metabolites. Profiles based on mass, retention time and signal amplitude provides the data required for filtering biomarkers. Typically a data processing pipeline can be divided into two steps: data processing (filtering, feature detection, alignment, and normalization) and data analysis (algorithm selection, training, evaluation, and model examination) (Witzel et al., 2015). To spread the use of metabolomics it is necessary to have public databases to compare the obtained results in the lab. There are two important characteristics to have a good metabolome analysis: the reproducibility to study the effect of different conditions in a concrete plant and the ease of quantitation and identification the number of metabolites to be measured (Verpoorte et al., 2007). Based on those characteristics Veerporte et al. (2007) established three different groups of methods: 1, Chromatographic methods; 2, MS; and 3, NMR, that can be reduced to two methods, the combination of 1 with 2 or 3. MS is now the most extended method but has a problem with the reproducibility due to the previous ionization step and the different mass spectrometers. NMR is a physical measurement of the resonances of the magnetic nuclei in a strong magnetic field. Each compound has a highly specific spectrum. This technique is highly reproducible which makes NMR the most suited method for a public domain metabolomics database. Chromatographic methods have shown to be very suitable for quantitative analysis. However, they require calibration curves for each compound as each one gives a different detector response. For qualitative analysis, chromatographic methods (HPLC with diode array detector, MS, and/or NMR, GC-MS) are the most powerful as they offer both retention behavior and physical characteristics as a tool for identification. MS allows the determination of the molecular weight, and in case of high resolution also of the elemental composition, but this is not always sufficient to determine the structure. Tandem MS might be of help to identify, in such a case, the compound through its fragmentation pattern, but in case of novel compounds further spectral data are required. The performance for NMR can be improved by using two-dimensional (2D) NMR methods, which even may enable structure elucidation of novel compounds in a mixture (Verpoorte et al., 2007). In any case, the metabolite identification in non-targeted approaches in the absence of reference compounds remains difficult (Witzel et al., 2015), but there are different platforms and spectral databases online and it is expected that these difficulties could be solved in the future, as the number of identified compounds is increasing.

Brassica crops are attacked by an important number of pathogens which compromise the development of the plants of this genus. Some works have evaluated the effect of glucosinolates (secondary metabolites specific of brassicas) on different plant diseases (Ma et al., 2015; Singh et al., 2015; Sotelo et al., 2015; Velasco et al., 2013) but, despite of the importance of these crops and the effect of this pathogens, little is known about the interactions of different diseases on the metabolome of *Brassica* spp. Dr. Verpoorte's lab (Leiden University) is the most active one working in *Brassica* metabolomics, using mainly NMR. So, till now, the only metabolome studies in *Brassica* crops, affected by diseases, were from or in collaboration with this lab. In 2008, Jahangir et al. (2008) evaluated the metabolomics response of *B. rapa* to different food borne bacteria such as *Bacillus subtilis, Staphylococcus aureus, Escherichia coli, Salmonella typhimurium and Shigella flexneri.* They used ^{1}H NMR and two-dimensional NMR spectra, coupled with principal component analysis (PCA) and partial least square-discriminant analysis (PLS-DA). The metabolic changes varied according to bacterial species. Gram-positive and Gram-negative bacteria had a different effect on the *Brassica* metabolome. While threonine and GABA were found to be the discriminating metabolites in Gram-positive bacteria treated plants, Gram-negative bacteria exhibited a significant increase in sinapoyl-malate, caffeoyl-malate and histidine. In general, amino acids, alcohols, carbohydrates and phenols were the discriminating metabolites. Using the same species, Simoh et al. (2009) evaluated the metabolomics changes that occur after infection with a disarmed and tumor-inducing strain of *Agrobacterium tumefaciens.* As in the previous report, the selected technique was ^{1}H NMR coupled with PLS. These authors observed differences between the two kinds of strains, affecting mostly to flavonoid, phenylpropanoid, sugar and free amino/organic acid content. As an interesting point, the infection with two tumor-inducing strains of *A. tumefaciens* provoke the suppression of some flavonoids and phenylpropanoids, compounds that are usually induced in the plants in response to biotic an abiotic stresses. In the same year, Abdel-Farid et al. (2009) evaluated the metabolite induction in *B. rapa* under the infection of several fungal pathogens (*L. maculans, Aspergillus niger, and F. oxysporum*). The results obtained from the ^{1}H NMR with PLS-DA, showed again that phenylpropanoids and phenols are associated with fungal infection. Besides, in this case, they found glucosinolates related to the infection response. Nevertheless, the metabolomic response was different for each fungal species, *F. oxysporum* infected plants accumulated more phenylpropanoids (sinapoyl-, feruloyl, and 5-hydroxyferuloyl malate), flavonoids (kaempferol and quercetin) and fumaric acid than plants infected with the other two species (Abdel-Farid et al., 2009).

In 2014, we started in our lab (Group of Genetics, Breeding and Biochemistry of Brassicas, MBG-CSIC) an experiment to evaluate the effect of *X. campestris* pv. *campestris* on the metabolome of *B. oleracea* at several times (1, 2, 3, 7, 14 days), using an LC-QTOF instrument. The preliminary results using the Welch test showed that several compounds are significantly affected by the infection and the time, both in negative and positive modes, from 94 compounds at day 1 to 300 at day 2, and decreasing during next days. Some of these compounds were present in the five times.

As far as we know, no other labs have investigated the metabolome of *Brassica* crops under the infection of different pathogens. In the model plants, *A. thaliana* and Thellungiella, several authors have reported the metabolite profile with different diseases (Botanga et al., 2012; Floerl et al., 2012; Huang et al., 2009; Pedras and Zheng, 2010). In this way, Botanga et al. (2012) inoculated Col-0 *A. thaliana* with *A. brassicicola.* They showed that almost half of the detected compounds were affected by the inoculation. Some of these compounds, e.g. ascorbate, affect disease severity when applied to *Alternaria* in following experiments. *Verticillium longisporium* is one of the most devastating diseases in oilseed crops of Brassicaceae and several authors have evaluated the changes in the metabolome under the infection of this fungus in *A. thaliana* (Floerl et al., 2012; Konig et al., 2014), by LC-QTOF or NMR. Several identified compounds (phytohormones, oxylipins, aminoacides, synapates, and lignans) and hundreds of potential biomarkers have been shown to be affected by infection of *V. longisporium.* The use of transformed *A. thaliana* to reduce the content of anti-nutritive compounds has led to discover of other metabolites implicated in disease response. In this way, Huang et al. (2009) transformed *A. thaliana* to reduce the content in sinapine. This transformation altered the metabolome resulting in changes of several sinapoyl derivatives, quercetin, salicylic acid and indolyl glucosinolates. The transformed plants were showed to be more susceptible to fungal infection indicating that these metabolites may have a role in the defense of *Brassicaceae* plants.

As showed above, few researches have been done to evaluate the role of the metabolome in the defense of *Brassica* diseases. From the data analyzed, it is clear that compounds from the phenylpropanoid route have an implication in the defense to several diseases but changes could occur in hundreds of metabolites and the response is plant and pathogen specific. Metabolomics is the newest 'omic' discipline and it has a broad field to develop. One of the key points will be the creation and improvement of compound databases with MS and MS/MS information to identify relevant metabolic markers.

Conclusion

Brassica crops are susceptible to infections produced by fungus, bacteria or virus. These pathogens affect seriously the productivity and quality of these crops which result in substantial economic losses every year worldwide. For this reason, during the last decades it has been paid a great research emphasis on the plant host responses and the mechanisms underlying the resistance. Nevertheless, plant-pathogen interactions are highly complex since multiple pathogen factors and plant-signaling events take place, which ultimately define the susceptibility or resistance of the plant exposed to the pathogen. The recent advances in 'omic' approaches including gene expression analysis as well as protein and metabolite quantification enable genome-scale capturing of complex biological processes at the molecular level in plant diseases. This opens up new possibilities for understanding the molecular complexity of plant-pathogen systems and thus gain a better understanding of the molecular mechanisms implicated in basal and specific plant defense responses against a particular infection. To date, different 'omic' tools have been employed to understand how brassicas respond to biotic stress conditions. In this regard, genomics and transcriptomics have progressed as expected along with the recent availabiliy of high-throughput sequence data generated from *Brassica* spp. but the other major 'omic' branches like proteomics and metabolomics are still lagging behind. In addition, the integration of 'omic'-scale information to address complex genetics and physiological questions is still a challenge. Despite these limitations, the detailed global comparison of *Brassica* vegetables responses under pathogen attack using 'omics' has allowed the identification of metabolic pathways, novel genes and proteins whose biological role warrants in-depth biochemical and cellular elucidation of resistance mechanism. Finally, a molecular-level understanding of biotic stress responses may identify promising novel targets for the development of *Brassica* cultivars with improved disease resistance.

References

Abdel-Farid, I.B., Jahangir, M., van den Hondel, C.A.M.J.J., Kim, H.K., Choi, Y.H., and Verpoorte, R. (2009). Fungal infection-induced metabolites in *Brassica rapa*. Plant Sci *176*, 608-615.

Ahn, Y.-K., Cho, Y.-I., Tripathi, S., Kim, D.-S., Kim, J.-H., and Lee, H.-E. (2014). Newly developed SNP markers related to genes for leaf morphological traits and disease resistance in Chinese cabbage (*Brassica rapa L. ssp. pekinensis*). Plant Genetic Resources-Characterization and Utilization *12*, 62-64.

Botanga, C.J., Bethke, G., Chen, Z., Gallie, D.R., Fiehn, O., and Glazebrook, J. (2012). Metabolite profiling of Arabidopsis inoculated with *Alternaria brassicicola* reveals that ascorbate reduces disease severity. Mol Plant-Microbe Interact *25*, 1628-1638.

Bradshaw, J.E., Gemmell, D.J., and Wilson, R.N. (1997). Transfer of resistance to clubroot (*Plasmodiophora brassicae*) to swedes (*Brassica napus* L. var. *napobrassica* Peterm) from *B. rapa*. Ann Appl Biol *130*, 337-348.

Brown, P.O., and Botstein, D. (1999). Exploring the new world of the genome with DNA microarrays. Nature *21*, 33-37.

Cao, T., Srivastava, S., Rahman, M.H., Kav, N.N.V., Hotte, N., Deyholos, M.K., and Strelkov, S.E. (2008). Proteome-level changes in the roots of *Brassica napus* as a result of *Plasmodiophora brassicae* infection. Plant Sci *174*, 97-115.

Chalhoub, B., Denoeud, F., Liu, S., Parkin, I.A.P., Tang, H., Wang, X., Chiquet, J., Belcram, H., Tong, C., Samans, B., *et al.* (2014). Early allopolyploid evolution in the post-Neolithic *Brassica napus* oilseed genome. Science *345*, 950-953.

Cho, Y.-I., Ahn, Y.-K., Tripathi, S., Kim, J.-H., Lee, H.-E., and Kim, D.-S. (2015). Comparative analysis of disease-linked single nucleotide polymorphic markers from *Brassica rapa* for their applicability to *Brassica oleracea*. Plos One *10*, e0120163.

Chu, M., Song, T., Falk, K.C., Zhang, X., Liu, X., Chang, A., Lahlali, R., McGregor, L., Gossen, B.D., Yu, F., *et al.* (2014). Fine mapping of Rcr1 and analyses of its effect on transcriptome patterns during infection by *Plasmodiophora brassicae*. BMC Genomics *15*, 1166-1185.

Dalton-Morgan, J., Hayward, A., Alamery, S., Tollenaere, R., Mason, A.S., Campbell, E., Patel, D., Lorenc, M.T., Yi, B., Long, Y., *et al.* (2014). A high-throughput SNP array in the amphidiploid species *Brassica napus* shows diversity in resistance genes. Functional and Integrative Genomics *14*, 643-655.

Diederichsen, E., Frauen, M., Linders, E.G.A., Hatakeyama, K., and Hirai, M. (2009). Status and perspectives of clubroot resistance breeding in crucifer crops. J Plant Growth Regul *28*, 265-281.

Eckardt, N.A. (2012). A MicroRNA Cascade in Plant Defense. Plant Cell *24*, 840-840.

Edwards, D., Batley, J., and Snowdon, R.J. (2013). Accessing complex crop genomes with next-generation sequencing. Theor Appl Genet *126*, 1-11.

Floerl, S., Majcherczyk, A., Possienke, M., Feussner, K., Tappe, H., Gatz, C., Feussner, I., Kües, U., and Polle, A. (2012). *Verticillium longisporum* infection affects the leaf apoplastic proteome, metabolome, and cell wall properties in *Arabidopsis thaliana*. Plos One *7*, e31435.

Fopa Fomeju, B., Falentin, C., Lassalle, G., Manzanares-Dauleux, M.J., and Delourme, R. (2014). Homoeologous duplicated regions are involved in quantitative resistance of *Brassica napus* to stem canker. BMC Genomics *15*, 498-498.

Hayward, A., McLanders, J., Campbell, E., Edwards, D., and Batley, J. (2012). Genomic advances will herald new insights into the Brassica: *Leptosphaeria maculans* pathosystem. Plant Biol *14*, 1-10.

Hill, C.B., and Roessner, U. (2013). Metabolic profiling of plants by GC-MS. In The handbook of plant metabolomics, W. Weckwerth, and G. Kahl, eds. (Wiley-Blackwell), p. 448.

Hirai, M. (2006). Genetic analysis of clubroot resistance in *Brassica* crops. Breed Sci *56*, 223-229.

Hu, T.T., Pattyn, P., Bakker, E.G., Cao, J., Cheng, J.-F., Clark, R.M., Fahlgren, N., Fawcett, J.A., Grimwood, J., Gundlach, H., *et al.* (2011). The *Arabidopsis lyrata* genome sequence and the basis of rapid genome size change. Nat Genet *43*, 476-481.

Huang, J., Bhinu, V.S., Li, X., Dallal Bashi, Z., Zhou, R., and Hannoufa, A. (2009). Pleiotropic changes in Arabidopsis f5h and sct mutants revealed by large-scale gene expression and metabolite analysis. Planta *230*, 1057-1069.

Hunter, P.J., Jones, J.E., and Walsh, J.A. (2002). Involvement of beet western yellows virus, cauliflower mosaic virus, and turnip mosaic virus in internal disorders of stored white cabbage. Phytopathology *92*, 816-826.

Jahangir, M., Kim, H.K., Choi, Y.H., and Verpoorte, R. (2008). Metabolomic response of *Brassica rapa* submitted to pre-harvest bacterial contamination. Food Chem *107*, 362-368.

Jung, Y.-J. (2013). Transcriptional profiling of soft-rot resistant transgenic Chinese cabbage (*Brassica rapa* L.) constitutively overexpressing a human cathelicidin antimicrobial peptide (hCAP18/LL-37). Plant Breed Biotech *1*, 80-90.

Kaido, M., Ishikawa, T., and Hori, H. (2007). Comparative proteomic analysis of the resistant response in *Brassica rapa* root culture to the clubroot disease agent *Plasmodiophora brassicae*. Bull Facul Agric Niigata Univ *60*, 67-71.

Kalischuk, M.L., Johnson, D., and Kawchuk, L.M. (2015). Priming with a double-stranded DNA virus alters *Brassica rapa* seed architecture and facilitates a defense response. Gene *557*, 130-137.

Kaul, S., Koo, H.L., Jenkins, J., Rizzo, M., Rooney, T., Tallon, L.J., Feldblyum, T., Nierman, W., Benito, M.I., Lin, X.Y., *et al.* (2000). Analysis of the genome sequence of the flowering plant *Arabidopsis thaliana*. Nature *408*, 796-815.

Kim, Y.-W., Jung, H.-J., Park, J.-I., Hur, Y., and Nou, I.-S. (2015). Response of NBS encoding resistance genes linked to both heat and fungal stress in *Brassica oleracea*. Plant Physiol Biochem *86*, 130-136.

Kodzius, R., Kojima, M., Nishiyori, H., Nakamura, M., Fukuda, S., Tagami, M., Sasaki, D., Imamura, K., Kai, C., Harbers, M., *et al.* (2006). CAGE: cap analysis of gene expression. Nat Meth *3*, 211-222.

Konig, S., Feussner, K., Kaever, A., Landesfeind, M., Thurow, C., Karlovsky, P., Gatz, C., Polle, A., and Feussner, I. (2014). Soluble phenylpropanoids are involved in the defense response of Arabidopsis against *Verticillium longisporum*. New Phytol *202*, 823-837.

Lee, J., Izzah, N.K., Jayakodi, M., Perumal, S., Joh, H.J., Lee, H.J., Lee, S.-C., Park, J.Y., Yang, K.-W., Nou, I.-S., *et al.* (2015). Genome-wide SNP identification and QTL mapping for black rot resistance in cabbage. BMC Plant Biol *15*, 32-43.

Lee, Y.H., and Hong, J.K. (2015). Differential defence responses of susceptible and resistant kimchi cabbage cultivars to anthracnose, black spot and black rot diseases. Plant Pathol *64*, 406-415.

Lema, M., Velasco, P., Soengas, P., Francisco, M., and Cartea, M.E. (2012). Screening for resistance to black rot in *Brassica oleracea* crops. Plant breeding *131*, 607-613.

Li, Y., Zhang, C., Song, X., Lv, S., and Hou, X. (2014). Using an Arabidopsis oligonucleotide microarray to analyze the secretory pathway genes' Response to TuMV infection in Brassica rapa. Physiol Mol Plant Pathol *88*, 85-93.

Liu, R., Zhao, J., Xiao, Y., and Meng, J. (2005). Identification of prior candidate genes for Sclerotinia local resistance in *Brassica napus* using *Arabidopsis* cDNA microarray and *Brassica-Arabidopsis* comparative mapping. Science in China Series C: Life Sciences *48*, 460-470.

Liu, S., Liu, Y., Yang, X., Tong, C., Edwards, D., Parkin, I.A.P., Zhao, M., Ma, J., Yu, J., Huang, S., *et al.* (2014). The *Brassica oleracea* genome reveals the asymmetrical evolution of polyploid genomes. Nature Communications *5*.

Liu, Z., Kong, L., Zhang, M., Lv, Y., Liu, Y., Zou, M., Lu, G., Cao, J., and Yu, X. (2013). Genome-wide identification, phylogeny, evolution and expression patterns of AP2/ERF genes and cytokinin response factors in *Brassica rapa* ssp pekinensis. PloS One *8*, e83444.

Love, C.G., Graham, N.S., S, O.L., Bowen, H.C., May, S.T., White, P.J., Broadley, M.R., Hammond, J.P., and King, G.J. (2010). A *Brassica* exon array for whole-transcript gene expression profiling. PLoS One *5*, e12812.

Lowe, R.G., Cassin, A., Grandaubert, J., Clark, B.L., Van de Wouw, A.P., Rouxel, T., and Howlett, B.J. (2014). Genomes and transcriptomes of partners in plant-fungal-interactions between canola (*Brassica napus*) and two *Leptosphaeria species*. PLoS One *9*, e103098.

Lv, H.-h., Yang, L.-m., Kang, J.-g., Wang, Q.-b., Wang, X.-w., Fang, Z.-y., Liu, Y.-m., Zhuang, M., Zhang, Y.-y., Lin, Y., *et al.* (2013). Development of InDel markers linked to *Fusarium* wilt resistance in cabbage. Mol Breed *32*, 961-967.

Lv, H., Fang, Z., Yang, L., Zhang, Y., Wang, Q., Liu, Y., Zhuang, M., Yang, Y., Xie, B., Liu, B., *et al.* (2014).

Mapping and analysis of a novel candidate *Fusarium* wilt resistance gene FOC1 in *Brassica oleracea*. BMC Genomics *15*, 1094.

Lloyd, S.R., Schoonbeek, H.J., Trick, M., Zipfel, C., and Ridout, C.J. (2014). Methods to study PAMP-triggered immunity in *Brassica* species. Mol Plant-Microbe Interact *27*, 286-295.

Ma, Y., Gentry, T., Hu, P., Pierson, E., Gu, M., and Yin, S. (2015). Impact of *Brassicaceous* seed meals on the composition of the soil fungal community and the incidence of *Fusarium* wilt on chili pepper. Appl Soil Ecol *90*, 41-48.

Marra, R., Li, H., Barbetti, M.J., Sivasithamparam, K., Vinale, F., Cavallo, P., and Lorito, M. (2010). Proteomic analysis of the interaction between *Brassica napus* cv. Surpass 400 and virulent or avirulent isolates of *Leptosphaeria maculans*. J Plant Pathol *92*, 89-101.

Matsumura, H., Yoshida, K., Luo, S., Kimura, E., Fujibe, T., Albertyn, Z., Barrero, R.A., Krüger, D.H., Kahl, G., Schroth, G.P., *et al.* (2010). High-throughput superSAGE for digital gene expression analysis of multiple samples using next generation sequencing. PLoS ONE *5*, e12010.

McSorley, R., and Frederick, J.J. (1995). Responses of some common *Cruciferae* to root-knot nematodes. J Nematol *27*, 550-554.

Meyers, B.C., Kozik, A., Griego, A., Kuang, H.H., and Michelmore, R.W. (2003). Genome-wide analysis of NBS-LRR-encoding genes in *Arabidopsis*. Plant Cell *15*, 809-834.

Mun, J.-H., Yu, H.-J., Park, S., and Park, B.-S. (2009). Genome-wide identification of NBS-encoding resistance genes in *Brassica rapa*. Mol Genet Genomics *282*, 617-631.

Narusaka, M., Abe, H., Kobayashi, M., Kubo, Y., and Narusaka, Y. (2006). Comparative analysis of expression profiles of counterpart gene sets between *Brassica rapa* and *Arabidopsis thaliana* during fungal pathogen *Colletotrichum higginsianum* infection. Plant Biotechnol *23*, 503-508.

Park, S., Yu, H.-J., Mun, J.-H., and Lee, S.-C. (2010). Genome-wide discovery of DNA polymorphism in *Brassica rapa*. Mol Genet Genomics *283*, 135-145.

Parkin, I.A.P., Koh, C., Tang, H., Robinson, S.J., Kagale, S., Clarke, W.E., Town, C.D., Nixon, J., Krishnakumar, V., Bidwell, S.L., *et al.* (2014). Transcriptome and methylome profiling reveals relics of genome dominance in the mesopolyploid *Brassica oleracea*. Genome Biology *15*, R77.

Pedras, M.S., and Zheng, Q.A. (2010). Metabolic responses of *Thellungiella halophila/salsuginea* to biotic and abiotic stresses: metabolite profiles and quantitative analyses. Phytochemistry *71*, 581-589.

Piao, Z., Ramchiary, N., and Lim, Y.P. (2009). Genetics of clubroot resistance in brassica species. J Plant Growth Regul *28*, 252-264.

Rabilloud, T., Chevallet, M., Luche, S., and Lelong, C. (2010). Two-dimensional gel electrophoresis in proteomics: Past, present and future. J Proteomics *73*, 2064-2077.

Raybould, A.F., Maskell, L.C., Edwards, M.L., Cooper, J.I., and Gray, A.J. (1999). The prevalence and spatial distribution of viruses in natural populations of *Brassica oleracea*. New Phytol *141*, 265-275.

Rennie, D.C., Holtz, M.D., Turkington, T.K., Leboldus, J.M., Hwang, S.-F., Howard, R.J., and Strelkov, S.E. (2015). Movement of *Plasmodiophora brassicae* resting spores in windblown dust. Can J Plant Pathol *37*, 188-196.

Saharan, G.S., and Mehta, N. (2008). Sclerotinia Diseases of Crop Plants: Biology, Ecology and Disease Management (Amsterdam: Springer).

Sardans, J., Peñuelas, J., and Rivas-Ubach, A. (2011). Ecological metabolomics: overview of current developments and future challenges. Chemoecology *21*, 34.

Sarosh, B.R., Danielsson, J., and Meijer, J. (2009). Transcript profiling of oilseed rape (*Brassica napus*) primed for biocontrol differentiate genes involved in microbial interactions with beneficial *Bacillus amyloliquefaciens* from pathogenic *Botrytis cinerea*. Plant Mol Biol *70*, 31-45.

Shen, D., Suhrkamp, I., Wang, Y., Liu, S., Menkhaus, J., Verreet, J.-A., Fan, L., and Cai, D. (2014). Identification and characterization of microRNAs in oilseed rape (*Brassica napus*) responsive to infection with the pathogenic fungus *Verticillium longisporum* using Brassica AA (*Brassica rapa*) and CC (*Brassica oleracea*) as reference genomes. New Phytol *204*, 577-594.

Shimizu, M., Fujimoto, R., Ying, H., Pu, Z.J., Ebe, Y., Kawanabe, T., Saeki, N., Taylor, J.M., Kaji, M., Dennis, E.S., *et al.* (2014). Identification of candidate genes for *Fusarium* yellows resistance in Chinese cabbage by differential expression analysis. Plant Mol Biol *85*, 247-257.

Simoh, S., Quintana, N., Kim, H.K., Choi, Y.H., and Verpoorte, R. (2009). Metabolic changes in *Agrobacterium tumefaciens*-infected *Brassica rapa*. J Plant Physiol *166*, 1005-1014.

Singh, A., Guest, D., and Copeland, L. (2015). Associations between glucosinolates, white rust, and plant defense activators in *Brassica* plants: A review. Int J Veg Sci *21*, 297-313.

Slotte, T., Hazzouri, K.M., Agren, J.A., Koenig, D., Maumus, F., Guo, Y.-L., Steige, K., Platts, A.E., Escobar, J.S., Newman, L.K., *et al.* (2013). The *Capsella rubella* genome and the genomic consequences of rapid mating system evolution. Nat Genet *45*, 831-U165.

Sotelo, T., Lema, M., Soengas, P., Cartea, M.E., and Velasco, P. (2015). *In vitro* activity of glucosinolates and their degradation products against *Brassica*-pathogenic bacteria and fungi. Appl Environ Microbiol *81*, 432-440.

Strehlow, B., de Mol, F., and Struck, C. (2014). Location and crop management affect clubroot severity in Germany. Gesunde Pflanzen *66*, 157-164.

Sun, C., Wang, L., Hu, D., Riquicho, A.R.M., Liu, T., Hou, X., and Li, Y. (2014). Proteomic analysis of non-heading Chinese cabbage infected with *Hyaloperonospora parasitica*. J Proteomics *98*, 15-30.

Sun, C.B. (2014). Characterization of the small RNA transcriptome in plant-microbe (*Brassica/Erwinia*) interactions by high-throughput sequencing. Biotechnol Lett *36*, 371-381.

Suwabe, K., Suzuki, G., Nunome, T., Hatakeyama, K., Mukai, Y., Fukuoka, H., and Matsumoto, S. (2012).

Microstructure of a *Brassica rapa* genome segment homoeologous to the resistance gene cluster on *Arabidopsis* chromosome 4. Breed Sci *62*, 170-177.

Tollenaere, R., Hayward, A., Dalton-Morgan, J., Campbell, E., Lee, J.R.M., Lorenc, M.T., Manoli, S., Stiller, J., Raman, R., Raman, H., *et al.* (2012). Identification and characterization of candidate Rlm4 blackleg resistance genes in *Brassica napus* using next-generation sequencing. Plant Biotechnol J *10*, 709-715.

Trick, M., Cheung, F., Drou, N., Fraser, F., Lobenhofer, E.K., Hurban, P., Magusin, A., Town, C.D., and Bancroft, I. (2009). A newly-developed community microarray resource for transcriptome profiling in *Brassica* species enables the confirmation of *Brassica*-specific expressed sequences. BMC Plant Biol *9*, 50.

Velasco, P., Lema, M., Francisco, M., Soengas, P., and Elena Cartea, M. (2013). In vivo and *in vitro* effects of secondary metabolites against *Xanthomonas campestris* pv. *campestris*. Molecules *18*, 11131-11143.

Verma, S.S., Rahman, M.H., Deyholos, M.K., Basu, U., and Kav, N.N. (2014). Differential expression of miRNAs in *Brassica napus* root following infection with *Plasmodiophora brassicae*. PLoS One *9*, e86648.

Verpoorte, R., Choi, Y.H., and Kim, H.K. (2007). NMR-based metabolomics at work in phytochemistry. Phytochem Rev *6*, 3-14.

Vicente, J.G., Conway, J., Roberts, S.J., and Taylor, J.D. (2001). Identification and origin of *Xanthomonas campestris* pv. *campestris* races and related pathovars. Phytopathology *91*, 492-499.

Vicente, J.G., Everett, B., and Roberts, S.J. (2006). Identification of isolates that cause a leaf spot disease of *Brassicas* as *Xanthomonas campestris* pv. *raphaniand* pathogenic and genetic comparison with related pathovars. Phytopathology *96*, 735-745.

Vicente, J.G., and Holub, E.B. (2013). *Xanthomonas campestris* pv. *campestris* (cause of black rot of crucifers) in the genomic era is still a worldwide threat to brassica crops. Mol Plant Pathol *14*, 2-18.

Wang, X., Wang, H., Wang, J., Sun, R., Wu, J., Liu, S., Bai, Y., Mun, J.-H., Bancroft, I., Cheng, F., *et al.* (2011). The genome of the mesopolyploid crop species *Brassica rapa*. Nat Genet *43*, 1035-U1157.

Wang, Y., Wu, H., Wei, G., Zhang, H., Zhang, Y., and Gao, X. (2015). Transcriptome profiling of transgenic rapeseed revealed genes related to the defense responses and signaling pathways. Crop Sci *55*, 800-810.

Witzel, K., Neugart, S., Ruppel, S., Schreiner, M., Wiesner, M., and Baldermann, S. (2015). Recent progress in the use of 'omics' technologies in brassicaceous vegetables. Front Plant Sci *6*, 244.

Yamamoto, M., Wakatsuki, T., Hada, A., and Ryo, A. (2001). Use of serial analysis of gene expression (SAGE) technology. J Immunol Methods *250*, 45-66.

Yang, B., Srivastava, S., Deyholos, M.K., and Kav, N.N.V. (2007). Transcriptional profiling of canola (*Brassica napus* L.) responses to the fungal pathogen *Sclerotinia sclerotiorum*. Plant Sci *173*, 156-171.

Yang, R., Jarvis, D.E., Chen, H., Beilstein, M.A., Grimwood, J., Jenkins, J., Shu, S.Q., Prochnik, S., Xin, M., Ma, C., *et al.* (2013). The reference genome of the halophytic plant *Eutrema salsugineum*. Front Plant Sci *4*, 46.

Yin, H., Li, S., Zhao, X., Du, Y., and Ma, X. (2006). cDNA microarray analysis of gene expression in *Brassica napus* treated with oligochitosan elicitor. Plant Physiol Biochem *44*, 910-916.

Yoshikawa, H. (1981). Breeding for clubroot resistance in Chinese cabbage. In Chinese Cabbage, N.S. Taleker, T.D. Griggs, and T.W. Shanhua, eds. (Taiwan), pp. 405-413.

Yu, J., Tehrim, S., Zhang, F., Tong, C., Huang, J., Cheng, X., Dong, C., Zhou, Y., Qin, R., Hua, W., *et al.* (2014). Genome-wide comparative analysis of NBS-encoding genes between *Brassica* species and *Arabidopsis thaliana*. BMC Genomics *15*, 3-3.

Zhao, J., Buchwaldt, L., Rimmer, S.R., Sharpe, A., McGregor, L., Bekkaoui, D., and Hegedus, D. (2009). Patterns of differential gene expression in *Brassica napus* cultivars infected with *Sclerotinia sclerotiorum*. Mol Plant Pathol *10*, 635-649.

Zhao, J., Wang, J., An, L., Doerge, R.W., Chen, Z.J., Grau, C., Meng, J., and Osborn, T. (2007). Analysis of gene expression profiles in response to *Sclerotinia sclerotiorum* in *Brassica napus*. Planta *227*, 13-24.

Other Publications of Interest

Xu, J.-P. (2014). Next-generation Sequencing: Current Technologies and Applications. Caister Academic Press, U.K. ISBN: 978-1-908230-33-1

He, Z. (2014). Microarrays: Current Technology, Innovations and Applications. Caister Academic Press, U.K. ISBN: 978-1-908230-49-2

Poptsova, M.S. (2014). Genome Analysis: Current Procedures and Applications. Caister Academic Press, U.K. ISBN: 978-1-908230-29-4

Fuentes, M. and LaBaer, J. (2014). Proteomics: Targeted Technology, Innovations and Applications. Caister Academic Press, U.K. ISBN: 978-1-908230-46-1

Murillo, J., Vinatzer, B.A., Jackson, R.W. and Arnold, D.L. (2015). Bacteria-Plant Interactions: Advanced Research and Future Trends. Caister Academic Press, U.K. ISBN: 978-1-908230-58-4

Caranta, C., Aranda, M.A., Tepfer, M. and Lopez-Moya, J.J. (2011). Recent Advances in Plant Virology. Caister Academic Press, U.K. ISBN: 978-1-904455-75-2

Nannipieri, P., Pietramellara, G. and Renella, G. (2014). Omics in Soil Science. Caister Academic Press, U.K. ISBN: 978-1-908230-32-4

Herold, K.E. and Rasooly, A. (2009). Lab-on-a-Chip Technology: Fabrication and Microfluidics. Caister Academic Press, U.K. ISBN: 978-1-904455-46-2

Rice Responses and Resistance to Planthopper-Borne Viruses at Transcriptomic and Proteomic Levels

Feng Cui§, Wan Zhao§, Lan Luo and Le Kang*

State Key Laboratory of Integrated Management of Pest Insects and Rodents, Institute of Zoology, Chinese Academy of Sciences, Beijing, China
§Contributed equally to this work
*Corresponding author. Email: lkang@ioz.ac.cn

Abstract
Rice (*Oryza sativa*) is one of the most important cereal crops in the world, especially in Asian areas. Rice virus diseases are considered as the most serious threat to rice yields. Most rice viruses are transmitted by hemipteran insects such as planthoppers and leafhoppers. In Asia five rice viruses are transmitted mainly by three planthopper species in a persistent manner: Rice stripe virus, Rice black-streaked dwarf virus, Rice ragged stunt virus, Rice grassy stunt virus, and Southern rice black-streaked dwarf virus. In rice antivirus studies, several individual genes have been shown to function in rice resistance to viruses. Since plant responses to viral infection are complex, system-level omic studies are required to fully understand the responses. Recently more and more omic studies have appeared in the literatures on relationships between planthoppers and viruses, employing microarray, RNA-Seq, small RNA deep sequencing, degradome sequencing, and proteomic analysis. In this paper, we review the current knowledge and progress of omic studies in rice plant responses and resistance to four planthopper-borned viruses. We also discuss progress in the omic study of the interactions of planthoppers and rice viruses. Future research directions and translational applications of fundamental knowledge of virus-vector-rice interactions are proposed.

Introduction
Rice (*Oryza sativa*) is one of the most important cereal crops in the world, second only to corn in worldwide crop production. Rice crop is planted in almost all Asian countries, in most of South and Central America, and in some of central and eastern Africa. More than 80% of the world's rice is cultivated in South, East and South-East Asian regions (Welch et al., 2010). The average yield of unhulled rice in Mainland China is 130 million metric tons (MMt), which account for about 36% of the world production. Other major rice production areas are located in India (95 MMt), Indonesia (37 MMt), and Bangladesh (30 MMt). Therefore, any disease that reduces rice yields significantly affects the well-being of the Asian population and beyond. Rice virus disease is known as rice cancer and causes a serious threat to rice production in South-East Asian countries (Ou, 1985). For example, from 2005 to 2006, more than 485,000 ha of rice plantation area in southern Vietnam were severely affected by viral diseases, resulting in the loss of 828,000 tons of rice valued at US $120 million (Du et al., 2007). To ensure global food supplies for meeting the challenge of continuous population growth, controlling the outbreaks of various rice viruses is crucial.

More than 30 viruses are reported to infect rice (Abo and Sy, 1998). Most rice viruses are transmitted by hemipteran insects (Huo et al., 2014), which are efficient vectors because they are able to transmit viruses into specific plant tissues using their piercing-sucking mouthparts. In Asia, eight rice viruses are transmitted by planthoppers or leafhoppers in a persistent manner: Rice stripe virus (RSV, *Tenuivirus*), Rice dwarf virus (RDV, Reoviridae), Rice gall dwarf virus (RGDV, Reoviridae), Rice ragged stunt virus (RRSV, Reoviridae), Rice yellow stunt virus (RYSV, Rhabdoviridae), Rice grassy stunt virus (RGSV, *Tenuivirus*), Rice black-streaked dwarf virus (RBSDV, Reoviridae), and Southern rice black-streaked dwarf virus (SRBSDV, Reoviridae) (Uehara-Ichiki et al., 2013). RSV and RBSDV are transmitted by the small brown planthopper *Laodelphax striatellus* (Fallen). RRSV and RGSV are transmitted by the brown planthopper *Nilaparvata lugens* (Stål). SRBSDV is transmitted by the whitebacked planthopper *Sogatella furcifera* (Horvath) (Hibino, 1996). In America, Rice hoja blanca virus (RHBV, *Tenuivirus*) is the most important rice virus and transmitted by the planthopper *Sogatodes orizicola* (Muir) in a persistent manner (Hibino, 1996). In Africa, Rice stripe necrosis virus (RSNV, *Benyvirus*) and Rice yellow mottle virus (RYMV, *Sobemovirus*) are transmitted by the soil-inhabiting fungal pathogen *Polymyxa graminis* and leaf beetles, respectively (Uehara-Ichiki et al., 2013).

Plants have developed elaborate and effective defense mechanisms to resist infection by viruses, including hypersensitive and necrotic resistance, systemic necrosis response, systemic acquired resistance, R gene-mediated response, ubiquitin proteasome system-mediated resistance, and RNA silencing (Soosaar et al., 2005; Mandadi and Scholthof, 2013; Palukaitis and Carr, 2008). More significant advances in the understanding of plant innate immunity resistance to viruses have been gained mainly from the model systems of dicotyledonous plants, such as tobacco *Nicotiana benthamiana*, *Arabidopsis thaliana*, and tomato *Solanum lycopersicum* (Mandadi and Scholthof, 2013). However, quite few studies focus on viruses that infect monocot plants such as rice and gramineous plants. Several individual genes have been shown to function in rice resistance to viruses. A recessive resistance gene, eukaryotic translation initiation factor eIF(iso)4G1, confers resistance against Rice yellow mottle virus (RYMV) (Albar et al., 2006) by a mutation disrupting its interaction with the viral genome-linked protein (VPg) (Hébrard et al., 2010). The defense genes *PR1a* (pathogenesis-related class 1a), *PAL* (phenylalanine ammonia-lyase), *CHS* (chalcone synthase), and *Gns1* (1,3; 1,4-β-glucanase) play roles in the defense responses to RSV (Hao et al., 2011). Recently a resistant allele of sulfotransferase *STV11* (*STV11-R*) is extensively proven to confer durable RSV resistance in rice (Wang et al., 2014). *STV11-R* encodes a sulfotransferase catalysing the conversion of salicylic acid (SA) into sulphonated SA, whereas the gene product encoded by the susceptible allele *STV11-S* loses this activity.

Since plant responses to viral infection are complex and relate to many kinds of physiological processes, system-level omic studies are required to fully understand the responses. Recently more and more omic studies appear in rice antivirus realm, employing microarray, next-generation deep-sequencing techniques such as RNA-Seq, small RNA deep sequencing, degradome sequencing, and proteomic analysis. In this paper, we review important progress in omic studies in rice responses and resistance to planthopper-borned viruses and the interaction between insect vectors and viruses.

Rice molecular responses to RSV

Rice stripe virus (RSV), a member of the genus *Tenuivirus*, causes rice stripe disease. The typical symptoms of this disease are the formation of stripes on leaves and necrosis of newly emerged leaf blades, eventually leading to complete stunt of rice growth (Takahashi et al., 1991). RSV has a thin, filamentous shape and no envelope. The genome consists of four single-stranded RNA segments, and the average length of each particle containing RNA1-4 is 510, 610, 840, and 2,110 nm, respectively, after the separation of RNAs by sucrose density gradient centrifugation (Ishikawa et al., 1989). RNA1 is negative-sense and encodes a 337-kDa protein referred to as the RNA-dependent RNA polymerase (RdRp). RNAs 2-4 are ambisense and encode NS2, NSvc2 (putative membrane glycoprotein), NS3 (gene silencing suppressor), CP (nucleocapsid protein), SP (nonstructural disease-specific protein), and NSvc4 (movement protein) (Kakutani et al., 1990, 1991; Zhu et al., 1992; Takahashi et al., 1993; Toriyama et al., 1994; Xiong et al., 2008, 2009).

Several genome-wide transcriptome analyses of rice in response to RSV infection have been performed. A 60-mer custom oligomicroarray containing 43,494 oligo probes representing 59,000 rice gene loci was used to examine time-course (3, 6, 9, and 12 d after infection) transcriptome changes in RSV-infected rice plants (Satoh et al., 2010). The results revealed that 9,788 genes were differentially expressed between RSV-infected and mock-inoculated plants at least at one of the time points. RSV infection led to the activation of protein-synthesis machineries and energy production in the mitochondrion while energy production in the chloroplast and synthesis of cell-structure components were suppressed. The transcription of genes related to host-defence systems under hormone signals and those for gene silencing were activated at the later infection phase (Satoh et al., 2010).

A systemic gene analysis method, combining gene expression profiles and known quantitative trait loci (QTL) information, was developed to identify putative rice genes associated with RSV infection and resistance (Kim et al., 2012). Using an Agilent-012106 Rice Oligo Microarray G4138A, 237 genes showed differential expression between the RSV-resistant and RSV-susceptible near-isogenic lines, of which 11 genes were selected based on known QTL. Among 417 genes showing differential expression between RSV-infected and uninfected plants, 20 genes were selected as RSV resistance factor candidates. For the 31 candidates, 21 genes were further analyzed with a network model to delineate their potential associations with RSV resistance-related processes. The network revealed several functional modules: gene silencing, NB-ARC-containing protein, protein folding, translation factor, and DEAD-box RNA helicase modules. Among the 31 potential resistance factor candidates, seven genes overlapped with the RSV infection-related genes previously reported by Satoh et al. (2010). These seven genes encode a NB-ARC containing protein, heparan-α-glucosaminide N-acetyltransferase, zinc knuckle domain containing protein, PMR5 homolog, translation initiation factor, and two DEAD-box RNA helicases (Kim et al., 2012).

Next-generation deep-sequencing techniques such as RNA-Seq have been applied to investigate the alteration in gene expression of rice in response to RSV infection. By *de novo* transcriptome assembling, 648 and 937 genes were differentially expressed from the disease-resistant and the susceptible varieties in early interaction with RSV (2 d after inoculation by RSV-borne *L. striatellus*), respectively (Zheng et al., 2013). Among these differently expressed genes, 12 peroxidase biosynthesis genes, 4 glycine-rich cell wall structural protein genes, and a cellulose synthase gene were upregulated, probably resulting in lignification and strengthening of cell walls. Moreover, 12 LRR receptor-like protein kinase genes and 6 genes coding pathogenesis-related proteins that were all related to disease resistance, signal transduction and the hypersensitive necrosis reaction were downregulated.

Deep-sequencing techniques are also used to characterize the small RNA profiles of rice plants infected by RSV. RNA silencing, mediated by small RNAs including microRNAs (miRNAs) and small interfering RNAs (siRNAs), is a widespread antiviral or antibacterial mechanism (Ding and Voinnet, 2007; Ding, 2010; Padmanabhan, et al., 2009). Du et al. (2011) reported that a number of the known miRNAs showed significant changes in expression in RSV-infected rice plants; expressions of miR156b, miR159a1, miR164a, miR166, miR167a, miR1884b, miR393b, miR396e and miR528 were downregulated, whereas miR535, miR390 and miR171 were upregulated. The expression levels of other known miRNAs were dramatically changed due to RSV infection (Guo et al., 2015). Furthermore, RSV infection induced the expression of novel miRNAs in a phased pattern from several conserved miRNA precursors, and enhanced the accumulation of some rice miRNA*s, but not their corresponding miRNAs. These include miRNA*s for some members of the families miR160, miR166, miR167, miR171, miR396 and miRNA* of miR1318, miR1425, miR159a, miR168, miR172d, miR390, miR444b. 2 and miR528. RSV infection also induced the accumulation of phased siRNAs from a particular precursor. The expression levels of the genes of RNA silencing machinery, *OsDCLs* and *OsAGOs*, were elevated by RSV infection (Du et al., 2011).

In RSV-infected rice, novel abundant miRNAs (with a minimum of 100 sequencing reads) were identified from deep-sequenced small RNA libraries (Guo et al., 2012). Seven novel miRNAs whose precursor sequences have not previously been described were recognized, five of which were over-expressed while the other two were down-expressed in RSV-infected rice after verified with real-time PCR. Predicted targets for the seven novel miRNAs were involved in gene expression, signal transduction, energy metabolism, and even resistance to disease. In addition, 23 putative novel miRNAs that were newly produced from known miRNA precursors were also identified, three of which were down-expressed and the other 20 were probably unchanged in RSV-infected rice. However, Guo et al. (2015) reported that no specific novel miRNA species were induced by RSV infection in both wild-type or transgenic rice plants.

Rice molecular responses to RGSV

Rice grassy stunt virus (RGSV) is another member of the genus *Tenuivirus* in a thin filamentous shape. The genome of RGSV is composed of six ambisense single-stranded RNA segments (RNA1-6) containing 12 open reading frames (Ramirez, 2008). RGSV RNA1, 2, 5, and 6 correspond to the four RNA segments of RSV, but RGSV RNA1 is ambisense while RSV RNA1 is negative sense. RGSV RNA3 and 4 are unique in this genus. RGSV forms a group distinct from the other tenuiviruses in phylogeny (Ramirez, 2008). Rice plants infected with RGSV show pronounced leaf yellowing (chlorosis), stunting, and excess tillering (branching) (Shikata et al., 1980). Chlorosis and stunting are also observed in plants infected with other tenuiviruses, whereas excess tillering is a specific symptom for RGSV infection. RGSV-infected rice cells contain masses of fibrils in the nuclei and cytoplasm, and membrane-bound bodies with fibrils in the cytoplasm (Hibino, 1986).

Fewer researches have explored the gene responses of rice to RGSV infection compared to the researches on RSV infection. Only one study reported the relationship between gene responses of rice and symptoms induced by RGSV (Satoh et al., 2013). A total of 8,203 genes were differently expressed between RGSV- and mock-inoculated plants. RGSV infection suppressed both cell wall formation and cell elongation in rice through downregulating genes of cellulose synthase (-like), arabinogalactan proteins, and expansin proteins, associating with the stunting symptom. Activation of *RCN1* and *OstiII1* alone may contribute to the development of excess tillers in RGSV-infected plants. RGSV infection inhibited chlorophyll synthesis and promoted degradation of chlorophyll, accounting for leaf yellowing symptom. For plant hormone metabolism and signaling, RGSV infection inhibited the accumulation of auxin in rice, but activated the IAA signaling process. RGSV activated genes were also involved in inactivation of gibberellic acid and indole-3-acetic acid. The genes for strigolactone signaling were suppressed by RGSV.

Rice molecular responses to RBSDV

Rice black-streaked dwarf virus (RBSDV) belongs to the genus *Fijivirus* within the family Reoviridae, and can infect rice and maize, leading to rice black streaked dwarf disease and maize rough dwarf diseases, respectively (Zhang *et al.*, 2001). The plants infected with RBSDV are characterized by dwarfing, leaf dark greening, white tumors or black streaked swellings along the veins on the back of leaf blades and stems (Kuribayashi and Shinkai, 1952). The genome of RBSDV consists of 10 double-stranded RNA (dsRNA) segments (S1-S10) (Milne *et al.*, 2005). Protein P1, encoded by S1, is a putative RNA-dependent RNA polymerase. P2 and P4 are the major core structural protein and outer shell B-like protein, respectively (Zhang *et al.*, 2001). S8 and S10 encode the major inner caspid and outer caspid proteins, respectively. Apart from serving as the inner caspid protein, P8 can enter into the nucleus and play a role in the inhibition of transcription (Liu *et al.*, 2007). P6 is likely to be a silencing suppressor (Zhang *et al.*, 2005). Both P7-1 and P9-1, encoded by S7 and S9, each of which contains two open reading frames, are involved in the formation of the tubular structure and viroplasm during infection, respectively (Isogai *et al.*, 1998; Zhang *et al.*, 2009).

A global proteome change of rice after a 50-day RBSDV infection was investigated (Xu et al., 2013). Of around 1800 protein spots detected on two-dimensional electrophoresis, 72 spots showed at least a 1.5-fold significant change in abundance, among which 69 spots were successfully identified by MALDI-TOF/TOF-MS and corresponded to 45 proteins. These proteins were ascribed to 12 functional clusters, including defense and stress related proteins, photosynthesis, redox homeostasis, energy pathway, amino acid metabolism, carbohydrate metabolism, nucleotide metabolism, transcription and translation, cell wall modifications, plant hormone responses, signal transduction and virus proteins. Most of the differently expressed proteins related to defense and stress response were upregulated while most proteins related to photosynthesis were downregulated, indicating that RBSDV infection triggers defense and stress responses while impairing photosynthesis.

RBSDV-responsive miRNAs in rice leaves and roots were explored using small RNA and degradome sequencing technology (Sun et al., 2015). In this study, five miRNAs in leaf tissue were up-regulated and nine were down-regulated, while in the roots, nine miRNAs were up-regulated and seven were downregulated as a result of virus infection. Five miRNAs (miR166abcdf, miR169 hijklm, miR408, miR827 and miR1428e) responded similarly in roots and leaves, but the other miRNAs had tissue-specific responses. Three miRNAs changed significantly in the roots but not in the leaves (i.e. miR394, miR397 and miR530), while three others changed in the leaves, but not in the roots (miR162, miR398 and miR1862d). miR159ab, miR171 and miR1432 changed significantly in both tissues but in opposite directions. Thus, a differential regulation of the accumulation of rice miRNAs appears in different tissues in response to RBSDV infection. High-throughput degradome sequencing technology has been used to characterize conserved and non-conserved miRNA targets from rice (Wu et al., 2009). Through this approach, a total of 104 target transcripts for 17 conserved and 16 non-conserved miRNAs were associated with RBSDV infection. These targets included protein kinase, zinc finger proteins, nuclear transcription factors, copper amine oxidase, IAA-alanine resistance protein, E3 ubiquitin ligase, etc. These miRNA targets probably play essential regulatory roles in various aspects of plant growth and development (Sun et al., 2015).

Rice molecular responses to SRBSDV

SRBSDV is a recently identified virus and causes Southern rice black-streaked dwarf disease throughout East and Southeast Asia since the mid-2000s. SRBSDV is currently ascribed to the genus *Fijivirus* of the family Reoviridae (Wang et al., 2010a). SRBSDV resembles RBSDV in disease symptoms, viral particles, and genome structure. The virions of SRBSDV are icosahedral particles 70-75 nm in diameter (Zhou et al., 2013). The viral genome contains ten linear dsRNA segments, named S1-S10, encoding 13 proteins. S5, S7, and S9 each encode two proteins, while the other segments each encode one. S1-S4, S8, and S10 of SRBSDV encode putative structural proteins: RNA-dependent RNA polymerase, major core structural protein, capping enzyme, outer shell B-spike protein, minor core protein, and major outer capsid protein, respectively (Zhang et al., 2008; Wang et al., 2010a). P6, P5-1 and P9-1 might participate in the genesis and maturation of the viroplasms induced by fijiviruses in insect vector cells (Mao et al., 2013). P7-1 is the major component of the tubules and maybe a virus movement protein (Liu et al., 2011). P9-1 is the major component of the viroplasm and is vital for viral proliferation in insect vectors (Jia et al., 2012).

Microarray profiling was performed to identify miRNA expression regulation patterns in SRBSDV-infected rice plants across four time points (3, 9, 15, and 20 days post-inoculation, dpi) (Xu et al., 2014). Of 309 investigated rice miRNAs, 56 miRNAs responded to SRBSDV infection at least at one time point. Fourteen miRNAs belonging to six families (miR164, miR396, miR530, miR1846, miR1858, and miR2097) were significantly regulated at 20 dpi. Some members of the miR164, miR396, miR530, and miR1846 families were positively or negatively correlated with their respective targets, whereas in more cases the rice miRNAs were not in correlation with their targets along the post-inoculation period. The reported functions of the target genes of the miR164, miR396, miR530, miR1846, and miR1858 families are associated with symptom development.

In conclusion, the viruses RSV, RGSV, or RBSDV regulate gene expression of various physiological processes in rice plants (Figure 1). Genes involved in translation, protein processing, nucleotide metabolism, amino acid metabolism, energy production in mitochondria, gene silencing, pathogenesis-related activity, and stress-response are all activated. Genes involved in photosynthesis in the chloroplast, lipid metabolism, cell wall, and cytoskeleton are repressed. Hormone synthesis is also perturbed. The miRNA-mediated RNA silencing response in rice is quite different towards the infection of RBSDV and SRBSDV although the two viruses share many characters in disease symptoms and genome structure. The miRNAs or miRNA*s of the families miR159, miR160, and miR535 are upregulated while members of the families miR396 and miR164 are downregulated in the rice infected by RSV or SRBSDV. The miRNAs or miRNA*s of the families miR166 and miR396 are upregulated while members of the family miR393 are downregulated in the rice infected by RSV or RBSDV (Figure 2). The majority of miRNAs shows a different response to different viruses.

Molecular responses of insect vectors to rice viruses

Owing to excellent mobility, insects are considered as outstanding vectors for plant pathogens. Much research on the transmission mechanisms of plant viruses in insect vectors has been undertaken to explore the reasons for the specificity of some insects as vectors and what molecules are required for viral transmission (Huo et al., 2014; Huang et al., 2015). Virus transmission consists of several essential processes in insect vectors: acquisition of virus from an infected plant, retention or replication of virus at specific organs, and release of virus from salivary glands upon salivation (Andret-Link and Fuchs, 2005). Based on the retention period in insect vectors, viruses are divided into three categories: non-persistent, semi-persistent, and persistent (Nault and Ammar, 1989). Persistent viruses are further classified as persistent-circulative (mostly non-propagative) or persistent-propagative based on their replication features (Nault, 1997). RSV, RGSV, RBSDV, and SRBSDV are transmitted by planthoppers in a persistent-propagative way. None of the four viruses can be transmitted by seeds or by introducing biologically active plant-derived viruses into plant cells through wounds or abrasions on the leave surface (Ling, 1972; Abo and Sy, 1998; Wang et al., 2010b). This fact indicates that persistent-propagative viruses are extremely dependent on and share complicated interactions with insect vectors. Here we highlight recent advances in the understanding of molecular interactions between rice viruses and their planthopper vectors at omic levels.

The first omic study on the molecular response of planthoppers to rice viruses was a comparative transcriptome analysis on the RSV-borne and naïve *L. striatellus* by 454-FLX high-throughput pyrosequencing (Zhang et al., 2010). A total of 16,885 contigs and 24,607 singletons (a total of 41,492 unigenes) were generated from viruliferous and naïve *L. striatellus*. Innate immunity regulatory genes of *L. striatellus* in antivirus related cascades, such as RNA interference (RNAi), JAK/STAT, and Imd cascade genes, were retrieved. However, there was no substantial difference in the transcript abundance of putative RNAi related genes between viruliferous and naïve samples of *L. striatellus*. This is in accordance with the results in *Drosophila melanogaster*, where insect RNAi pathway genes are constitutively expressed and are not

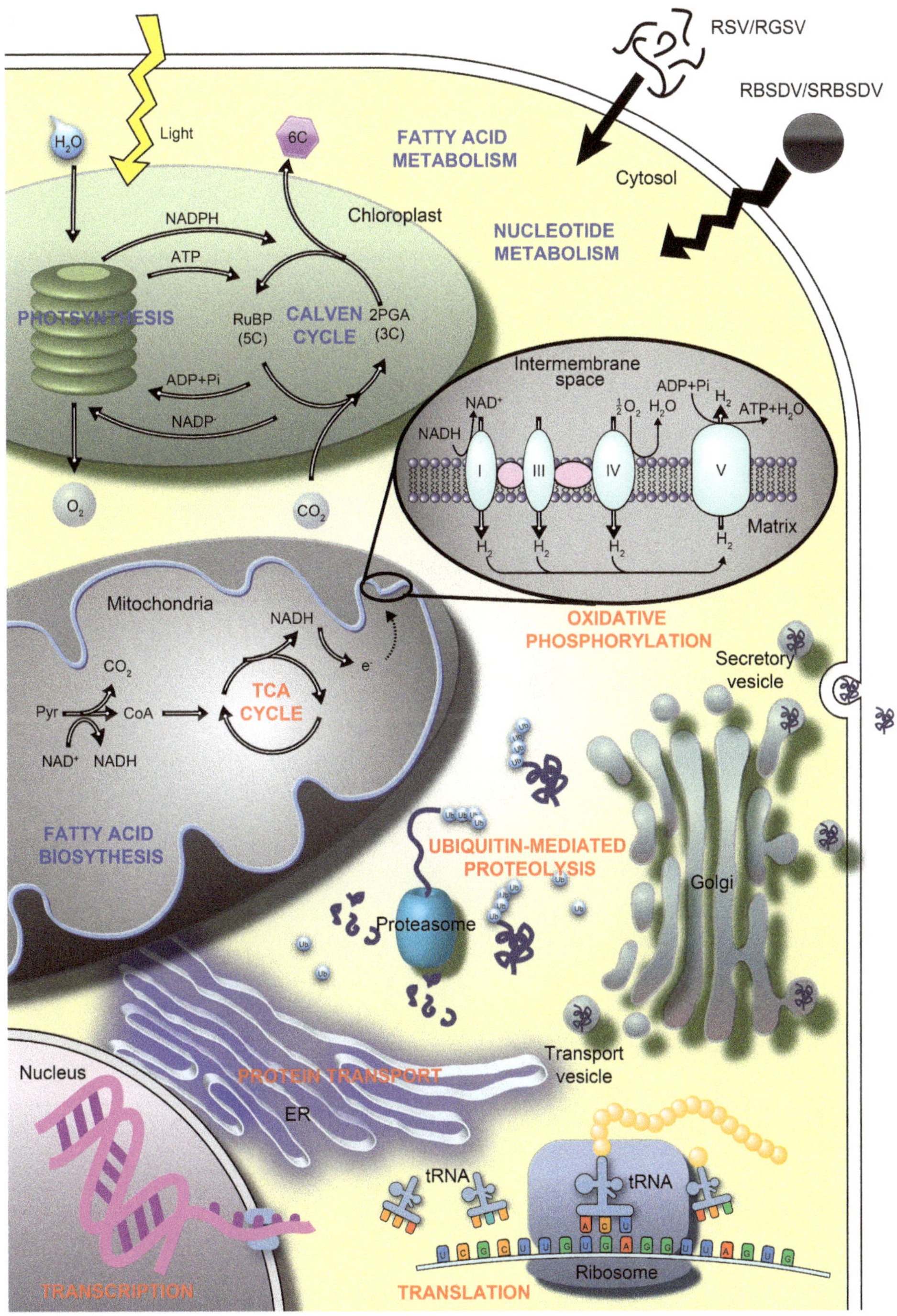

Figure 1. The physiological processes of rice plants affected by viruses RSV, RGSV, or RBSDV. The process in red is activated and the process in blue is suppressed by viruses. The information is from Satoh et al., 2010, 2013, Kim et al., 2012, and Xu et al., 2013.

upregulated by viral infection (Carpenter et al., 2009). Genes encoding vitellogenin, transferrin, ribosomal proteins L4 and L15, malic enzyme, G-protein, ATP-citrate synthase, and mitochondrial ATP synthase alpha subunit were upregulated in the viruliferous samples.

Another study on the interaction between RSV and *L. striatellus* by 454-pyrosequencing obtained a total of 2,555 contigs and 33,808 singletons (Lee et al., 2013). The gene expression pattern revealed that 453 and 428 contigs were upregulated and downregulated, respectively, in the viruliferous *L. striatellus*, Genes related to folding, sorting and degradation, and cellular process were significantly upregulated while genes related to translation and xenobiotics metabolism were downregulated when compared with those of the naïve samples. The

Figure 2. The known miRNA families of rice plants responding to the infection of RSV, RBSDV, or SRBSDV. Members of these families include miRNAs or miRNA*s. The family in red is upregulated and the family in blue is downregulated by viruses. The information is from Du et al., 2011, Guo et al., 2012, Xu et al., 2014, and Sun et al., 2015.

upregulation of genes encoding vitellogenin, ribosomal protein 1, 2, 3 and 4, and heat shock protein 1 and 2, and the downregulation of genes encoding translation initiation factor 1, 2 and 3 in the viruliferous *L. striatellus* were further confirmed by real-time quantitative PCR. These RSV-regulated genes possibly have important roles in the physiology of *L. striatellus*, transmission of RSV, and RSV and *L. striatellus* interaction.

The interaction between *S. furcifera* and SRBSDV has also been investigated by *de novo* transcriptome assembling and massive parallel pyrosequencing using the Illumina Hiseq2000 platform (Xu et al., 2012). A total of 81,388 unigenes were obtained from two transcriptomes of SRBSDV-infected and naïve *S. furcifera*, and 58% of the differentially expressed genes were upregulated in the viruliferous samples. Genes involved in protein synthesis and amino acid metabolism were downregulated by SRBSDV infection, while a large percentage of the genes involved in lipid metabolism and lipogenic compound metabolism were upregulated. SRBSDV infection perturbed the cell cycle, cytoskeleton organization, and ubiquitin-proteasome pathways. A majority of ubiquitin-proteasome pathway genes were downregulated. Furthermore, SRBSDV infection activated the immunity regulatory systems of *S. furcifera*, such as RNAi, autophagy, lysosome, melanization, and antimicrobial peptide production. Dicer 1 and Dicer 2 of *S. furcifera* were upregulated by SRBSDV infection, while no substantial difference in the expression levels of putative RNAi related genes was found in the *L. striatellus* with RSV infection (Zhang et al., 2010). These differences may be caused by the diversity among the replication cycles of the two viruses or the vector defense systems.

Future prospects

A better understanding of the molecular interactions among virus, insect vector, and rice plant will require the integration of multiple approaches and collaboration among plant pathologists, molecular biologists, and entomologists. Exploitation of natural genetic sources of resistance is one of more promising approaches available to protect rice from virus infection. Unfortunately, very few naturally-occurring rice genes have been identified that confer resistance against rice viruses.

Genetically engineering rice is a potential strategy to control viral diseases. One of the most promising methods to confer resistance against rice viruses is the use of RNAi through the expression of dsRNA, enabling us to apply this technology to silence the expression of both specific endogenous genes and genes of invading pathogens. However, not all RNAi constructs against viral RNAs are equally effective in preventing virus infection. It is important to target viral genes that play critical roles in viral infection and proliferation at an early stage of viral replication. Transgenic rice plants that harbour RNAi constructs against pC1 (RdRp), pC3 (CP) or pC4 (NSvc4, movement protein) are highly resistant to RSV infection, whereas RNAi constructs against pC2 (NSvc2, membrane glycoprotein) or p4 (SP) did not result in resistance to RSV infection. Transgenic rice plants with the p2 (NS2) or p3 (NS3, gene silencing suppressor) constructs exhibit moderate resistance against RSV infection (Shimizu et al., 2011a). However, studies from other two groups showed that SP RNAi transgenic rice showed resistance to RSV infection (Ma et al., 2011; Zhou et al., 2012). Transgenic rice plants that express dsRNAs of the RGSV genes for the nucleocapsid protein pC5 or movement protein pC6 are strongly resistant to RGSV infection and did not allow the virus proliferation (Shimizu et al., 2013). The P9-1 is a viroplasm component protein of RBSDV, playing an important role in viral proliferation. Rice plants with an RNA interference construct against the P9-1 encoding gene are completely resistant to RBSDV infection and no virus was detectable after challenge with RBSDV (Shimizu et al., 2011b).

Based on these related researches, we predict that interfering with the expression of the outer capsid protein S10, the movement protein P7-1, or the major viroplasm component P9-1 should also be effective for controlling SRBSDV infection in rice plants. In addition, the expression of the RNAi constructs against virus genes does not affect the morphology or growth of the rice plants, such as plant height, number of tillers, and rice grain yields. Thus, the use of RNAi to genetically engineer rice plants with improved resistance against RSV, RGSV, RBSDV, and SRBSDV is one of the most promising approaches to solve these virus problems.

Blocking virus transmission within insect vectors would be a new and effective strategy for control of rice virus disease outbreak in fields. The interference can be effective at any step during the virus transmission process in insect vectors, i.e. acquisition, retention or replication, and release of virus. Much work has focused on each process at molecular or cellular levels (Hayakawa et al., 1992). However, most persistent-propagative plant viruses are extremely dependent on insect vectors to become pathogenic. For example, RSV extracted from viruliferous rice cannot infect healthy rice through mechanical inoculation despite the various inoculation buffers, conditions, techniques, and inoculum sources that have been utilized (Ling, 1972). Our group recently successfully inoculated RSV into rice leaves through midrib micro-injection and found that insect-derived RSV induced a typical stripe symptom, whereas plant-derived RSV only produced chlorosis in rice leaves (unpublished data). The non-pathogenicity of plant-derived RSV probably is caused by the lack of required modifications in insect vectors. Clarification of the mechanism by which persistent plant viruses acquire pathogenicity from insect vectors will be an outstanding breakthrough in unraveling the enigma of the insect vector-dependency transmission mode, and also provides a fascinating way to disarm plant viruses before the battle with host plants. RNA-guided gene drives based on the CRISPR nuclease Cas9 would be a promising technique for spreading the virus-incompatibility trait through wild populations of vector insects (Esvelt et al., 2014). In conclusion, omic studies on interactions among plant, virus, and insect vector will provide more new cues to explore molecular mechanisms and control approaches for rice virus diseases.

Acknowledgments

This work was supported by the Major State Basic Research Development Program of China (973 Program) (No. 2014CB13840402) and the Strategic Priority Research Program of the Chinese Academy of Sciences (No. XDB11040200).

References

Abo, M.E., Sy, A.A. (1998). Rice virus diseases. J. Sustain. Agric. *11*, 113-134.

Albar, L., Bangratz-Reyser, M., Hébrard, E., Ndjiondjop, M.N., Jones, M., and Ghesquière, A. (2006). Mutations in the eIF(iso)4G translation initiation factor confer high resistance of rice to Rice yellow mottle virus. Plant J. *47*, 417-426.

Andret-Link, P., and Fuchs, M. (2005). Transmission specificity of plant viruses by vectors. Eur. J. Plant Pathol. *87*, 153-165.

Carpenter, J., Hutter, S., Baines, J.F., Roller, J., Saminadin-Peter, S.S., Parsch, J., and Jiggins, F.M. (2009). The transcriptional response of *Drosophila melanogaster* to infection with the sigma virus (*Rhabdoviridae*). PLoS One *4*, e6838.

Ding, S.W., Li, H., Lu, R., Li, F., and Li, W.X. (2004). RNA silencing: a conserved antiviral immunity of plants and animals. Virus Res. *102*, 109-115.

Ding, S.W., and Voinnet, O. (2007). Antiviral immunity directed by small RNAs. Cell *130*, 413-426.

Ding, S.W. (2010). RNA-based antiviral immunity. Nat. Rev. Immunol. *10*, 632-644.

Du, P., Wu, J.G., Zhang, J.Y., Zhao, S.Q., Zheng H, Gao, G., Wei, L.P., and Li, Y. (2011). Viral infection induces expression of novel phased microRNAs from conserved cellular microRNA precursors. PLoS Pathog. *7*, e1002176.

Du, P.V., Cabunagan, R.C., Cabauatan, P.Q., Choi, H.S., Choi, I.R., Chien, H.V., and Huan, N.H. (2007). Yellowing syndrome of rice: etiology, current status, and future challenges. Omonrice *15*, 94-101.

Esvelt, K.M., Smidler, A.L., Catteruccia, F., and Church, G.M. (2014). Concerning RNA-guided gene drives for the alteration of wild populations. eLife doi: http://dx.doi.org/10.7554/eLife.03401.

Guo, C., Li, L., Wang, X.F., and Liang, C. (2015). Alterations in siRNA and miRNA expression profiles detected by deep sequencing of transgenic rice with siRNA-mediated viral resistance. PLoS ONE *10*, e0116175.

Guo, X.X., Wu, G.T., Yan, F., Lu, Y.W., Zheng, H.Y., Lin, L., Chen, H.R., and Chen, J.P. (2012). Identification of novel *Oryza sativa* miRNAs in deep sequencing-based small RNA libraries of rice infected with rice stripe virus. PLoS ONE 7(10), e46443.

Hao, Z., Wang, L., He, Y., Liang, J., and Tao, R. (2011). Expression of defense genes and activities of antioxidant enzymes in rice resistance to Rice stripe virus and small brown planthopper. Plant Physiol. Biochem. *49*, 744-751.

Hayakawa, T., Zhu, Y., Itoh, K., Kimura, Y., Izawa, T., Shimamoto, K., and Toriyama, S. (1992). Genetically engineered rice resistant to rice stripe virus, an insect-transmitted virus. Proc. Natl. Acad. Sci. USA. *89*, 9865-9869.

Hébrard, E., Poulicard, N., Ge´rard, C., Traore´, O., Wu, H. C., Albar, L., Fargette, D., Bessin, Y., and Vignols, F. (2010). Direct interaction between the Rice yellow mottle virus (RYMV) VPg and the central domain of the rice eIF(iso)4G1 factor correlates with rice susceptibility and RYMV virulence. Mol. Plant Microbe. Interact. *23*, 1506-1513.

Hibino, H. (1986). Rice grassy stunt virus. Tropical Agriculture Research Series *19*, 165-172.

Hibino, H. (1996). Biology and epidemiology of rice viruses. Annu. Rev. Phytopathol. *34*, 249-274.

Huang, H.J., Bao, Y.Y., Lao, S.H., Huang, X.H., Ye, Y.Z., Wu, J.X., Xu, H.J., Zhou, X.P., and Zhang, C.X. (2015). Rice ragged stunt virus-induced apoptosis affects virus transmission from its insect vector, the brown planthopper to the rice plant. Sci. Rep. *5*, doi:10.1038/srep11413.

Ishikawa, K., Omura, T., and Hibino, H. (1989). Morphological characteristics of rice stripe virus. J. Gen. Virol. *70*, 3465-3468.

Isogai, M., Uyeda, I., and Lee, B. (1998). Detection and assignment of proteins encoded by rice black-streaked dwarf *fijivirus* S7, S8, S9 and S10. J. Gen. Virol. *79*, 1487-1494.

Jia, D., Chen, H., Zheng, A., Chen, Q., Liu, Q., Xie L., Wu, Z., and Wei, T. (2012). Development of an insect vector cell culture and RNA interference system to investigate the functional role of *fijivirus* replication protein. J. Virol. *86*, 5800-5807.

Kakutani, T., Hayano, Y., Hayashi, T., and Minobe, Y. (1990). Ambisense segment 4 of rice stripe virus: possible evolutionary relationship with phleboviruses and uukuviruses (*Bunyaviridae*). J. Gen. Virol. *71*, 1427-1432.

Kakutani, T., Kayano, Y., Hayashi, T., and Minobe, Y. (1991). Ambisense segment 3 of rice stripe virus: the first instance of a virus containing two ambisense segments. J. Gen. Virol. *72*, 465-468.

Kim, K., Choi, D., Kim, S.M., Kwak, D.Y., Choi, J., Lee, S., Lee, B.C., Hwang, D., and Hwang, I. (2012). A systems approach for identifying resistance factors to Rice stripe virus. Mol. Plant Microbe. Interact. *25*, 534-545.

Kuribayashi, K., and Shinkai, A. (1952). On the rice black-streaked dwarf disease, a new disease of rice in Japan. Ann. Phytopath. Soc. Japan *16*, 41.

Lee, J.H., Choi, J.Y., Tao, X.Y., Kim, J.S., Kim, W.J., Je, Y.H. (2013). Transcriptome analysis of the small brown planthopper, *Laodelphax striatellus* carrying rice stripe virus. Plant Pathol. J. *29*, 330-337.

Ling, K.C. (1972). Rice virus diseases: International Rice Research Institute.

Liu Y., Jia, D., Chen, H., Chen, Q., Xie, L., Wu, Z., and Wei, T.Y. (2011). The P7-1 protein of Southern rice black-streaked dwarf virus, a *fijivirus*, induces the formation of tubular structures in insect cells. Arch. Virol. *156*, 1729-1736.

Liu, H., Wei, C., Zhong, Y., and Li, Y. (2007). Rice black-streaked dwarf virus minor core protein P8 is a nuclear dimeric protein and represses transcription in tobacco protoplasts. FEBS Lett. *581*, 2534-2540.

Ma, J., Song, Y., Wu, B., Jiang, M., Li, K., Zhu, C., and Wen, F. (2011). Production of transgenic rice new permplasm with strong resistance against two isolations of Rice stripe virus by RNA interference. Transgenic Res. *20*, 1367-1377.

Mandadi K.K., and Scholthof K.-B.G. (2012). Characterization of a viral synergism in the monocot *Brachypodium distachyon* reveals distinctly altered host molecular processes associated with disease. Plant Physiol. *160*, 1432-1452.

Mandadi K.K., and Scholthof K.-B.G. (2013). Plant immune responses against viruses: How does a virus cause disease? Plant Cell *25*, 1489-1505.

Mao Q., Zheng S., Han Q., Chen H., Ma Y., Jia D., and Wei, T. (2013). New model for the genesis and maturation of viroplasms induced by fijiviruses in insect vector cells. J. Virol. *87*, 6819-6828.

Milne, R.G., del Vas, M., Harding, R.M., Marzachi, R., and Mertens, P.P.C. (2005). Genus *Fijivirus*. In: Virus taxonomy: eighth report of the international committee on taxonomy of viruses, Fauquet, C.M., Mayo, M.A., Maniloff, J., Desselberger, U., and Ball, L.A., eds. (London: Elsevier/Academic Press), pp. 534-542.

Nault, L.R., and Ammar, E.D. (1989). Leafhopper and planthopper transmission of plant viruses. Ann. Rev. Entomol. *34*, 503-2.

Nault, L.R. (1997). Arthropod transmission of plant viruses: A new synthesis. Ann. Entomol. Soc. Am. *90*, 522-541.

Ou, S.H. (1985). Rice Diseases, 2nd ed: Commonwealth Mycological Institute.

Padmanabhan, C., Zhang, X., and Jin, H. (2009). Host small RNAs are big contributors to plant innate immunity. Curr. Opin. Plant Biol. *12*, 465-472.

Palukaitis, P., and Carr, J.P. (2008). Plant resistance responses to viruses. J. Plant Pathol. *90*, 153-171.

Ramirez, B.C. (2008). *Tenuivirus*, in encyclopedia of virology (Third Edition), Mahy B.W.J., van Regenmortel M.H.V., eds. (Oxford, UK: Academic Press), pp. 24-27.

Satoh, K., Yoneyama, K., Kondoh, H., Shimizu, T., Sasaya, T., Choi, I.-R., Yoneyama, K, Omura, T., and Kikuchi, S. (2013). Relationship between gene responses and symptoms induced by Rice grassy stunt virus. Front. Microbiol. *4*, 313.

Satoh, K., Kondoh, H., Sasaya, T., Shimizu, T., Choi, I.R., Omura, T., and Kikuchi, S. (2010). Selective modification of rice (*Oryza sativa*) gene expression by Rice stripe virus infection. J. Gen. Virol. *91*, 294-305.

Shikata, E., Senboku, T., and Ishimizu, T. (1980). The causal agent of rice grassy stunt disease. Proc. Jpn. Acad. Series B *56*, 89-94.

Shimizu, T., Nakazono-Nagaoka, E., Akita, F., Uehara-Ichiki, T., Omura, T., and Sasaya,T. (2011b). Immunity to Rice black streaked dwarf virus, a plant reovirus, can be achieved in rice plants by RNA silencing against the gene for the viroplasm matrix protein. Virus Res. *160*, 400-403.

Shimizu, T., Nakazono-Nagaoka, E., Uehara-Ichiki, T., Sasaya, T., and Omura, T. (2011a). Targeting specific genes for RNA interference is crucial to the development of strong resistance to Rice stripe virus. Plant Biotechnol. J. *9*, 503-512.

Shimizu, T., Ogamino, T., Hiraguri, A., Nakazono-Nagaoka, E., Uehara-Ichiki, T., Nakajima, M., Akutsu, K., Omura, T., and Sasaya, T. (2013). Strong resistance against Rice grassy stunt virus is induced in transgenic rice plants expressing dsRNA of the viral genes for nucleocapsid or movement proteins as targets for RNA interference. Phytopathology *103*, 513-519.

Soosaar J.L.M., Burch-Smith, T.M., and Dinesh-Kumar, S.P. (2005). Mechanisms of plant resistence to viruses. Nat. Rev. Microbiol. *3*, 789-798.

Sun, W., Xu, X.H., Wang, Y., Lu, X.B., Sun, H.W., and X, X.Z. (2015). Genome-wide identification of microRNAs and their targets in wild type and phyB mutant provides a key link between microRNAs and the phyB-mediated light signaling pathway in rice. Front. Plant Sci. *6*, 372.

Takahashi, M., Toriyama, S., Hamamatsu, C., and Ishihama, A. (1993). Nucleotide sequence and possible ambisense coding strategy of rice stripe virus RNA segment 2. J. Gen. Virol. *74*, 769-773.

Takahashi, Y., Omura, T., Shohara, K., and Tsuchizaki, T. (1991). Comparison of four serological methods for practical detection of ten viruses of rice in plants and insects. Plant Dis. *75*, 458-461.

Toriyama, S., Takahashi, M., Sano, Y., Shimizu, T., and Ishihama, A. (1994). Nucleotide sequence of RNA1, the largest genomic segment of rice stripe virus, the prototype of the *tenuiviruses*. J. Gen. Virol. *75*, 3569-3579.

Uehara-Ichiki, T., Shiba, T., Matsukura, K., Ueno, T., Hirae, M., and Sasaya, T. (2013). Detection and diagnosis of rice-infecting viruses. Front. Microbiol. *4*, 289.

Wang, Q., Yang, J.A., Zhou, G.H., Zhang, H. M., Chen, J.P., and Adams, M.J. (2010a). The complete genome sequence of two isolates of Southern rice black-streaked dwarf virus, a new *Fijivirus*. J. Phytopathol. *158*, 733-737.

Wang, K., Zheng, J.J., Zhang, S.G., and Zhou, G.H. (2010b). Seedling test confirms Southern rice black-streaked dwarf virus cannot be transmitted via rice seed. Guangdong Agric. Sci. *37*, 95-96.

Wang, Q. Liu, Y.Q., He, J., Zheng, X.M., Hu, J.L., Liu, Y.L., Dai, H.M., Zhang, Y.X., Wang, B.X., Wu, W.X., Gao, H., Zhang, Y.H., Tao, X.R., Deng, H.F., Yuan, D.Y., Jiang, L., Zhang, X., Guo, X.P., Cheng, X.N., Wu, C.Y., Wang, H.Y., Yuan, L.P., and Wan, J.M. (2014). STV11 encodes a sulphotransferase and confers durable resistance to rice stripe virus. Nat. Commun. *5*, 4768.

Welch, J.R., Vincent, J.R, Auffhammer, M., Moyae, P.F, Dobermann, A., and Dawe, D. (2010). Rice yields in tropical/subtropical Asia exhibit large but opposing sensitivities to minimum and maximum temperatures. Proc. Natl. Acad. Sci. USA. *107*, 14562-14567.

Wu, L., Zhang, Q., Zhou, H., Ni, F., Wu, X., and Qi, Y. (2009). Rice microRNA effector complexes and targets. Plant Cell *21*, 3421-3435.

Xiong, R., Wu, J., Zhou, Y., and Zhou, X. (2008). Identification of a movement protein of the *tenuivirus* rice stripe virus. J. Virol. *82*(24), 12304-12311.

Xiong, R., Wu, J., Zhou, Y., and Zhou, X. (2009). Characterization and subcellular localization of an RNA silencing suppressor encoded by Rice stripe *tenuivirus*. Virology *387*, 29-40.

Xu, D.L., Mou, G.P., Wang, K., and Zhou, G.H. (2014). MicroRNAs responding to southern rice black-streaked dwarf virus infection and their target genes associated with symptom development in rice. Virus Res. *190*, 60-68.

Xu, Q.F., Ni, H.P., Chen, Q.Q., Sun, F., Zhou, T., and Lan, Y. (2013). Comparative proteomic analysis reveals the cross-talk between the responses induced by H_2O_2 and by long-term rice black-streaked dwarf virus infection in rice. PLoS ONE *8*, e81640.

Xu, Y., Zhou, W.W., Zhou, Y.J., Wu, J.X., and Zhou, X.P. (2012). Transcriptome and comparative gene expression analysis of *Sogatella furcifera* (Horváth) in response to southern rice black-streaked dwarf virus. PLoS One *7*, e36238.

Zhang, H.M., Yang, J., Chen, J.P., Adams, M.J. (2008). A black-streaked dwarf disease on rice in China is caused by a novel *Fijivirus*. Arch. Virol. 153 1893-1189.

Zhang, C., Liu, Y., Liu, L., Lou, Z., Zhang, H., Miao, H., Hu, X., Pang, Y., and Qiu, B. (2009). Rice black streaked dwarf virus P9-1, an a-helical protein, self-interacts and forms viroplasms *in vivo*. J. Gen. Virol. *89*, 1770-1776.

Zhang, F.J., Guo, H.Y., Zheng, H.J., Zhou, T., Zhou, Y.J., Wang, S.Y., Fang, R.X., Qian, W., and Chen, X.Y. (2010). Massively parallel pyrosequencing-based transcriptome analyses of small brown planthopper (*Laodelphax striatellus*), a vector insect transmitting rice stripe virus (RSV). BMC Genomics *11*, 303-316.

Zhang, H.M., Chen, J.P., and Adams, M.J. (2001). Molecular characterization of segments 1 to 6 of Rice black-streaked dwarf virus from China provides the complete genome. Arch. Virol. *146*, 2331-2339.

Zhang, L., Wang, Z., Wang, X., Zhang, W., Li, D., Han, C., Qu, Y., and Yu, J. (2005). Two virus-encoded RNA silencing suppressors, P14 of Beet necrotic yellow vein virus and S6 of Rice black-streaked dwarf virus. Chin. Sci. Bull. *50*, 305-310.

Zheng, W.J., Ma, L., Zhao, J.M., Li, Z.Q., Sun, F.Y., and Lu, X.C. (2013). Comparative transcriptome analysis of two rice varieties in response to rice stripe virus and small brown planthoppers during early interaction. PLoS One *8*, e82126.

Zhou, G.H., Wen, J.J., Cai, D.J., Li, P., Xu, D.L., and Zhang, S.G. (2008). Southern rice black-streaked dwarf virus: a new proposed *Fijivirus* species in the family Reoviridae. Chin. Sci. Bull. *53*, 3677-3685.

Zhou, G.H., Xu, D.L., Xu, D.G., and Zhang, M.X. (2013). Southern rice black-streaked dwarf virus: a white-backed planthopper-transmitted *fijivirus* threatening rice production in Asia. Front. Microbiol. *4*, 270.

Zhou, Y., Yuan, Y., Yuan, F., Wang, M., Zhong, H., Gu, M., and Liang, G. (2012). RNAi-directed down-regulation of RSV results in increased resistance in rice (*Oryza sativa* L.). Biotechnol. Lett. *34*, 965-972.

Zhu, Y.F., Hayakawa, T., and Toriyama, S. (1992). Complete nucleotide sequence of RNA 4 of rice stripe virus isolate T, and comparison with another isolate and with maize stripe virus. J. Gen. Virol. *73*, 1309-1312.

Other Publications of Interest

Xu, J.-P. (2014). Next-generation Sequencing: Current Technologies and Applications. Caister Academic Press, U.K. ISBN: 978-1-908230-33-1

He, Z. (2014). Microarrays: Current Technology, Innovations and Applications. Caister Academic Press, U.K. ISBN: 978-1-908230-49-2

Poptsova, M.S. (2014). Genome Analysis: Current Procedures and Applications. Caister Academic Press, U.K. ISBN: 978-1-908230-29-4

Fuentes, M. and LaBaer, J. (2014). Proteomics: Targeted Technology, Innovations and Applications. Caister Academic Press, U.K. ISBN: 978-1-908230-46-1

Murillo, J., Vinatzer, B.A., Jackson, R.W. and Arnold, D.L. (2015). Bacteria-Plant Interactions: Advanced Research and Future Trends. Caister Academic Press, U.K. ISBN: 978-1-908230-58-4

Caranta, C., Aranda, M.A., Tepfer, M. and Lopez-Moya, J.J. (2011). Recent Advances in Plant Virology. Caister Academic Press, U.K. ISBN: 978-1-904455-75-2

Nannipieri, P., Pietramellara, G. and Renella, G. (2014). Omics in Soil Science. Caister Academic Press, U.K. ISBN: 978-1-908230-32-4

Herold, K.E. and Rasooly, A. (2009). Lab-on-a-Chip Technology: Fabrication and Microfluidics. Caister Academic Press, U.K. ISBN: 978-1-904455-46-2

The Power of Omics to Identify Plant Susceptibility Factors and to Study Resistance to Root-knot Nematodes

Javier Cabrera, Marta Barcala, Carmen Fenoll and Carolina Escobar*

Universidad de Castilla-La Mancha, Facultad de Ciencias Ambientales y Bioquímica, Avda. Carlos III s/n, Toledo, Spain
*Corresponding Author: Carolina.Escobar@uclm.es

Abstract
Technology has contributed to the advances on the genomic, transcriptomic, metabolomic and proteomic analyses of the plant-root-knot nematode (RKN) interaction. Holistic approaches to obtain expression profiles, such as cDNA libraries, differential display, q-PCR, microarray hybridization, massive sequencing, etc., have increased our knowledge on the molecular aspects of the interaction and have triggered the development of biotechnological tools to control this plague. An important limitation, however, has been the difficulty of cross-comparative analysis of these data. The construction of a database, NEMATIC, compiling microarray data available in Arabidopsis of the interaction with plant endoparasitic nematodes facilitated the *in silico* analysis, but is not sufficient for the handling of "omic" information of different plant species. Omics combined with cell isolation techniques have shed some light on the heterogeneous expression signatures of nematode induced gall tissues, i.e., plant defences are specifically inhibited in giant cells within the gall aiding the nematode for a successful establishment. The natural resistance against RKNs varies from an early hypersensitive reaction before the establishment of the nematode, to the arrest of gall growth. The molecular bases of these mechanisms, not fully understood yet, could disclose powerful targets for the development of biotechnology based tools for nematode control.

Introduction
Plant parasitic nematodes (PPNs) constitute a major pest for crop plants worldwide, causing economic and yield losses estimated in $118 billion and 11%, respectively (McCarter et al., 2008). To these numbers, losses in non-food crops including ornamentals, turf and forest trees should be added, increasing even further the global impact of these parasites on society and on the environment (McCarter et al., 2008). Root- knot nematodes (RKNs; *Meloidogyne* spp.) are sedentary endoparasites considered one of the most harmful group of PPNs as they can infect a wide range of plant species (Escobar et al., 2015). RKNs penetrate into the roots through the elongation zone and migrate intercellulary towards the root apex entering into the vascular cylinder where they establish and inducing a feeding site necessary for life cycle completion (Escobar et al., 2015). RKNs secretions interfere with the plant molecular pathways to differentiate specialized cell types into their feeding sites (Truong et al., 2015), called giant cells (GCs) as they are far larger than the surrounding cells in the vascular cylinder (Cabrera et al., 2015). During the development of the GCs, surrounding vascular cells divide profusely and cortical cells hypertrophy which, together with GC growth, lead to the formation of a knot or gall in the root that is the typical structure that characterizes the infection by these nematodes (Escobar et al., 2015). GCs inside the galls act as transfer cells that sink the nutrients from the plant to help the nematode complete its life cycle (Rodiuc et al., 2014). The differentiation of vascular cells into highly specialized GCs requires massive transcriptional reprogramming to allow changes in the cell cycle machinery (de Almeida-Engler et al., 2015), lipid, carbohydrate and amino acid metabolisms (Brown et al., 2015; Siddique and Grundler, 2015) or cell wall composition (Wieczorek, 2015) among others.

In the last 10-15 years, technology has assisted the advances in the genomic, transcriptomic, metabolomic and proteomic analyses which have contributed to the huge increase in the research on the molecular aspects of the plant-RKNs interaction. Hence, studies based on different omics have tried to decipher those molecular pathways leading to the successful establishment of the nematode in the root and to GCs differentiation, as well as those pathways governing the plant resistance that lead to incompatible plant-RKN interactions. Expectations on the development of biotechnological tools based on this molecular knowledge to control this plague were, at least partly, the fundament of the research. A quick online search in the Pubmed library using the keyword "*Meloidogyne*" retrieves a higher number of publications with this word in the abstract within the last 15 years than in all the 20th century (1132 and 1041, respectively). Moreover, 246 of the publications since 2000 contain in the abstract the words "gene expression". These data reflect the intense research on this field in recent years which has resulted from the development of the omics techniques that might open a new horizon not only for the basic understanding of the plant-nematode interaction, but also to design effective and economically viable tools for its control. In this review we describe the main studies based on different omics analyses focusing mainly in those with relevant data on plant defence responses and resistance.

Transcriptomic analyses of plant-RKN interactions
cDNA libraries and differential display
Differential screening of cDNA libraries from infected root tissue compared to the cDNA of uninfected roots has been broadly used to uncover those genes responsive to the infection of different plant species when microarrays or

massive sequencing techniques were not available (Figure 1; Table S1). First attempts to achieve a global gene expression analysis were performed by Wilson et al., in 1994 by constructing a cDNA library from RNA of 1-2 months-old hand dissected GCs induced by *M. incognita* in tomato roots (Figure 1; Table S1). 58 cDNA clones were more abundant in GCs compared to the uninfected roots (Bird and Wilson, 1994; Wilson et al., 1994). Among them, some were not expressed in uninfected roots but in other parts of the plant, suggesting that the pathways interfered during GC maintenance were not necessarily specific for roots. Only 12 cDNA clones could be associated to a putative homolog using a Blast algorithm, including a putative MYB transcription factor (Bird and Wilson, 1994). More recently, several publications have confirmed the involvement of the *MYB* family in the regulation of the plant responses to different abiotic and biotic stresses (Ambawat et al., 2013) including plant-RKN interaction, with several MYB transcription factors resulting differentially regulated in different transcriptomic analyses (e.g. Jammes et al., 2005; Barcala et al., 2010; see next section).

Based also on a cDNA library from 5 weeks-old *M. incognita* tomato galls, Van der Eycken et al. (1996), found a gene named *LEMMI9*, in addition to other genes such as those coding for two extensins. *LEMMI9* encodes a late embryogenesis related protein whose induction was confirmed in GCs by *in situ* hybridization. It shows a high homology to cotton LEA14, a hydrophilic protein belonging to the late embryogenesis abundant (LEA) class of proteins that accumulate both during embryogenesis and in vegetative tissues under water stress (Dure, 1993), and whose function has been proposed to be ion sequestering during water deficit. Hence, it was proposed that the accumulation in GCs of *LEMMI9* might be related to osmotic changes during their development and functioning (Van der Eycken et al., 1996). *LEMMI9* promoter was also active in potato, and a nematode responsive *cis*-element was identified by deletions and EMSA analysis (Escobar et al., 1999). This gene was one of the firsts to be obtained from transcriptomic studies, and demonstrated to play a role in the plant-nematode interaction as evidenced by the

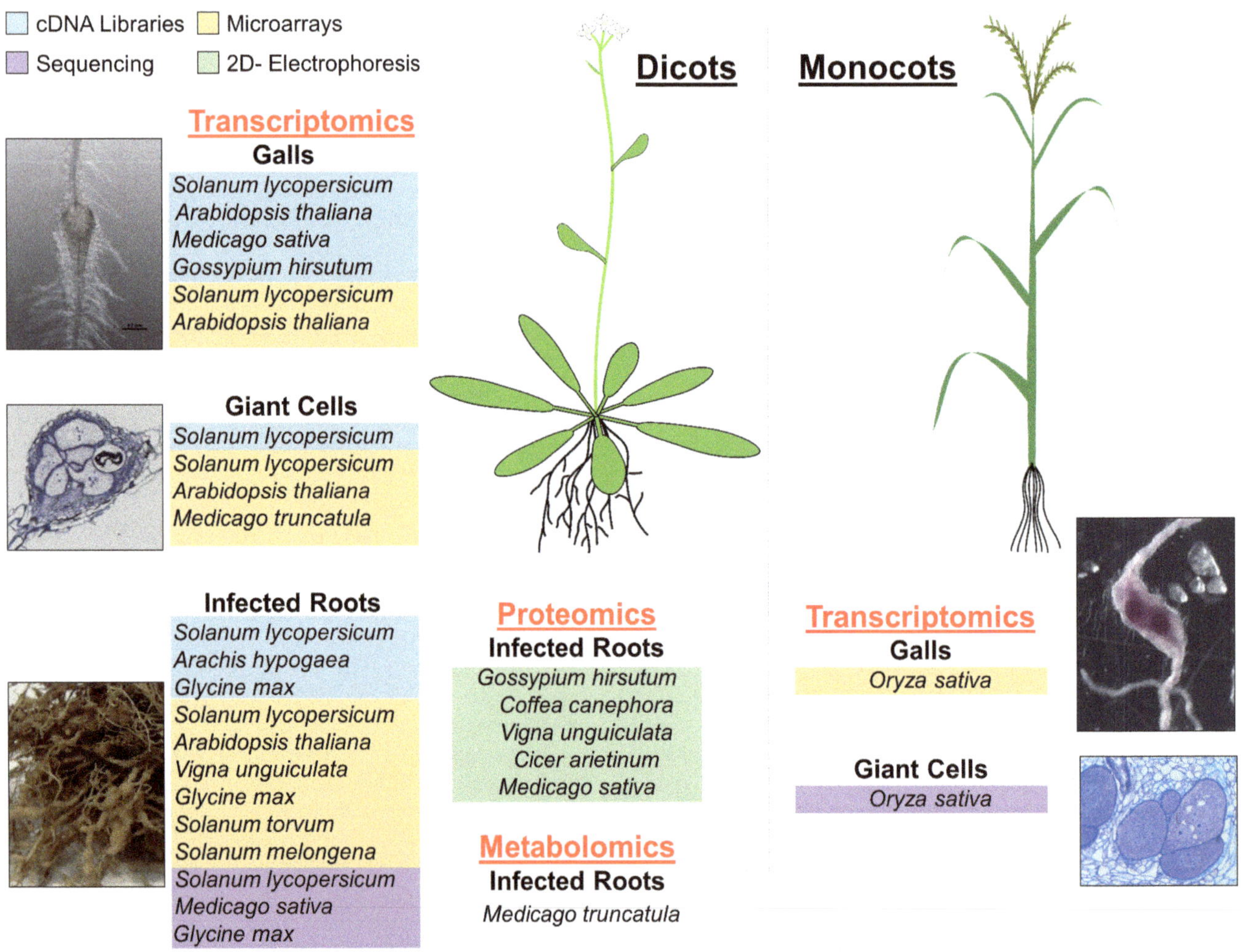

Figure 1. Different omics techniques used to decipher the molecular mechanisms underlying the interaction of *Meloidogyne* spp. with plants. The use of giant cells, galls and/or infected root tissues from dicotyledonous and monocotyledonous plant species are indicated. We acknowledge Prof. Dr. Godelieve Gheysen and Ji Hongli and Kamrun Nahar from Ghent University for kindly providing us with the pictures of galls and giant cells from rice.

fact that its silencing in tomato hairy roots produced a decrease in nematode reproduction (Plovie et al., 2003).

Most of the early attempts to disclose the gall/GC transcriptomes were made in well-developed GCs at 1-2 months after infection, therefore retrieving no information on the establishment and early development of the GCs. The first cDNA library constructed from very early stages of the plant-nematode interaction during the penetration-migration phase was performed using 12 hours post inoculation (hpi) nematode-resistant tomato root tips (carrying *Mi-1* gene) infected with *M. javanica* (Lambert et al., 1999). Eight genes from the cDNA library were characterized, some of them associated with defence or stress responses, such as those encoding an ascorbate reductase or a peroxidase. However, these 8 genes were expressed in both the compatible and incompatible interaction (Lambert et al., 1999). This is in accordance with the production of reactive oxygen species (ROS) at the early stages, not only in the incompatible but also in the compatible interaction as confirmed experimentally (Melillo et al., 2011). However, the H_2O_2 levels were lower in the compatible than in the incompatible interaction, which suggests a role of ROS in nematode resistance. Accordingly, a cDNA library of *M. incognita* galls at 3 days post inoculation (dpi) in *Medicago sativa* roots revealed 6 induced cDNAs. Two of them, an isoflavone-like protein and a metallothionein-like protein have a potential role in ROS scavenging (Potenza et al., 2001). In this respect, it was recently shown that the cyst nematode *Heterodera schachtii* stimulates the formation of ROS by an NADPH-oxidase to modulate the distribution of dead cells and to allow the proper formation of their feeding site (Siddique et al., 2014). Differential display was also used to identify alterations of gene expression of *A. thaliana* root galls caused by *M. incognita* at early infection stages (2, 3, 4, 5 and 7 dpi; Vercauteren et al., 2001). The differential expression of six genes was confirmed by DNA gel blot hybridization. A general defence mechanism operating during this compatible interaction was suggested because of the enhanced expression of genes encoding proteinase inhibitors and peroxidases, both described during other plant-pathogen interactions (e.g. Koiwa et al., 1997; Habib and Fazili, 2007). Interestingly, most of the induced genes were also localized by *in situ* hybridisation within the gall tissues, among them, two genes encoding a trypsin inhibitor and a peroxidase were expressed mainly in the endodermal cells surrounding the expanding GCs, which can be considered as part of a host wound reaction (Vercauteren et al., 2001). Their expression pattern in the outer layers of the galls, but not within the GCs, is in accordance, essentially, with the data obtained in a highly sensitive analysis performed later by combining microarray analysis with cell biology techniques. This analysis involved GCs isolation from the rest of the gall tissues in Arabidopsis, tomato and Medicago (Barcala et al., 2010; Damiani et al., 2012; Portillo et al., 2013). In general, it was observed that defence genes were repressed within the GCs, whereas defence responses were observed in the rest of the gall.

As a result of a proteomic study (see next section), a cotton protein correlated with RKN resistance was identified. It was DE at 8 dpi when nematode development is arrested in the incompatible interaction (Callahan et al., 1997). Analysis with a specific antiserum recognized the 14 kDa protein and revealed a negative correlation between the presence of this protein in root tissues and root galling from a large group of RKN resistant breeding lines, and susceptible cultivars of cotton (Callahan et al., 2004). Additionally, it helped to identify a gene named *MIC-3* (*Meloidogyne Induced Cotton3*) from a cDNA library of 10 dpi *M. incognita* galls that was induced in a nematode-resistant line of cotton (*Gossypium hirsutum*; Zhang et al., 2002) as compared to the expression in a susceptible line (Callahan et al., 1997; Zhang et al., 2002) corresponding to this protein. More recently, the authors demonstrated that the over-expression of the protein in a susceptible line reduces egg production by 60-75% compared to the non-transgenic lines (Wubben et al., 2015). At least 15 *MIC*-like cDNAs are expressed in the roots of susceptible and resistant cotton lines (Wubben et al., 2008), differing in the timing of induction. In resistant lines their maximum expression occurs before the appearance of visible signs of resistance, while susceptible lines showed a delayed expression. This suggests that the *MIC* gene family is part of a root-specific defence response mechanism in cotton that operates independently from other known pathways such as gossypol biosynthesis, lipid peroxidation, and *PR10* expression (Wubben et al., 2008). *MIC-3* constitutes a clear example of how proteomic and transcriptomic studies could complement each other to assist the development of biotechnological tools based on modifications of the plant defence systems against the RKNs.

Although some of the previously mentioned papers used root material enriched in GCs for the construction of the cDNA library, e.g. hand dissected galls, there was a serious technical difficulty to isolate the GCs from the rest of the gall where they were immersed. GCs are the nourishing cells and their differentiation and continuous maintenance is strictly necessary for the nematode to complete its life cycle (Hussey et al., 1989). It was not until 2003 when the first transcriptomic study from isolated GCs was carried out (Wang et al., 2003). In this work, 25 dpi GCs from tomato roots inoculated with *M. javanica* were isolated using a pressure probe system (Wang et al., 2001; 2003). The mRNA extracted was further used for differential display RT-PCR, finding that 73 genes were up-regulated and 8 genes down-regulated (Wang et al., 2003). Many of those genes were classified into functional categories that reflect the high metabolic rate of the GCs more than the defence response or process related to the GCs development. These results were not surprising due to the age of the GCs, 25 dpi, when they are fully developed nourishing the nematode. However, the authors demonstrated the importance of GCs isolation preceding the transcriptomic analysis, as the fold change of the up- regulated genes was much lower using GC-enriched root segments than that from isolated GCs. This confirms the strong dilution effect of GC mRNAs when whole infected tissues are used (Wang et al., 2003). Later, 87 and 54 transcripts were identified from a cDNA library of laser captured isolated GCs from tomato galls at 4 and 7 dpi respectively (Fosu-Nyarko et al., 2009). It was suggested that the presence and initial movement of the nematode juveniles (J2s) between cells may activate defence and stress transduction pathways in the host root, as transcripts encoding pathogenesis-related proteins (e.g. TSI-1 and stress related proteins such as superoxide dismutases) were induced at 4 and 7 dpi as compared to uninfected cells surrounding the vascular cylinder (Fosu-Nyarko et al., 2009). In contrast, most of the later transcriptomic analyses based on microarrays or massive sequencing combined with laser microdissection of GCs pointed to gene repression of these groups of genes within GCs, not only in Arabidopsis but in tomato and Medicago as well (Barcala et al., 2010; Daimiani et al., 2012; Portillo et al., 2013).

Even though in the last decade researchers preferred holistic transcriptomic techniques such as microarrays and high throughput sequencing analyses, cDNA libraries have been used in other studies when no microarray chips were available, i.e. in the interactions of *M. arenaria* with *Arachis hypogaea* and *M. javanica* with *Glycine max* (Tirumalaraju et al., 2011; de Sa et al., 2012). cDNA libraries from resistant and susceptible peanut lines confirmed the higher number of genes related to stress categories in the resistant line than in the susceptible one, although both of them presented genes from this category. The resistant cultivar (NemaTAM)-specific transcripts included those encoding putative PR proteins, patatin-like proteins and universal stress proteins (USP). The expression of most PR proteins is contingent upon pathogen infection and is mediated through salicylic acid (SA), jasmonic acid (JA) and ethylene signal transduction pathways (Tirumalaraju et al., 2011). In contrast, the number of genes related to transcriptional regulation was higher in the susceptible than in the resistant lines, probably reflecting an intense reprogramming needed for GC establishment (Tirumalaraju et al., 2011).

Microarrays

The first studies reporting DNA microarrays for the study of the plant- *Meloidogyne* interaction constituted a prologue for a veritable explosion of analyses based on this technique (Figure 1; Table S1). The sequencing of the Arabidopsis genome coupled to the valuable information from other research fields, made this plant a suitable system to study the transcriptional profiles of galls and GCs. The first study was centred in the regulation of membrane transporters in Arabidopsis roots infected with *M. incognita* at 1, 2 and 4 weeks post inoculation (wpi; Hammes et al., 2005). Using the GeneChip® Arabidopsis ATH1 Genome Array from Affymetrix, the authors focused their study in the expression of the 805 membrane-proteins represented in the chip (Hammes et al., 2005). 50 transporters including members of 19 of the approximately 80 transporter gene families identified in Arabidopsis showed differential expression at 1, 2 and 4 wpi in whole infected roots compared to the uninfected control (Hammes et al., 2005). Some of them were specifically expressed in the galls and not in the rest of the root, as demonstrated by qPCR analysis and GUS assays. Consistent with the role of GCs as transfer cells to feed the nematode (Cabrera et al., 2014a; Rodiuc et al., 2014), an up- regulation of auxin, amino acid and sucrose transporters was observed (Hammes et al., 2005). The second microarray study reported for RKNs was a full time-course along different infection stages, from establishment to maturation (7, 14 and 21 dpi) using CATMA microarrays representing 22089 Arabidopsis genes (Jammes et al., 2005). A total of 3373 genes were differentially expressed (DE) at some of the selected infection times, increasing greatly the number of genes found DE in the galls as compared to those found in the previous cDNA library screens. Such a number of DE genes indicated a vast molecular reprogramming of the cells forming the gall, including the GCs. The number of genes up and down- regulated were similar (1606 and 1742, respectively), highlighting the importance of gene repression in the compatible interaction (Jammes et al., 2005). Moreover, the number of repressed genes related to plant defence was remarkable, suggesting a necessity for the down- regulation of defence pathways in the gall for a proper establishment during the compatible interaction (Jammes et al., 2005). The gene repression in galls was further confirmed in later microarray analysis of galls from Arabidopsis infected with *M. incognita* (Fuller et al., 2007) where two-thirds of the 259 DE were down-regulated. Those repressed genes were mainly classified in the response to abiotic or biotic stresses (Fuller et al., 2007).

The first transcriptome combining cell isolation techniques and DNA-microarrays was performed with RNA extracted from GCs at 3 dpi isolated by laser capture micro-dissection from frozen sections of hand dissected Arabidopsis-*M. javanica* galls (Barcala et al., 2010). It detected 1161 DE genes, of which 851 were down-regulated and 310 were up-regulated. Consistently, most of the stress related genes were down-regulated in GCs. Hence, a distinctive feature of the GCs transcriptome was the general down-regulation of genes classified in the secondary metabolism and stress groups. In particular, most genes involved in the phenylpropanoid pathway, typically part of a defence mechanism against pathogen attack, including *PAL1*, *PAL2*, *4CL3* and *CAD* (Dixon et al., 2002) were down-regulated. Furthermore, many *WRKY*, *MYB*, *AP2/EREBP* and *bZIP* transcription factor family members, induced by biotic or abiotic stresses (Bakshi and Oelmüller, 2014; Mizoi et al., 2012; Liu, et al., 2015) were down-regulated in GCs, but not in the transcriptome of whole galls. However, defence and PRP genes directly or indirectly regulated by SA and JA were not DE in 3 dpi GCs (Barcala et al., 2010). The study by Barcala et al. (2010) allowed comparison of the GC transcriptome to that of the whole gall at the same developmental stage (3 dpi), revealing that their transcriptional profiles are different, with only 120 shared genes out of a total of 1161 DE in GCs. A group of distinctive stress and defence related genes were clearly induced in galls, mainly related to biotic stress and secondary metabolism. In contrast, most genes of those categories were repressed in GCs. This study also confirmed the high dilution of GC mRNAs when the whole gall is analysed, as the fold changes of the common DE genes were much higher in GCs than in galls. Therefore, it served to characterize the transcriptional changes associated to GCs formation, which were proved to be distinct to that of the rest of gall tissues (Barcala et al., 2010).

Vast amount of transcriptomic data for the plant-nematode interaction was generated in the last 10 years. However, that quantity of data was not directly translated into a gain of knowledge about gene functions or metabolic pathways within the feeding sites, neither to the development of genetic tools for biotechnological nematode control. One reason for this lag was the difficulty of cross-comparative analysis of these data. A first attempt to overcome this limitation was the construction of a database called NEMATIC (Nematode-Arabidopsis Transcriptomic Interactive Compendium) that includes most of the transcriptomic data generated in Arabidopsis related to feeding sites induced by endoparasitic nematodes (Cabrera et al., 2014b). This tool was designed for the selection of putative target genes, for the development of biotechnological tools and to contribute to the basic knowledge of the plant-nematode interaction. NEMATIC is completed with transcriptomic data from several biological processes: plant biotic and abiotic interactions, hormonal regulation, specific transcriptomes for different root cell types, etc. (Cabrera et al., 2014b). It allows to compare lists of genes commonly or distinctly regulated throughout gall development or in syncytia (induced by cyst nematodes), and/or GCs (Table 1; Table S2). A direct example is the use of NEMATIC for the comparison of all stress- related genes contained in Mapman (Thimm et al., 2004) across all the GC and gall transcriptomes compiled in the tool. Applying this filter, a list of 200 genes related to stress/disease resistance that were DE at some of the developmental stages of GCs or galls was retrieved (Table S2). From a biotechnological point of view, the most interesting defence-related genes are those down-regulated at the early stages of GCs or gall development (3 and 7 dpi). By adding these criteria, the number of genes in the list was reduced to 88 (Table 1). Some of them (Table 1) are strongly down-regulated, as *KTI1,* involved in cell death modulation during plant-pathogen interactions (Li et al., 2008), or *PR1*, a typical gene induced by different pathogens and a marker for the systemic acquired resistance (Asai et al., 2014). Other highly down-regulated genes are *PR4*, a pathogenesis-related protein that participates in the resistance to *Fusarium* head blight disease in wheat (Zhuang et al., 2012), and *SPX1*, a Pi-dependent inhibitor of *PHR1* and a master regulator of Pi starvation responses (Puga et al., 2014). Other genes obtained in this selection that have been already related with disease resistance are *HSP90* (Takahashi et al., 2003), *TGA1*, a redox-controlled regulator of systemic acquired resistance (Fobert and Després, 2005), or *HS1*, induced by exposure to SA and JA and with antimicrobial and antifungal activity (Park et al., 2007). Several disease resistance proteins (*AT1G58170, AT2G21100, AT4G19920, AT5G40910, AT5G51630*), pathogenesis related proteins (*AT4G25780, AT4G33720, AT4G36010, AT5G40020*) and germin-like proteins (*AT1G10460, AT1G18980, AT3G05950, AT3G62020, AT4G14630, AT5G39100, AT5G39150, AT5G39180*) are also down regulated in GCs and galls at early and medium stages of development (Table 1). This list of 88 stress and resistance related genes is an example of the vast information that transcriptomics produced in the last decade and the usefulness of tools like NEMATIC, which allows easy cross comparison through several databases and transcriptomes. Genes listed in Table 1 could be useful for basic studies and for biotechnology based strategies for the RKN control.

Microarray based transcriptomic has been successfully applied not only in Arabidopsis but also in the crop model tomato (*Solanum lycopersicum*; Figure 1; Table S1). Pathogenesis-related genes, as defensins, were DE in galls induced by *M. javanica* in tomato roots (Bar-Or et al., 2005). The up-regulation of *HIN1* (harpin-induced gene 1), induced by bacterial harpins (a class of virulent/avirulent proteins) in tomato and potato, was suggested as a putative virulence/avirulence mechanism in the RKN-tomato interaction (Bar-Or et al., 2005). Differences between the compatible and the incompatible interaction was studied in more detail using microarray technology with resistant and susceptible tomato plants infected with *M. incognita* (*Mi*-susceptible) and *M. hapla* (*Mi*- resistant; Schaff et al., 2007). During a time course study at 12, 36 and 72 hpi and 2 wpi, they found 58 genes that exhibited differential regulation in resistant roots compared to uninfected roots (Schaff et al., 2007). 31 genes were induced in the incompatible interaction, encoding proteins such as a glycosyltransferase, a peroxidase, and an ethylene-responsive protein that were not DE in the susceptible plants at 12 to 36 hpi. In absence of nematodes, the same glycosyltransferase corresponded to the only gene which was induced in the transcriptome from resistant (*Mi-1/Mi-1*) Motelle as compared to the susceptible (*Mi-*) Moneymaker cultivar. Furthermore, the silencing of this gene restored the susceptibility to nematodes in the resistant cultivar Motelle, suggesting that this gene participates in the resistance mechanism. A similar approach, but focusing in the roles of JA and SA, was used by Bhattarai et al., 2008. RNA was extracted from root tips infected at 24 hpi. in cultivar Motelle (resistant) and cultivar Moneymaker (susceptible). After microarray hybridisation, resistant root tips presented twice as many DE genes than the susceptible line (1497 and 750, respectively). Several genes involved in JA biosynthesis and genes regulated by JA were expressed in both compatible and incompatible interactions, inferring that the JA signalling pathway had a role in basal defence but not in *Mi-1*-mediated resistance to RKNs (Bhattarai et al., 2008). Two of the transcription factors found, *SlWRKY72a* and *SlWRKY72b,* were up-regulated in resistant but not in susceptible roots after RKN infection (Bhattarai et al., 2010). The silencing of these genes in resistant plants allowed the infestation by RKN in contrast to non-transgenic resistant plants, showing that *SlWRKY72a* and *SlWRKY72b* are required for nematode resistance in Motelle (Bhattarai et al., 2010). Furthermore, their silencing in susceptible Moneymaker plants increased nematode reproduction, indicating that they have a role in the basal defence response of the plant as well (Bhattarai et al., 2010).

Table 1. List of genes from the stress Mapman category that appear as down- regulated at early and/or medium stages (3 and 7 dpi) of giant cells (GCs) and gall development in the compatible interaction between *Arabidopsis thaliana* and *Meloidogyne javanica* or *M. incognita*. M values for each gene in each of the experiments (Jammes et al., 2005; Barcala et al., 2010) are indicated. "- -" indicates "Non-differentially expressed".

	M value			
Gene ID	GC 3dpi	Gall 3dpi	Gall 7dpi	Description
At1g04430	--	--	-1,07	dehydration-responsive protein-related
At1g05850	-1,48	--	--	POM1: Encodes an endo chitinase-like protein AtCTL1.
At1g10460	-0,99	--	--	GLP7: germin-like protein (GLP7)
At1g11960	--	--	-0,98	early-responsive to dehydration protein-related
At1g13930	-1,79	--	-1,18	Involved in response to salt stress.
At1g18980	--	--	-0,71	germin-like protein, putative
At1g20440	--	--	-2,45	COR47: Belongs to the dehydrin protein family
At1g21080	--	--	-0,91	DNAJ heat shock N-terminal domain-containing protein
At1g23120	-2,63	--	--	major latex protein-related
At1g30360	--	--	-1,61	ERD4 (early-responsive to dehydration 4)
At1g31850	-1,67	--	--	dehydration-responsive protein, putative
At1g35260	-2,63	--	--	MLP165 (MLP-LIKE PROTEIN 165)
At1g52560	-1,54	--	--	26.5 kDa class I small heat shock protein-like
At1g58170	-1,55	--	--	disease resistance-responsive protein-related
At1g64140	--	--	-1,11	loricrin-related
At1g65390	-1,39	--	--	ATPP2-A5 (ARABIDOPSIS THALIANA PHLOEM PROTEIN 2 A5)
At1g70850	-2,52	--	-0,79	MLP34 (MLP-LIKE PROTEIN 34)
At1g70880	--	-2,62	--	Bet v I allergen family protein
At1g70890	-1,79	--	--	MLP43 | MLP43 (MLP-LIKE PROTEIN 43)
At1g72500	--	--	-0,71	inter-alpha-trypsin inhibitor heavy chain-related
At1g73260	-3,05	--	--	KTI1: Encodes a trypsin inhibitor involved in modulating programmed cell death in plant-pathogen interactions.
At1g73330	--	--	-1,17	DR4: encodes a plant-specific protease inhibitor-like protein
At1g75030	-1,17	--	--	TLP-3: encodes a PR5-like protein
At1g76180	--	--	-1,85	ERD14: Encodes a dehydrin protein whose expression is induced early on in response to dehydration stress.
At1g77260	-2,15	--	--	dehydration-responsive protein-related
At1g78040	-1,23	--	--	pollen Ole e 1 allergen and extensin family protein
At2g01520	-2,09	--	--	MLP328 (MLP-LIKE PROTEIN 328); copper ion binding
At2g01530	-1,42	--	--	MLP329 (MLP-LIKE PROTEIN 329); copper ion binding
At2g14610	-4,07	--	--	PR1: PR1 gene expression is induced in response to a variety of pathogens.
At2g15970	-1,53	--	-1,59	COR413-PM1: encodes an alpha form of a protein similar to the cold acclimation protein WCOR413 in wheat.
At2g19990	--	--	-0,75	PR-1-like protein homolog differentially expressed in resistant cultivars by powdery mildew infection.
At2g21100	-1,13	--	--	disease resistance-responsive protein-related
At2g21620	-1,87	--	--	RD2: Encodes gene that is induced in response to dessication
At2g24550	--	--	-0,83	unknown protein
At2g33790	--	-2,26	--	AGP30: pollen Ole e 1 allergen protein
At2g39200	--	--	-0,88	MLO12: homolog of the barley mildew resistance locus o (MLO) protein.
At2g45130	--	--	-0,69	ATSPX3, SPX3 | SPX3 (SPX DOMAIN GENE 3)
At2g47770	--	--	-0,89	TSPO: Encodes a membrane-bound protein designated AtTSPO
At3g01420	--	--	-1,97	DOX1: Encodes an alpha-dioxygenase involved in protection against oxidative stress and cell death.
At3g04720	-4,15	--	--	PR4: Encodes a protein similar to the antifungal chitin-binding protein hevein from rubber tree latex.
At3g05880	-2,12	--	-1,32	RCI2A: Induced by low temperatures, dehydration and salt stress and ABA.
At3g05950	-1,23	--	1,22	germin-like protein, putative
At3g10985	-1,82	-1,14	-1,59	SAG20: A senescence-associated gene induced in response to a fungal protein that causes necrosis.
At3g17020	-0,88	--	--	universal stress protein (USP) family protein
At3g17210	-1,21	--	--	HS1: Encodes a heat stable protein with antimicrobial and antifungal activity.
At3g23300	-1,41	--	--	dehydration-responsive protein-related
At3g28940	-1,92	--	--	avirulence induced gene (AIG) protein, putative
At3g44670	-1,32	--	--	ATP binding / nucleoside-triphosphatase/ nucleotide binding / protein binding / transmembrane receptor
At3g47540	-1,40	-1,02	0,69	chitinase, putative
At3g48450	-2,29	--	1,00	nitrate-responsive NOI protein, putative
At3g50460	-0,97	--	--	HR2: Homolog of RPW8
At3g50970	--	--	-2,05	LTI30: Belongs to the dehydrin protein family
At3g59930	-4,03	--	--	Encodes a defensin-like (DEFL) family protein.
At3g62020	--	-1,10	--	GLP10: germin-like protein (GLP10)

At3g62190	-1,39	--	--	DNAJ heat shock N-terminal domain-containing protein
At4g09940	-1,43	--	--	avirulence induced gene (AIG1) family protein
At4g11650	-2,99	--	--	OSM34: osmotin-like protein
At4g14630	--	--	-1,90	GLP9: germin-like protein
At4g19920	--	-1,16	--	disease resistance protein (TIR class), putative
At4g22214	--	-1,27	--	Encodes a defensin-like (DEFL) family protein.
At4g23680	-1,41	--	--	major latex protein-related
At4g24190	-0,77	--	--	SHD: an ortholog of GRP94, involved in regulation of meristem size and organization.
At4g25780	--	--	-0,81	pathogenesis-related protein, putative
At4g28240	-1,19	--	-0,68	wound-responsive protein-related
At4g28480	-1,13	--	--	DNAJ heat shock family protein
At4g30650	-2,85	--	--	low temperature and salt responsive protein, putative
At4g33300	--	--	-0,74	ADR1-L1 (ADR1-like 1); ATP binding / protein binding
At4g33720	-0,80	--	--	pathogenesis-related protein, putative
At4g36010	--	--	-0,82	pathogenesis-related thaumatin family protein
At4g36040	--	--	-0,68	J11: DNAJ heat shock N-terminal domain-containing protein (J11)
At5g20150	-3,40	--	--	ATSPX1, SPX1 | SPX1 (SPX DOMAIN GENE 1)
At5g33355	--	-1,05	--	Encodes a defensin-like (DEFL) family protein.
At5g39100	--	-1,40	--	GLP6: germin-like protein (GLP6)
At5g39150	-2,33	-1,51	--	germin-like protein, putative
At5g39180	-1,77	-1,46	--	germin-like protein, putative
At5g39730	-1,81	--	--	avirulence induced gene (AIG) protein-related
At5g40020	--	-1,77	--	pathogenesis-related thaumatin family protein
At5g40910	-1,18	--	--	disease resistance protein (TIR-NBS-LRR class), putative
At5g45050	-1,59	--	--	TTR1: Encodes a member of WRKY Transcription Factor that confers resistance to tobacco ringspot nepovirus.
At5g45490	--	--	-1,14	disease resistance protein-related
At5g49910	-1,25	--	--	cpHsc70-2: Stromal heat shock protein involved in protein import into chloroplast.
At5g51630	-1,25	--	--	disease resistance protein (TIR-NBS-LRR class), putative
At5g52060	--	--	-1,05	BAG1: A member of Arabidopsis BAG proteins, plant homologs of mammalian regulators of apoptosis.
At5g52640	--	--	-0,99	HSP90.1: Encodes a cytosolic heat shock protein AtHSP90.1.
At5g54855	-1,19	--	--	pollen Ole e 1 allergen and extensin family protein
At5g64890	--	--	-0,81	PROPEP2 (Elicitor peptide 2 precursor)
At5g64930	-0,81	--	--	CPR5: Regulator of expression of pathogenesis-related (PR) genes.
At5g65210	--	--	-0,79	TGA1: Encodes TGA1, a redox-controlled regulator of systemic acquired resistance.

The first microarray study performed with RNA extracted from isolated GCs induced by *M. javanica* in a model crop, tomato (*S. lycopersicum*) was carried out by laser capture microdissection of 3 and 7 dpi GCs (Portillo et al., 2013). The transcriptomes of the isolated GCs were compared with that of the complete galls at 1, 3, 7 and 14 dpi, showing a higher repression state in 3 dpi GCs (86% of the DE genes) than at 7 dpi when the GCs are already well developed (56% of the DE genes). This repression trend in GCs at early developing stages was already observed in Arabidopsis (Barcala et al., 2010). Moreover, genes down-regulated in early developing GCs were robustly conserved between Arabidopsis and tomato (Portillo et al., 2014). The phenylpropanoid pathway was significantly over-represented among the repressed genes in both tomato and Arabidopsis GCs, particularly genes involved in lignin biosynthesis, such as those coding for a group of peroxidases, together with genes from a biotic stress subcategory encoding protease inhibitors (Portillo et al., 2013). Infection tests with a tomato line overexpressing the *TPX-1* peroxidase, highly repressed in GCs from Arabidopsis and tomato, showed a 35% of reduction in the number of galls formed. In contrast, a remarkable induction (above nine-fold) was observed for this peroxidase during the incompatible interaction with a resistant cultivar carrying the *Mi-1* gene, *S. lycopersicum* cv Motelle (*Mi-1/Mi-1*) as compared to the susceptible near-isogenic line *S. lycopersicum* cv Moneymaker (Portillo et al., 2013). Transcriptomic studies during the plant-nematode interaction have been also performed on a model plant for legume genetics, *Medicago truncatula* (Damiani et al., 2012) challenged with *M. incognita*. Laser microdissected GCs were compared with laser microdissected neighbouring cells in 7 dpi galls from Medicago using Affymetrix GeneChip®_Medicago genome (Damiani et al., 2012). 36% of the down-regulated genes and 78% of those up-regulated in GCs were not expressed in neighbouring cells, reinforcing the importance of the genomic studies which focus specifically in the feeding cells (Damiani et al., 2012). Medicago GCs also showed down-regulation of secondary metabolism and defence related genes as compared to the neighbouring cells, similar to that occurring in Arabidopsis and tomato (Barcala et al., 2010; Portillo et al., 2013). All these data demonstrate that the down-regulation of stress-defence related genes is a hallmark of GCs during the compatible interaction.

Microarray technology has been successfully used as well with other species for which genomic information is scarce. Some examples are *Glycine max* infected with *M. incognita* during a compatible interaction (Ibrahim et al., 2011). At 12 dpi, six of seven members of the lipoxygenase (*LOX*) gene family were up-regulated as compared to non-infected plants. *LOX* genes are essential for the biosynthesis of oxylipins, which have an important function in the plant defence response against wounding and pathogen attack. Interestingly, some members have been recently reported to have significant roles during plant-nematode interactions. *LOX4* has a key role controlling plant defence in Arabidopsis when infected with *M. javanica* and *H. schachtii,* as proved by the fact that loss-of-function mutants were more susceptible to nematodes than control plants, and showed increased levels of JA and transcripts from the defence related genes encoding allene oxide synthase, allene oxide cyclase, ethylene-responsive transcription factor 4 and phenylalanine ammonia lyase (*AOS, AOC2, ERF4, PAL1*) (Ozalvo et al., 2014). Genes encoding PR proteins were induced at 12 dpi in soybean galls formed by *M. incognita* after microarray analysis with Affymetrix Soybean GeneChip containing 37,500 probe sets (Ibrahim et al., 2011). Additionally, although its precise role on the plant-nematode interaction is not yet known, it has been reported that SA induced the expression of *PR-1* and SA treatment of tomato plants inoculated with *M. incognita* enhanced the synthesis of *PR-1*, which resulted in a significant increase in resistance to the nematode (Nandi et al., 2003).

Interestingly, the *Rk* locus from cowpea (*Vigna unguiculata*) confers resistance against several species of RKNs, but the mechanism of resistance is substantially different to that of the characterized *Mi-1* gene in tomato (Williamson, 1999; Mantelin et al., 2013). *Rk* confers a delayed but strong resistance mechanism without a hypersensitive reaction-mediated cell death process, which allows nematode development but blocks reproduction. Gene ontology based functional categorization of genes DE in *RK* plants, as compared to a susceptible nearly isogenic line, revealed that the typical defence response was partially suppressed in resistant roots, even at 9 dpi, allowing nematode juvenile development. However, differences in ROS concentrations, induction of toxins and other defence related genes seem to play a role in this unique resistance mechanism (Das et al., 2010). Another example of transcriptome analysis on a non-model species is that of *Solanum torvum*, used worldwide as rootstock for eggplant cultivation because of its vigour and resistance/ tolerance to bacterial, fungal wilts and RKNs. The transcriptome was obtained by combining 454 pyrosequencing and microarray technology and it was compared to that of the phylogenetically-related nematode-susceptible eggplant species *S. melongena*. The expression profiling of *S. torvum* challenged with *M. incognita* points to sesquiterpenoids and chitinases as major effectors of nematode resistance. In addition, 16 transcripts of *S. torvum* were annotated as disease resistance genes (*R* genes) and only one of these induced transcripts has a validated counterpart in eggplant, pointing to major sequence divergence and/or lack of counterparts in eggplants for this group of induced *S. torvum* candidate *R* genes (Bagnaresi et al., 2013).

High throughput sequencing

In the last years, the development of high throughput sequencing techniques opened a new field for the transcriptomic analyses of the RKN- plant interaction (Figure 1). Interesting examples are those from monocots as *Oryza sativa* (Kyndt et al., 2012a; 2014; Ji et al., 2013). Consistently with the findings described in previously discussed microarray experiments, the galls induced by *M. graminicola* in rice roots at 3 dpi showed down-regulation of genes related to biotic stresses and to the phenylpropanoids pathway, responsible for the biosynthesis of different metabolites, such as lignin precursors, flavonoids and hydroxycinnamic acid esters and SA (Kyndt et al., 2012). However, genes related to metabolic processes such as protein and sucrose biosynthesis, were up-regulated (Kyndt et al., 2012a). The sequencing of the mRNAs obtained from laser captured microdissected isolated GCs from *M. graminicola* galls in rice showed the same tendency, with the defence and secondary metabolism related genes being strongly down-regulated (Ji et al., 2013). Moreover, the gene regulation trend in GCs was almost the same at 7 and 14 dpi. However, genes related with epigenomic changes as those coding for *AGO* or *DCL* proteins, showed up-regulation at 7 dpi but not at 14 dpi (Ji et al., 2013).

In order to achieve novel insights into incompatible interactions, comprehensive gene expression profiling was also obtained by pyrosequencing (Beneventi et al., 2013). Transcriptomic studies in soybean lines resistant to specific strains and races of RKNs identified DE transcripts for several members of the lipid transfer protein (*LTPs*) family. For instance, *PR-14*, an important player in the general plant stress response, was induced only in the incompatible interaction whereas *PR-1*, *PR-2* and *PR-5* protein families showed increased expression in the compatible interaction. The data also suggest a key role of glycosyltransferases, auxins and components of gibberellin related genes in the resistance reaction. In addition, some overlaps between salt stress responses and this particular interaction were also observed, such as the induction of annexins and of negative regulators of JA signalling (*JAZ/ TIFY*). A complex model integrating putative crosstalk mechanisms between plant hormones, mainly gibberellins and auxins, which can be crucial to modulate the levels of ROS in the resistance reaction to nematode invasion, has been proposed (Beneventi et al., 2013).

Recently, high throughput sequencing has also been used to study the transcriptome of the interaction between *M. javanica* and tomato hairy roots (Iberkleid et al., 2015). This is one of the few experiments in which the transcriptome of a wild type plant is compared with a transgenic plant overexpressing a nematode-secreted protein. Local transcriptional changes were monitored in tomato hairy root lines with constitutive expression of the *M. javanica* fatty acid- and retinol-binding protein (Mj-FAR-1), compared with control roots without inoculation and after 2, 5 and 15 dpi (Iberkleid et al., 2015). Mj-FAR-1 is involved in nematode development and reproduction in tomato roots and plants over-expressing Mj-FAR-1 showed increased susceptibility to the infection by *M. javanica* (Iberkleid et al., 2013). A total of 3970 genes were DE in over-expressing and control lines (2069 up-regulated and 2205 down-regulated), further confirming the down-regulation of the phenylpropanoids and secondary metabolism pathways (Iberkleid et al., 2015), which might facilitate nematode infection.

mRNA high throughput sequencing was also used to unravel the mechanism of nematode resistance by sequencing the transcriptome of roots from *M. sativa* susceptible and resistant lines at 10 dpi after infection with *M. incognita* (Postnikova et al., 2015). Most of the genes DE were unique to either resistant or compatible cultivars, including a high number of *R* genes (36 and 121, respectively). These results suggest that the presence of *R* genes in the susceptible interaction may be part of a general or passive defence reaction while those DE in the incompatible interaction may be responsible for the resistance (Postnikova et al., 2015). Moreover, they corroborate as well the significant repression of plant defences in the susceptible line but not in the resistant one (Postnikova et al., 2015).

Small RNAs (sRNAs) DE in the systemic response of the plants to RKN have been also analysed by high throughput sequencing (Zhao et al., 2015). Sequencing of sRNAs was performed on tomato phloem tissue after RKN inoculation in WT and *spr2* (suppressor of prosystemin-mediated response 2, JA-deficient). Interestingly, a negative correlation was observed between miR319 and its target *TCP4* in leaf, stem, and root under RKN stress, implying that the miR319/TCP4 module is involved in the systemic defensive response (Zhao et al., 2015). Reverse genetics also demonstrated that the miR319/*TCP4* module affected JA synthetic genes and the endogenous JA level in leaves, thereby mediating RKN resistance (Zhao et al., 2015).

Other approaches and tools

Apart from the holistic techniques previously described, other molecular tools have been used to study gene expression during gall formation. This is the case for real-time quantitative PCR (qPCR) to measure specific transcripts, GUS reporter lines or promoter trap lines to check promoter activity in feeding sites, or *in situ* RT-PCR, which have been successfully used and reported in many articles to show the regulation of genes or to validate those results obtained by holistic studies. In this section, due to the vast amount of publications related to the plant- RKN interaction using these techniques, we will only discuss some of those which describe the regulation of defence- or stress- related groups or families of genes. One of the pioneer analysis using genomic tools to uncover the regulation of gene expression from a holistic point of view was the large-scale screening of 1472 Arabidopsis promoter tag lines, harbouring a promoter-less *GUS* gene, challenged with cyst or with RKNs (Barthels et al., 1997). From this screen, the line Att0001 (*Arabidopsis thaliana* tag 0001) with *GUS* expression was clearly activated in galls and GCs (Barthels et al., 1997) was selected. Further Southern analysis and inverse-PCR were used to identify the promoter of the gene *WRKY23*, as the tagged locus and its induction by nematodes was confirmed by the same expression pattern of a *pWRKY23::GUS* line under RKN infection (Grunewald et al., 2008). Grunewald et al. (2008) hypothesised that the induction of *WRKY23* in the feeding cells could led to a reduction in the plant basal defences. The *GUS* expression in Att0728, another line showing distinct induction in galls, has been shown to be driven by the promoter of the transcription factor *JAZ1/TIFY10A* (Grunewald et al., 2009), a negative regulator of JA-responsive genes (Thines et al., 2007).

Hormone mediated systemic defence against RKNs has been broadly studied in rice infested with *M. graminicola* (Nahar et al., 2011; Kyndt et al., 2012b; Nahar et al., 2013). Defence induced genes, as shown by qPCR, were found to be down-regulated in infected roots at early stages of development, while they were mostly up-regulated in infected roots under treatment with the systemic defence-inducing compounds JA and ethephon (ethylene), indicating the ability of the nematodes to supress the defence pathways (Nahar et al., 2011). This suppression was further confirmed by qPCR at 3 dpi, as general defence genes (*OsPR10* and *OsPR1a*), as well as *OsPAL1*, *OsPR1b*, the JA carboxyl methyl transferase gene (*OsJMT1*) and ethylene biosynthesis and signalling genes (*OsAC07* and *OsEIN2a*) were down-regulated in shoots of infected plants while the *PR* genes were up- regulated in roots (Kyndt et al., 2012b). Accordingly, in Arabidopsis plants infected with *M. incognita*, the genes *PR1*, *PR2*, *PR3* and *PR5* were induced in roots while *PR1* to *PR5* were down-regulated in leaves (Hamamouch et al., 2011). Similar results were found in tomato, although *PR* genes were described as being down-regulated in roots in addition to shoots, but a strong up-regulation of *PR* genes was found in the shoots of resistant tomato genotypes (Molinari et al., 2014).

Genome-wide analysis of the *HSP20* gene family in soybean after *M. incognita* infection revealed six responsive *GmHSP20* genes by qRT-PCR (Lopes-Caitar et al., 2013). Some of them were down-regulated in a susceptible line but up- regulated in the resistant genotype. The proteins encoded by these genes are promising targets for developing crop varieties that are better adapted to biotic and abiotic stresses. QPCR analysis has also pointed to the down-regulation of thionins, peptides involved in plant defence, in galls induced by *M. graminicola* at early and medium stages of development (Ji et al., 2015). Addtionally, the expression of six PAL genes related to the phenylpropanoids pathway, in three maize genotypes that were good, moderate, and poor hosts for *M. incognita* showed that *ZmPAL4* was most strongly expressed in the most-resistant maize line (Starr et al., 2014), suggesting a role for this pathway in the defence against the RKNs as previously commented.

The above compilation represents only a brief selection of the successful use of techniques as qPCR and GUS reporter lines to study the defence mechanisms of the plants against the root-knot nematodes. There are a multitude of studies, however, which are based on the use of these techniques to decipher the expression of individual genes that could be related with the plant- nematode interaction.

Proteomic and Metabolomic analyses on plant-RKN interaction

Proteomic and metabolomic profiles of plants infected with RKN are very scarce as compared to the literature regarding transcriptomic studies. Moreover, traditional model plants as tomato and Arabidopsis have not been used for this purpose, which somehow limits the possibility to combine the data with the genomic information available in different databases to gain further molecular knowledge of the interaction. One of the first proteomic studies was made in a *Medicago sativa* cultivar resistant to *M. incognita*, Moapa 69, which lacks the typical hypersensitive response identified in tomato and tobacco (Potenza et al, 1996). Differences in J2 behaviour between the susceptible and resistant cultivars first appeared 48 to 72 hours after infection, coinciding with the entry of J2 into the vascular cylinder of the susceptible line Lahontan. Bidimensional electrophoresis (2D electrophoresis) of translation products of RNA from infected roots of both cultivars revealed no specific transcript variation in the resistant cultivar as compared to that of the susceptible one and to a general mechanical wounding response (Potenza et al., 1996). In contrast, another pioneer study was performed via one-and two dimensional polyacrylamide gel electrophoresis (PAGE) on a resistant cotton line that also showed nematode development arrest soon after infection (Callahan et al.,1997). From this study, a highly abundant protein that temporally correlated with the resistance response was detected and later identified as *MIC-3* (*Meloidogyne* Induced Cotton3) whose over-expression reduces egg production by 60-75% (Wubben et al., 2015).

Franco et al. (2010) studied the proteome of resistant cotton and coffee cultivars after *M. paranaensis* and *M. incognita* infection using 2D electrophoresis. They found 4 up and 5 down-regulated proteins in cotton infected roots at 6 dpi, and 13 up and 2 down-regulated at 10 dpi together with 7 induced / 3 repressed proteins at 6 dpi in coffee (Franco et al., 2010). Two of those proteins from coffee and one from cotton were identified as a chitinase, a pathogenesis related protein and a quinone reductase 2, respectively, all related to disease resistance (Franco et al., 2010). In this respect, overexpression of quinone reductases in Arabidopsis caused hypersensitive reaction to necrotrophic fungi (Heyno et al., 2013). Likewise, when comparing a highly resistant and a slightly resistant cowpea line at 3, 6 and 9 dpi after challenging both with *M. incognita,* significant differential expression profiles were observed. Interestingly, 13 unique proteins were identified in the highly resistant cultivar compared to the slightly resistant line (Villeth et al., 2015). Several of the induced proteins are associated to the oxidative stress responses in the highly resistant cultivar, including a multicatalytic endopeptidase complex (proteasome), hydroxyacid oxidase, gamma-type carbonic anhydrase family protein, ferredoxin-NADP reductase isozyme 2, and glutathione S-transferase (Villeth et al., 2015). Some of them, like the hydroxyacid oxidase and the gamma-type carbonic anhydrase, probably participate in the hypersensitive response of the highly resistant cultivar. The authors also found resistance related proteins such as patatin or the resistance associated protein RPP13 (Villeth et al., 2015).
At later stages of development (35-40 dpi), 2D electrophoresis has been also used to unravel the proteome of the tripartite interaction *M. artiellia-Fusarium oxysporum*, in fungus resistant cultivars of *Cicer arietinum* (chickpea) (Palomares-Rius et al., 2011). One of the cultivars lost the resistance to *F. oxysporum* after *M. artiellia* interaction. Class I quitinases were significantly induced in the cultivar maintaining *F. oxysporum* (Foc-5) resistance after *M. artiellia* infection. This differential response could thus be interpreted as a key determinant of the maintenance of Foc-5-resistance after infection with both pathogens. Additionally, a class III peroxidase and a R protein of the NBS-LRR class increased 5.5-fold and 4 fold, respectively in the cultivar that lost the *F. oxysporum* resistance, after *Meloidogyne artiellia* infection compared with the control, but did not change after infection by either pathogen alone. The class III peroxidase and catalase level changes reinforce the idea that redox perturbation is associated with the chickpea response to *M. artiella* infection.

In this respect, an antioxidant molecule involved in the cellular redox homeostasis is glutathione and homoglutathione, (h)GSH. Its crucial role in plant signalling under biotic stress is well known (Dubreuil-Maurizi and Poinssot, 2012). The (h)GSH metabolism is modified in nematode-induced root galls as the (h)GSH content is significantly higher in galls than in roots at late infection stages. A detailed analysis of glutathione (GSH) and homoglutathione in galls, (hGSH), demonstrated the importance of these compounds for the success of *M. incognita* infection in *Medicago truncatula*. Moreover, the impairment of nematode development and reproduction in (h)GSH-depleted galls was linked to a modified carbon metabolism at the feeding site (Baldacci-Cresp et al., 2012). Hence, a substantial modification of starch and γ-aminobutyrate metabolism and of malate and glucose content in (h)GSH-depleted galls compared to control roots was detected by untargeted proton Nuclear Magnetic Resonance. Eighteen of the 37 quantified metabolites were related to sugar, organic acid and amino acid metabolism. GABA was specifically detected in (h)GSH-depleted galls (Baldacci-Cresp et al., 2012) and it is known to accumulate in the plant in response to abiotic and biotic stresses (Kinnersley and Turano, 2000). The reproduction of *M. hapla* in transgenic plants accumulating GABA is lower than in control plants (McLean et al., 2003). Although details regarding how (h)GSH content can modulate metabolism are still lacking, there is evidence suggesting that the control of the plant cell redox status through the modification of the (h)GSH content may be a key regulator of the GC potential to sustain nematode feeding.

Concluding Remarks

Meloidogyne spp. is one of the most devastating pathogens for plants. Technology has contributed massively to the advances on genomic, transcriptomic, metabolomic and proteomic analyses of plant-*Meloidogyne* interactions, increasing the knowledge on the molecular aspects of the RKNs interaction in the last 10-15 years. Studies based on different omics tried to decipher the molecular pathways leading to the successful establishment of the nematode in the root and to GC differentiation, as well as those governing the plant resistance leading to incompatible plant-RKN interactions. Different holistic approaches to obtain gene expression profiles, such as cDNA libraries, differential display, q-PCR, hybridization of microarrays, massive sequencing etc., anticipated the development of biotechnological tools based on this molecular knowledge to control this plague. However, the vast amount of transcriptomic data for the plant-nematode interaction has not always correlated directly with an increase in the identification of genes functionality or metabolic pathways within the feeding sites, neither with an enhancement of genetic tools for biotechnological purposes leading to nematode control. One of the limitations has typically been the difficulty of cross- comparative analysis of these data. A first attempt to achieve this objective was the construction of a database, NEMATIC, compiling microarray data available in Arabidopsis for the interaction with plant endoparasitic nematodes. It also integrates transcriptomic data from several biological processes, plant biotic and abiotic interactions, hormonal regulation, specific transcriptomes for different root cell types, etc., but is not sufficient for the handling of "omic" information of different plant species, thus it should be expanded to other plant species.

The main information obtained from the RKN interaction based on omics is that plant defence shut down at early infection stages within GCs should serve the nematode for its successful establishment and development. Omics combined with cell isolation techniques have also shed some light on the heterogeneous expression signatures of nematode induced gall tissues, showing contrasted expression profiles in the GCs compared to the rest of the tissues. The natural resistance gene *Mi-1* and their homologues, are based in the generation of a typical hypersensitive reaction, modulating plant defence pathways before the establishment of the nematode. However, there are also other resistance modes in other crops that allow the arrest of gall formation at early stages of development without a typical hypersensitive reaction. Omics are contributing to increase the knowledge on the molecular bases of nematode resistance. This is an open field still to be fully uncovered that could help generating powerful targets to the development of biotechnology based tools for nematode control.

Acknowledgements

This work was supported by the Spanish Government (AGL2013-48787 to C. Escobar, and CSD2007-057 and PCIN-2013-053 to C. Fenoll) and by the Castilla-la Mancha Government (PEII-2014-020-P to CF). Javier Cabrera was supported by a fellowship from the Ministry of Education, Spain. We acknowledge Prof. Dr. Godelieve Gheysen and Ji Hongli and Kamrun Nahar from Ghent University for kindly provide us with the pictures of galls and giant cells from rice in Figure 1, Alberto de Marcos from University of Castilla- La Mancha for the plant designs in Figure 1 and Celia Martin for the reading of the manuscript.

References

Ambawat, S., Sharma, P., Yadav, N.R., and Yadav, R.C. (2013). MYB transcription factor genes as regulators for plant responses: an overview. Physiol Mol Biol Plants *19*, 307-321.

Asai, S., Rallapalli, G., Piquerez, S.J.M., Caillaud, M.-C., Furzer, O.J., Ishaque, N., Wirthmueller, L., Fabro, G., Shirasu, K., and Jones, J.D.G. (2014). Expression Profiling during Arabidopsis/Downy Mildew Interaction Reveals a Highly-Expressed Effector That Attenuates Responses to Salicylic Acid. PLoS Pathog *10*, e1004443.

Bagnaresi, P., Sala, T., Irdani, T., Scotto, C., Lamontanara, A., Beretta, M., Rotino, G.L., Sestili, S., Cattivelli, L., and Sabatini, E. (2013). Solanum torvum responses to the root-knot nematode Meloidogyne incognita. BMC Genomics *14*, 540.

Bakshi, M., and Oelmüller, R. (2014). WRKY transcription factors. Plant Signaling and Behavior *9*, e27700.

Baldacci-Cresp, F., Chang, C., Maucourt, M., Deborde, C., Hopkins, J., Lecomte, P., Bernillon, S., Brouquisse, R., Moing, A., Abad, P., *et al.* (2012). (Homo)glutathione deficiency impairs root-knot nematode development in Medicago truncatula. PLoS Pathog *8*, e1002471.

Bar-Or, C., Kapulnik, Y., and Koltai, H. (2005). A broad characterization of the transcriptional profile of the compatible tomato response to the plant parasitic root knot nematode Meloidogyne javanica. European Journal of Plant Pathology *111*, 181-192.

Barcala, M., Garcia, A., Cabrera, J., Casson, S., Lindsey, K., Favery, B., Garcia-Casado, G., Solano, R., Fenoll, C., and Escobar, C. (2010). Early transcriptomic events in microdissected Arabidopsis nematode-induced giant cells. Plant J *61*, 698-712.

Barthels, N., van der Lee, F.M., Klap, J., Goddijn, O.J., Karimi, M., Puzio, P., Grundler, F.M., Ohl, S.A., Lindsey, K., Robertson, L., *et al.* (1997). Regulatory sequences of Arabidopsis drive reporter gene expression in nematode feeding structures. Plant Cell *9*, 2119-2134.

Beneventi, M.A., da Silva, O.B., Jr., de Sa, M.E., Firmino, A.A., de Amorim, R.M., Albuquerque, E.V., da Silva, M.C., da Silva, J.P., Campos Mde, A., Lopes, M.J., *et al.* (2013). Transcription profile of soybean-root-knot nematode interaction reveals a key role of phythormones in the resistance reaction. BMC Genomics *14*, 322.

Bhattarai, K.K., Atamian, H.S., Kaloshian, I., and Eulgem, T. (2010). WRKY72-type transcription factors contribute to basal immunity in tomato and Arabidopsis as well as gene-for-gene resistance mediated by the tomato R gene Mi-1. Plant J *63*, 229-240.

Bhattarai, K.K., Xie, Q.G., Mantelin, S., Bishnoi, U., Girke, T., Navarre, D.A., and Kaloshian, I. (2008). Tomato susceptibility to root-knot nematodes requires an intact jasmonic acid signaling pathway. Mol Plant Microbe Interact *21*, 1205-1214.

Bird, D.M., and Wilson, M.A. (1994). DNA sequence and expression analysis of root-knot nematode-elicited giant cell transcripts. Mol Plant Microbe Interact *7*, 419-424.

Brown, S., Ionit, I., Buki, P., and Kolomiets, M. (2015). The Role of Lipid Signalling in Regulating Plant-Nematode Interactions. In Advances in Botanical Research Volume 73, C. Escobar and C. Fenoll, eds. (Oxford, UK: Elsevier Academic Press), pp. 139-166.

Cabrera, J., Barcala, M., Fenoll, C., and Escobar, C. (2014a). Transcriptomic signatures of transfer cells in early developing nematode feeding cells of Arabidopsis focused on auxin and ethylene signaling. Front Plant Sci *5*, 107.

Cabrera, J., Bustos, R., Favery, B., Fenoll, C., and Escobar, C. (2014b). NEMATIC: a simple and versatile tool for the in ilico analysis of plant-nematode interactions. Molecular Plant Pathology *15*, 627-636.

Cabrera, J., Diaz-Manzano, F.E., Barcala, M., Arganda-Carreras, I., de Almeida-Engler, J., Engler, G., Fenoll, C., and Escobar, C. (2015). Phenotyping nematode feeding sites: three-dimensional reconstruction and volumetric measurements of giant cells induced by root-knot nematodes in Arabidopsis. New Phytol *206*, 868-880.

Callahan, F.E., Jenkins, J.N., Creech, R.G., and Lawrence, G.W. (1997). Changes in root cotton proteins correlated with resistance to root knot nematode development. J Cot Sci *1*, 38-47.

Callahan, F.E., Zhang, X.D., Ma, D.P., Jenkins, J.N., Hayes, R.W., and Tucker, M.L. (2004). Comparison of MIC-3 protein accumulation in response to root-knot nematode infection in cotton lines displaying a range of resistance levels. J Cot Sci *8*, 186-190.

Damiani, I., Baldacci-Cresp, F., Hopkins, J., Andrio, E., Balzergue, S., Lecomte, P., Puppo, A., Abad, P., Favery, B., and Herouart, D. (2012). Plant genes involved in harbouring symbiotic rhizobia or pathogenic nematodes. New Phytol *194*, 511-522.

Das, S., Ehlers, J.D., Close, T.J., and Roberts, P.A. (2010). Transcriptional profiling of root-knot nematode induced feeding sites in cowpea (Vigna unguiculata L. Walp.) using a soybean genome array. BMC Genomics *11*, 480.

de Almeida-Engler, J., Vieira, P., Rodiuc, N., Grossi de Sa, M.F., and Engler, G. (2015). The Plant Cell Cycle Machinery: Usurped and Modulated by Plant-Parasitic Nematodes. In Advances in Botanical Research Volume 73, C. Escobar and C. Fenoll, eds. (Oxford, UK: Elsevier Academic Press), pp. 91-118.

de Sa, M.E., Conceicao Lopes, M.J., de Araujo Campos, M., Paiva, L.V., Dos Santos, R.M., Beneventi, M.A., Firmino, A.A., and de Sa, M.F. (2012). Transcriptome analysis of resistant soybean roots infected by Meloidogyne javanica. Genet Mol Biol *35*, 272-282.

Dixon, R.A., Achnine, L., Kota, P., Liu, C.-J., Reddy, M.S.S., and Wang, L. (2002). The phenylpropanoid pathway and plant defence—a genomics perspective. Molecular Plant Pathology *3*, 371-390.

Dubreuil-Maurizi, C., and Poinssot, B. (2012). Role of glutathione in plant signaling under biotic stress. Plant Signal Behav *7*, 210-212.

Dure, L. (1993). The LEA proteins of higher plants. In Control of Plant Gene Expression. D. P. S. Verma, ed. (Cyty, Country: CRC Press), pp. 325-335.

Escobar, C., Barcala, M., Cabrera, J., and Fenoll C. (2015). Overview of Root-Knot Nematodes and Giant Cells. In Advances in Botanical Research Volume 73, C. Escobar and C. Fenoll, eds. (Oxford, UK: Elsevier Academic Press), pp. 1-32.

Escobar, C., De Meutter, J., Aristizábal, F.A., Sanz-Alférez, S., del Campo, F.F., Barthels, N., Van der Eycken, W., Seurinck, J., van Montagu, M., Gheysen, G., *et al.* (1999). Isolation of the LEMMI9 Gene and Promoter Analysis During a Compatible Plant-Nematode Interaction. Molecular Plant-Microbe Interactions *12*, 440-449.

Fobert, P.R., and Després, C. (2005). Redox control of systemic acquired resistance. Current Opinion in Plant Biology *8*, 378-382.

Fosu-Nyarko, J., Jones, M.G., and Wang, Z. (2009). Functional characterization of transcripts expressed in early-stage Meloidogyne javanica-induced giant cells isolated by laser microdissection. Mol Plant Pathol *10*, 237-248.

Franco, O.L., Pereira, J.L., Costa, P.H., Rocha, T.L., Albuquerque, E.V., Grossi-de-Sa, M.F., Carneiro, R.M., Carneiro, R.G., and Mehta, A. (2010). Methodological evaluation of 2-DE to study root proteomics during nematode infection in cotton and coffee plants. Prep Biochem Biotechnol *40*, 152-163.

Fuller, V.L., Lilley, C.J., Atkinson, H.J., and Urwin, P.E. (2007). Differential gene expression in Arabidopsis following infection by plant-parasitic nematodes Meloidogyne incognita and Heterodera schachtii. Mol Plant Pathol *8*, 595-609.

Grunewald, W., Karimi, M., Wieczorek, K., Van de Cappelle, E., Wischnitzki, E., Grundler, F., Inze, D., Beeckman, T., and Gheysen, G. (2008). A role for AtWRKY23 in feeding site establishment of plant-parasitic nematodes. Plant Physiol *148*, 358-368.

Grunewald, W., Vanholme, B., Pauwels, L., Plovie, E., Inze, D., Gheysen, G., and Goossens, A. (2009). Expression of the Arabidopsis jasmonate signalling repressor JAZ1/TIFY10A is stimulated by auxin. EMBO Rep *10*, 923-928.

Habib, H., and Fazili, K.M. (2007). Plant protease inhibitors: a defense strategy in plants. Biotechnology and Molecular Biology Reviews. *2*, 068-085.

Hamamouch, N., Li, C., Seo, P.J., Park, C.M., and Davis, E.L. (2011). Expression of Arabidopsis pathogenesis-related genes during nematode infection. Mol Plant Pathol *12*, 355-364.

Hammes, U.Z., Schachtman, D.P., Berg, R.H., Nielsen, E., Koch, W., McIntyre, L.M., and Taylor, C.G. (2005). Nematode-induced changes of transporter gene expression in Arabidopsis roots. Mol Plant Microbe Interact *18*, 1247-1257.

Heyno, E., Alkan, N., and Fluhr, R. (2013). A dual role for plant quinone reductases in host-fungus interaction. Physiol Plant *149*, 340-353.

Iberkleid, I., Sela, N., and Brown Miyara, S. (2015). Meloidogyne javanica fatty acid- and retinol-binding protein (Mj-FAR-1) regulates expression of lipid-, cell wall-, stress- and phenylpropanoid-related genes during nematode infection of tomato. BMC Genomics *16*, 272.

Iberkleid, I., Vieira, P., de Almeida Engler, J., Firester, K., Spiegel, Y., and Horowitz, S.B. (2013). Fatty acid-and retinol-binding protein, Mj-FAR-1 induces tomato host susceptibility to root-knot nematodes. PLoS One *8*, e64586.

Ibrahim, H.M., Hosseini, P., Alkharouf, N.W., Hussein, E.H., Gamal El-Din Ael, K., Aly, M.A., and Matthews, B.F. (2011). Analysis of gene expression in soybean (Glycine max) roots in response to the root knot nematode Meloidogyne incognita using microarrays and KEGG pathways. BMC Genomics *12*, 220.

Jammes, F., Lecomte, P., de Almeida-Engler, J., Bitton, F., Martin-Magniette, M.L., Renou, J.P., Abad, P., and Favery, B. (2005). Genome-wide expression profiling of the host response to root-knot nematode infection in Arabidopsis. Plant J *44*, 447-458.

Ji, H., Gheysen, G., Denil, S., Lindsey, K., Topping, J.F., Nahar, K., Haegeman, A., De Vos, W.H., Trooskens, G., Van Criekinge, W., *et al. (*2013). Transcriptional analysis through RNA sequencing of giant cells induced by Meloidogyne graminicola in rice roots. J Exp Bot *64*, 3885-3898.

Ji, H., Gheysen, G., Ullah, C., Verbeek, R., Shang, C., De Vleesschauwer, D., Hofte, M., and Kyndt, T. (2015). The role of thionins in rice defence against root pathogens. Mol Plant Pathol.

Kinnersley, A.M., and Turano, F.J. (2000). Gamma Aminobutyric Acid (GABA) and Plant Responses to Stress. Critical Reviews in Plant Sciences *19*, 479-509.

Koiwa, H., Bressan, R.A., and Hasegawa, P.M. (1997). Regulation of protease inhibitors and plant defense. Trends in Plant Science 2, 379-384.

Kyndt, T., Denil, S., Haegeman, A., Trooskens, G., Bauters, L., Van Criekinge, W., De Meyer, T., and Gheysen, G. (2012a). Transcriptional reprogramming by root knot and migratory nematode infection in rice. New Phytol *196*, 887-900.

Kyndt, T., Fernandez, D., and Gheysen, G. (2014). Plant-parasitic nematode infections in rice: molecular and cellular insights. Annu Rev Phytopathol *52*, 135-153.

Kyndt, T., Nahar, K., Haegeman, A., De Vleesschauwer, D., Hofte, M., and Gheysen, G. (2012b). Comparing systemic defence-related gene expression changes upon migratory and sedentary nematode attack in rice. Plant Biol (Stuttg) *14 Suppl 1*, 73-82.

Hussey, R.S. (1989). Disease-inducing secretions of plant-parasitic nematodes. Annual Review of Phytopathology. 27,123-141.

Lambert, K.N., Ferrie, B.J., Nombela, G., Brenner, E.D., and Williamson, V.M. (1999). Identification of genes whose transcripts accumulate rapidly in tomato after root-knot nematode infection. Physiological and Molecular Plant Pathology *55*, 341-348.

Li, J., Brader, G., and Palva, E.T. (2008). Kunitz Trypsin Inhibitor: An Antagonist of Cell Death Triggered by Phytopathogens and Fumonisin B1 in Arabidopsis. Molecular Plant *1*, 482-495.

Liu, J., Osbourn, A., and Ma, P. (2015). MYB Transcription Factors as Regulators of Phenylpropanoid Metabolism in Plants. Molecular Plant *8*, 689-708.

Lopes-Caitar, V.S., de Carvalho, M.C., Darben, L.M., Kuwahara, M.K., Nepomuceno, A.L., Dias, W.P., Abdelnoor, R.V., and Marcelino-Guimaraes, F.C. (2013). Genome-wide analysis of the Hsp20 gene family in soybean: comprehensive sequence, genomic organization and expression profile analysis under abiotic and biotic stresses. BMC Genomics *14*, 577.

Mantelin, S., Bhattarai, K.K., Jhaveri, T.Z., and Kaloshian, I. (2013). *Mi-1*-Mediated Resistance to *Meloidogyne incognita* in Tomato May Not Rely on Ethylene but Hormone Perception through ETR3 Participates in Limiting Nematode Infection in a Susceptible Host. PLoS ONE *8*, e63281.

McCarter, J.P. (2008). Molecular approaches toward resistance to plant-parasitic nematodes. In Cell Biology of Plant Nematode Parasitism, R. H. Berg and C. G. Taylor, eds. (Berlin, Germany: Springer), pp. 239-267.

McLean, M.D., Yevtushenko, D.P., Deschene, D., Van Cauwenberghe, O.R., Makhmoudova, A., Potter, J.W., Bown, A.W., and Shelp BJ (2003) Overexpression of glutamate decarboxylase in transgenic tobacco plants confers resistance to the northern root-knot nematode. Mol. Breed. *11*, 277-285

Melillo, M.T., Leonetti, P., Leone, A., Veronico, P., and Bleve-Zacheo, T. (2011) ROS and NO production in compatible and incompatible tomato-Meloidogyne incognita interactions. Eur J Plant Pathol *130*, 489-502.

Mizoi, J., Shinozaki, K., and Yamaguchi-Shinozaki, K. (2012). AP2/ERF family transcription factors in plant abiotic stress responses. Biochimica et Biophysica Acta (BBA) - Gene Regulatory Mechanisms *1819*, 86-96.

Molinari, S., Fanelli, E., and Leonetti, P. (2014). Expression of tomato salicylic acid (SA)-responsive pathogenesis-related genes in Mi-1-mediated and SA-induced resistance to root-knot nematodes. Mol Plant Pathol *15*, 255-264.

Nahar, K., Kyndt, T., De Vleesschauwer, D., Hofte, M., and Gheysen, G. (2011). The jasmonate pathway is a key player in systemically induced defense against root knot nematodes in rice. Plant Physiol *157*, 305-316.

Nahar, K., Kyndt, T., Hause, B., Hofte, M., and Gheysen, G. (2013). Brassinosteroids suppress rice defense against root-knot nematodes through antagonism with the jasmonate pathway. Mol Plant Microbe Interact *26*, 106-115.

Nandi, B., Sukul, N.C., Banewee, N., Sengupta, S., Das, P., and Sinha BSP (2003). Induction of pathogenesis-related protein by salicylic acid and resistance to root-knot nematode in tomato. Indian J. Nematology. *33*,111-116.

Ozalvo, R., Cabrera, J., Escobar, C., Christensen, S.A., Borrego, E.J., Kolomiets, M.V., Castresana, C., Iberkleid, I., and Brown Horowitz, S. (2014). Two closely related members of Arabidopsis 13-lipoxygenases (13-LOXs), LOX3 and LOX4, reveal distinct functions in response to plant-parasitic nematode infection. Mol Plant Pathol *15*, 319-332.

Palomares-Rius, J.E., Castillo, P., Navas-Cortes, J.A., Jimenez-Diaz, R.M., and Tena, M. (2011). A proteomic study of in-root interactions between chickpea pathogens: the root-knot nematode Meloidogyne artiellia and the soil-borne fungus Fusarium oxysporum f. sp. ciceris race 5. J Proteomics *74*, 2034-2051.

Park, S.-C., Lee, J.R., Shin, S.-O., Park, Y., Lee, S.Y., and Hahm, K.-S. (2007). Characterization of a heat-stable protein with antimicrobial activity from Arabidopsis thaliana. Biochemical and Biophysical Research Communications *362*, 562-567.

Paulson, R.E., and Webster, J.M. (1972). Ultrastructure of the hypersensitive reaction in roots of tomato, Lycopersicon esculentum L., to infection by the root-knot nematode, Meloidogyne incognita. Physiological Plant Pathology *2*, 227-234.

Plovie, E., De Buck, S., Goeleven, E., Tanghe, M., Vercauteren, I., and Gheysen, G. (2003). Hairy roots to test for transgenic nematode resistance: think twice. Nematology *5*, 831-841.

Portillo, M., Cabrera, J., Lindsey, K., Topping, J., Andres, M.F., Emiliozzi, M., Oliveros, J.C., Garcia-Casado, G., Solano, R., Koltai, H., *et al.* (2013). Distinct and conserved transcriptomic changes during nematode-induced giant cell development in tomato compared with Arabidopsis: a functional role for gene repression. New Phytol *197*, 1276-1290.

Postnikova, O.A., Hult, M., Shao, J., Skantar, A., and Nemchinov, L.G. (2015). Transcriptome analysis of resistant and susceptible alfalfa cultivars infected with root-knot nematode Meloidogyne incognita. PLoS One *10*, e0118269.

Potenza, C., Thomas, S.H., and Sengupta-Gopalan, C. (2001). Genes induced during early response to Meloidogyne incognita in roots of resistant and susceptible alfalfa cultivars. Plant Sci *161*, 289-299.

Potenza, C.L., Thomas, S.H., Higgins, E.A., and Sengupta-Gopalan, C. (1996). Early Root Response to Meloidogyne incognita in Resistant and Susceptible Alfalfa Cultivars. J Nematol *28*, 475-484.

Puga, M.I., Mateos, I., Charukesi, R., Wang, Z., Franco-Zorrilla, J.M., de Lorenzo, L., Irigoyen, M.L., Masiero, S., Bustos, R., Rodríguez, J., *et al.* (2014). SPX1 is a phosphate-dependent inhibitor of PHOSPHATE STARVATION RESPONSE 1 in Arabidopsis. Proceedings of the National Academy of Sciences *111*, 14947-14952.

Rodiuc, N., Vieira, P., Banora, M.Y., and de Almeida Engler, J. (2014). On the track of transfer cell formation by specialized plant-parasitic nematodes. Front Plant Sci *5*, 160.

Schaff, J.E., Nielsen, D.M., Smith, C.P., Scholl, E.H., and Bird, D.M. (2007). Comprehensive transcriptome profiling in tomato reveals a role for glycosyltransferase in Mi-mediated nematode resistance. Plant Physiol *144*, 1079-1092.

Siddique, S., and Grundler, F.M.W. (2015). Metabolism in Nematode Feeding Sites. In Advances in Botanical Research Volume 73, C. Escobar and C. Fenoll, eds. (Oxford, UK: Elsevier Academic Press), pp. 119-138.

Siddique, S., Matera, C., Radakovic, Z.S., Shamim Hasan, M., Gutbrod, P., Rozanska, E., Sobczak, M., Torres, M.A., and Grundler, F.M.W. (2014). Parasitic Worms Stimulate Host NADPH Oxidases to Produce Reactive Oxygen Species That Limit Plant Cell Death and Promote Infection. *7*, ra33

Starr, J.L., Yang, W., Yan, Y., Crutcher, F., and Kolomiets, M. (2014). Expression of Phenylalanine Ammonia Lyase Genes in Maize Lines Differing in Susceptibility to Meloidogyne incognita. J Nematol *46*, 360-364.

Takahashi, A., Casais, C., Ichimura, K., and Shirasu, K. (2003). HSP90 interacts with RAR1 and SGT1 and is essential for RPS2-mediated disease resistance in Arabidopsis. Proceedings of the National Academy of Sciences of the United States of America *100*, 11777-11782.

Thimm, O., Blasing, O., Gibon, Y., Nagel, A., Meyer, S., Kruger, P., Selbig, J., Muller, L.A., Rhee, S.Y., and Stitt, M. (2004). MAPMAN: a user-driven tool to display genomics data sets onto diagrams of metabolic pathways and other biological processes. Plant J *37*, 914-939.

Thines, B., Katsir, L., Melotto, M., Niu, Y., Mandaokar, A., Liu, G., Nomura, K., He, S.Y., Howe, G.A., and Browse, J. (2007). JAZ repressor proteins are targets of the SCF(COI1) complex during jasmonate signalling. Nature *448*, 661-665.

Tirumalaraju, S.V., Jain, M., and Gallo, M. (2011). Differential gene expression in roots of nematode-resistant and -susceptible peanut (Arachis hypogaea) cultivars in response to early stages of peanut root-knot nematode (Meloidogyne arenaria) parasitization. J Plant Physiol *168*, 481-492.

Truong, N.M., Nguyen, C.N., Abad, P., Quentin, M., and Favery, B. (2015). Function of Root-Knot Nematode Effectors and Their Targets in Plant Parasitism. In Advances in Botanical Research Volume 73, C. Escobar and C. Fenoll, eds. (Oxford, UK: Elsevier Academic Press), pp. 293-324.

Van der Eycken, W., de Almeida Engler, J., Inze, D., Van Montagu, M., and Gheysen, G. (1996). A molecular study of root-knot nematode-induced feeding sites. Plant J *9*, 45-54.

Vercauteren, I., Van Der Schueren, E., Van Montagu, M., and Gheysen, G. (2001). Arabidopsis thaliana Genes Expressed in the Early Compatible Interaction with Root-Knot Nematodes. Molecular Plant-Microbe Interactions *14*, 288-299.

Villeth, G.R., Carmo, L.S., Silva, L.P., Fontes, W., Grynberg, P., Saraiva, M., Brasileiro, A.C., Carneiro, R.M., Oliveira, J.T., Grossi-de-Sa, M.F., *et al.* (2015). Cowpea-Meloidogyne incognita interaction: Root proteomic analysis during early stages of nematode infection. Proteomics *15*, 1746-1759.

Wang, Z., Potter, R.H. and Jones, M.G.K. (2001). A novel approach to extract and analyse cytoplasmic contents from individual giant cells in tomato roots induced by *Meloidogyne javanica*. Int. J. Nematol. *11*, 219-225.

Wang, Z., Potter, R.H., and Jones, M.G.K. (2003). Differential display analysis of gene expression in the cytoplasm of giant cells induced in tomato roots by Meloidogyne javanica. Molecular Plant Pathology *4*, 361-371.

Wieczorek, K. (2015). Cell Wall Alterations in Nematode-Infected Roots. In Advances in Botanical Research Volume 73, C. Escobar and C. Fenoll, eds. (Oxford, UK: Elsevier Academic Press), pp. 61-90.

Williamson, V.M. (1999). Plant nematode resistance genes. Curr Opinion Plant Biol 2,327-331.

Wilson, M.A., Bird, D.M., and van der Knaap, E. (1994). A Comprehensive Subtractive cDNA Cloning Approach to

Identify Nematode-Induced Transcripts in Tomato. Phytopathology *84*, 299-303.

Wubben, M., Callahan, F., Hayes, R., and Jenkins, J. (2008). Molecular characterization and temporal expression analyses indicate that the MIC (Meloidogyne Induced Cotton) gene family represents a novel group of root-specific defense-related genes in upland cotton (Gossypium hirsutum L.). Planta *228*, 111-123.

Wubben, M., Callahan, F., Velten, J., Burke, J., and Jenkins, J. (2015a). Overexpression of MIC-3 indicates a direct role for the MIC gene family in mediating Upland cotton (Gossypium hirsutum) resistance to root-knot nematode (Meloidogyne incognita). Theoretical and Applied Genetics *128*, 199-209.

Zhang, X.D., Callahan, F.E., Jenkins, J.N., Ma, D.P., Karaca, M., Saha, S., and Creech, R.G. (2002). A novel root-specific gene, MIC-3, with increased expression in nematode-resistant cotton (Gossypium hirsutum L.) after root-knot nematode infection. Biochim Biophys Acta *1576*, 214-218.

Zhao, W., Li, Z., Fan, J., Hu, C., Yang, R., Qi, X., Chen, H., Zhao, F., and Wang, S. (2015). Identification of jasmonic acid-associated microRNAs and characterization of the regulatory roles of the miR319/TCP4 module under root-knot nematode stress in tomato. J Exp Bot.

Zhuang, Y., Gala, A., and Yen, Y. (2012). Identification of Functional Genic Components of Major Fusarium Head Blight Resistance Quantitative Trait Loci in Wheat Cultivar Sumai 3. Molecular Plant-Microbe Interactions *26*, 442-450.

Other Publications of Interest

Xu, J.-P. (2014). Next-generation Sequencing: Current Technologies and Applications. Caister Academic Press, U.K. ISBN: 978-1-908230-33-1

He, Z. (2014). Microarrays: Current Technology, Innovations and Applications. Caister Academic Press, U.K. ISBN: 978-1-908230-49-2

Poptsova, M.S. (2014). Genome Analysis: Current Procedures and Applications. Caister Academic Press, U.K. ISBN: 978-1-908230-29-4

Fuentes, M. and LaBaer, J. (2014). Proteomics: Targeted Technology, Innovations and Applications. Caister Academic Press, U.K. ISBN: 978-1-908230-46-1

Murillo, J., Vinatzer, B.A., Jackson, R.W. and Arnold, D.L. (2015). Bacteria-Plant Interactions: Advanced Research and Future Trends. Caister Academic Press, U.K. ISBN: 978-1-908230-58-4

Caranta, C., Aranda, M.A., Tepfer, M. and Lopez-Moya, J.J. (2011). Recent Advances in Plant Virology. Caister Academic Press, U.K. ISBN: 978-1-904455-75-2

Nannipieri, P., Pietramellara, G. and Renella, G. (2014). Omics in Soil Science. Caister Academic Press, U.K. ISBN: 978-1-908230-32-4

Herold, K.E. and Rasooly, A. (2009). Lab-on-a-Chip Technology: Fabrication and Microfluidics. Caister Academic Press, U.K. ISBN: 978-1-904455-46-2

Table S1. Chronologically-ordered list of articles that made use of different "omics" techniques to study the molecular mechanisms altered during the plant-root-knot nematode interaction. The authors, techniques, type of interaction, nematode and plant species and the biological material studied in each article are indicated in different columns.

Year	Reference	Omics Type	Techique	Interaction	Nematode Species	Plant Species	Biological material
1994	Wilson et al.,	Transcriptomic	cDNA Library	Susceptible	*M. incognita*	*S. lycopersicum*	Hand dissected GCs from dark grown 2 months old galls Vs dark grown uninfected root segments
1996	Van der Eycken et al.,	Transcriptomic	cDNA Library	Susceptible	*M. incognita*	*S. lycopersicum*	Hand dissected 5 weeks old galls Vs uninfected root segments
1996	Potenza et al.,	Proteomic	Bidimensional electrophoresis	Susceptible/ Resistant	*M. incognita*	*M. sativa*	3 days post inoculation roots
1997	Callahan et al.,	Proteomic	Bidimensional electrophoresis	Susceptible/ Resistant	*M. incognita*	*G. hirsutum*	Infected roots at 8 days after inoculation
1999	Lambert et al.,	Transcriptomic	cDNA Library	Resistant	*M. javanica*	*S. lycopersicum*	12 hours infected root tips
2001	Potenza et al.,	Transcriptomic	cDNA Library	Susceptible	*M. incognita*	*M. sativa*	3 days old galls Vs uninfected root segments
2001	Vercauteren et al.,	Transcriptomic	Differential display	Susceptible	*M. incognita*	*A. thaliana*	Hand dissected 2, 3, 4, 5 and 7 days post infection galls Vs uninfected root segments
2002	Zhang et al.,	Transcriptomic	cDNA Library	Resistant	*M. incognita*	*G. hirsutum*	10 days after infection galls
2003	Wang et al.,	Transcriptomic	Differential display	Susceptible	*M. javanica*	*S. lycopersicum*	25 days after inoculation GCs
2005	Bar-Or et al.,	Transcriptomic	Microarray	Susceptible	*M. javanica*	*S. lycopersicum*	5 and 10 days after inoculation galls
2005	Hammes et al.,	Transcriptomic	Microarray	Susceptible	*M. incognita*	*A. thaliana*	Roots containing at least 10 galls at 1, 2 and 4 weeks after inoculation
2005	Jammes et al.,	Transcriptomic	Microarray	Susceptible	*M. incognita*	*A. thaliana*	7, 14 and 21 days after inoculation galls
2007	Fuller et a.,	Transcriptomic	Microarray	Susceptible	*M. incognita*	*A. thaliana*	galls at 21 days after inoculation
2007	Schaff et al.,	Transcriptomic	Microarray	Susceptible/ Resistant	*M. incognita/M. hapla*	*S. lycopersicum*	Root segments containing galls at 12, 36 and 72 hours after inoculation and galls at 4 weeks after inoculation
2008	Bhattarai et al.,	Transcriptomic	Microarray	Susceptible/ Resistant	*M. incognita/M. javanica*	*S. lycopersicum*	24 h after inoculation, 1 cm of the infected root tips
2009	Fosu-Nyarko et al.,	Transcriptomic	cDNA Library	Susceptible	*M. javanica*	*S. lycopersicum*	GCs at 4 and 7 days post-infection
2010	Barcala et al.,	Transcriptomic	Microarray	Susceptible	*M. javanica*	*A. thaliana*	Isolated GCs and galls at 3 days post-infection
2010	Das et al.,	Transcriptomic	Microarray	Susceptible/ Resistant	*M. incognita*	V. unguiculata	Nematode infected root tissue at 3 and 9 days post-inoculation
2010	Franco et al.,	Proteomic	Bidimensional electrophoresis	Resistant	*M. paranaensis/M. incognita*	*G. hirsutum and C. canephora*	Roots were collected at 6 and 10 days after inoculation
2011	Ibrahim et al.,	Transcriptomic	Microarray	Susceptible	*M. incognita*	*G. max*	Infected roots at 12 days after inoculation and 10 weeks after inoculation
2011	Palomares-Rius et al.,	Proteomic	Bidimensional electrophoresis	Susceptible	*M. artiellia*	C. arietinum	Root segments at 35 to 40 days after inoculation
2011	Tirumalaraju et al.,	Transcriptomic	cDNA Library	Susceptible/ Resistant	M. arenaria	A. hypogaea	Infected roots at 12, 24, 48, and 72h post-inoculation
2012	Baldacci-Cresp et al.,	Metabolomic		Susceptible	*M. incognita*	M. truncatula	
2012	Damiani et al.,	Transcriptomic	Microarray	Susceptible	*M. incognita*	M. truncatula	Isolated GCs and sourranding cells at 7dpi
2012	de Sa et al.,	Transcriptomic	cDNA Library	Resistant	*M. javanica*	*G. max*	Infected roots at 6, 12, 24, 48, 96, 144 and 192 h post inoculation
2012	Kyndt et al.,	Transcriptomic	mRNA sequencing	Susceptible	*M. graminicola*	*O. sativa*	Galls at 23 and 7 days after inoculation
2013	Bagnaresi et al.,	Transcriptomic	Microarray	Susceptible	*M. incognita*	*S. torvum/S. melongena*	Infected roots at 14 days after inoculation
2013	Beneventi et al.,	Transcriptomic	mRNA sequencing	Resistant	*M. javanica*	*G. max*	Root sections at 0, 6, 12 h, 1, 2, 4, 6 and 8 days post inoculation
2013	Ji et al.,	Transcriptomic	mRNA sequencing	Susceptible	*M. graminicola*	*O. sativa*	Isolated GCs at 7 and 14 days after inoculation
2013	Portillo et al.,	Transcriptomic	Microarray	Susceptible	*M. javanica*	*S. lycopersicum*	3 and 7 days after infection isolated GCs and 1, 3, 7 and 14 days after infection galls
2015	Iberkleid et al.,	Transcriptomic	mRNA sequencing	Susceptible	*M. javanica*	*S. lycopersicum*	Infected roots at 2, 5 and 15 days after inoculation
2015	Postnikova et al.,	Transcriptomic	mRNA sequencing	Susceptible/ Resistant	*M. incognita*	*M. sativa*	Infected roots at 10 days after inoculation
2015	Villeth et al.,	Proteomic	Bidimensional electrophoresis	Resistant	*M. incognita*	V. unguiculata	Infected roots at 3, 6 and 9 days after inoculation
2015	Zhao et al.	Transcriptomic	mRNA sequencing	Susceptible	*M. incognita*	*S. lycopersicum*	tomato phloem tissue at 20 days post infection

Table S2. List of 200 genes from the stress Mapman category differentially expressed during giant cell (GC) and gall development in the compatible interaction between *Arabidopsis thaliana* and *Meloidogyne javanica* or *M. incognita*. M values for each gene in each of the experiments (Jammes et al., 2005; Barcala et al., 2010) are indicated. NDE stands for "Non-differentially expressed".

	M value					
Gene ID	GC 3dpi	Gall 3dpi	Gall 7dpi	Gall 14dpi	Gall 21dpi	Description
AT1G01170	NDE	NDE	0.82	NDE	NDE	ozone-responsive stress-related protein, putative | chr1:73931-74737 REVERSE
AT1G04430	NDE	NDE	-1.07	NDE	-1.06	dehydration-responsive protein-related | chr1:1198118-1201527 FORWARD
AT1G05850	-1.48	NDE	NDE	1.12	0.89	POM1: Encodes an endo chitinase-like protein AtCTL1. Essential for tolerance to heat, salt and drought stresses. Also involved in root hair development, cell expansion and response to cytokinin. Allelic to erh2. 11 alleles described in Hauser (1995). Mutant is defective in acquired thermotolerance, appears semidwarf throughout its life cycle and has extra lateral branches. There are two EMS alleles. Expression of AtHSP101 is not affected in the mutants.
AT1G07400	NDE	NDE	NDE	NDE	-2.10	17.8 kDa class I heat shock protein (HSP17.8-CI) | chr1:2274943-2275758 FORWARD
AT1G09560	NDE	NDE	NDE	-1.37	-0.94	GLP5: germin-like protein (GLP5)
AT1G10460	-0.99	NDE	NDE	NDE	NDE	GLP7: germin-like protein (GLP7)
AT1G11960	NDE	NDE	-0.98	-0.73	-0.68	FUNCTIONS IN: molecular_function unknown; INVOLVED IN: biological_process unknown; LOCATED IN: plasma membrane; EXPRESSED IN: 22 plant structures; EXPRESSED DURING: 13 growth stages; CONTAINS InterPro DOMAIN/s: Protein of unknown function DUF221 (InterPro:IPR003864); BEST Arabidopsis thaliana protein match is: early-responsive to dehydration protein-related / ERD protein-related (TAIR:AT1G62320.1); Has 746 Blast hits to 706 proteins in 111 species: Archae - 0; Bacteria - 4; Metazoa - 131; Fungi - 370; Plants - 224; Viruses - 0; Other Eukaryotes - 17 (source: NCBI BLink). | chr1:4039711-4043880 REVERSE
AT1G12270	NDE	NDE	0.89	NDE	NDE	Hop1: stress-inducible protein, putative | chr1:4172075-4174775 FORWARD
AT1G13930	-1.79	NDE	-1.18	NDE	NDE	Involved in response to salt stress. Knockout mutants are hypersensitive to salt stress.
AT1G18250	NDE	NDE	1.13	1.36	1.83	ATLP-1: encodes a thaumatin-like protein
AT1G18980	NDE	NDE	-0.71	-0.98	NDE	germin-like protein, putative | chr1:6557254-6558045 REVERSE
AT1G20440	NDE	NDE	-2.45	NDE	-1.45	COR47: Belongs to the dehydrin protein family, which contains highly conserved stretches of 7-17 residues that are repetitively scattered in their sequences, the K-, S-, Y- and lysine rich segments. Cold regulated gene, amino acid sequence homology with Group II LEA (late embryogenesis abundant) proteins. Also responds to osmotic stress, ABA, dehydration and inhibits e.coli growth while overexpressed. COR47 and RAB18 double overexpressor plants are cold tolerant.
AT1G21080	NDE	NDE	-0.91	NDE	NDE	DNAJ heat shock N-terminal domain-containing protein | chr1:7378589-7382462 REVERSE
AT1G23120	-2.63	NDE	NDE	NDE	-0.65	major latex protein-related / MLP-related | chr1:8198720-8199557 FORWARD
AT1G24020	3.24	2.57	NDE	NDE	NDE	MLP423: Symbols: MLP423 | MLP423 (MLP-LIKE PROTEIN 423) | chr1:8500451-8501520 REVERSE
AT1G28290	NDE	NDE	1.60	2.55	1.77	AGP31: Encodes an atypical arabinogalactan protein that is localized to the plasma membrange. AGP31 is highly expressed in flowers and vascular tissue and is repressed by jasmonic acid. AGP31 may play a role in vascular tissue function during defense and development.
AT1G30360	NDE	NDE	-1.61	NDE	-0.82	ERD4: Symbols: ERD4 | ERD4 (early-responsive to dehydration 4) | chr1:10715665-10718997 FORWARD
AT1G31850	-1.67	NDE	NDE	NDE	NDE	dehydration-responsive protein, putative | chr1:11430222-11433699 FORWARD
AT1G32090	NDE	NDE	NDE	-0.81	NDE	early-responsive to dehydration protein-related / ERD protein-related | chr1:11540025-11544131 REVERSE
AT1G35260	-2.63	NDE	NDE	NDE	NDE	MLP165: Symbols: MLP165 | MLP165 (MLP-LIKE PROTEIN 165) | chr1:12936849-12937787 REVERSE
AT1G47540	NDE	1.84	NDE	NDE	NDE	trypsin inhibitor, putative | chr1:17455533-17456156 REVERSE
AT1G48605	1.32	NDE	NDE	NDE	NDE	ATHAL3B: Encodes a protein similar to yeast HAL3, which regulates the cell cycle and tolerance to salt stress through inhibition of the PPZ1 type-1 protein phosphatase. AtHAL3b mRNA levels are induced by salt stress. HAL3B presumably encodes for phosphopantothenoylcysteine decarboxylase being involved in Coenzyme A biosynthesis as indicated by functional complementation of a double mutant hal3 aaBb.
AT1G52560	-1.54	NDE	NDE	NDE	NDE	26.5 kDa class I small heat shock protein-like (HSP26.5-P) | chr1:19574760-19575866 REVERSE
AT1G54050	NDE	NDE	NDE	0.79	NDE	17.4 kDa class III heat shock protein (HSP17.4-CIII) | chr1:20179422-20180281 REVERSE
AT1G56300	NDE	NDE	NDE	-1.18	-1.01	DNAJ heat shock N-terminal domain-containing protein | chr1:21078820-21080423 REVERSE
AT1G58170	-1.55	NDE	NDE	1.41	0.70	disease resistance-responsive protein-related / dirigent protein-related | chr1:21536128-21536836 FORWARD
AT1G59860	NDE	NDE	NDE	NDE	-1.22	17.6 kDa class I heat shock protein (HSP17.6A-CI) | chr1:22031413-22032132 FORWARD
AT1G64140	NDE	NDE	-1.11	NDE	NDE	FUNCTIONS IN: molecular_function unknown; INVOLVED IN: biological_process unknown; EXPRESSED IN: sperm cell, male gametophyte, pollen tube; EXPRESSED DURING: L mature pollen stage, M germinated pollen stage; BEST Arabidopsis thaliana protein match is: loricrin-related (TAIR:AT5G64550.1); Has 2298 Blast hits to 1429 proteins in 121 species: Archae - 0; Bacteria - 42; Metazoa - 1626; Fungi - 29; Plants - 243; Viruses - 9; Other Eukaryotes - 349 (source: NCBI BLink). | chr1:23803793-23807159 REVERSE
AT1G64160	NDE	NDE	NDE	NDE	-1.02	DIR5: disease resistance-responsive family protein / dirigent family protein | chr1:23814010-23814839 FORWARD
AT1G65280	NDE	NDE	NDE	NDE	0.93	heat shock protein binding | chr1:24245296-24248762 FORWARD
AT1G65390	-1.39	NDE	NDE	NDE	-0.70	PP2-A5: Symbols: ATPP2-A5 | ATPP2-A5 (ARABIDOPSIS THALIANA PHLOEM PROTEIN 2 A5); carbohydrate binding | chr1:24292478-24294631 FORWARD
AT1G69450	NDE	NDE	NDE	-1.01	-1.05	LOCATED IN: endomembrane system, membrane; EXPRESSED IN: 23 plant structures; EXPRESSED DURING: 13 growth stages; CONTAINS InterPro DOMAIN/s: Protein of unknown function DUF221 (InterPro:IPR003864); BEST Arabidopsis thaliana protein match is: HYP1 (HYPOTHETICAL PROTEIN 1) (TAIR:AT3G01100.1); Has 911 Blast hits to 825 proteins in 137 species: Archae - 0; Bacteria - 0; Metazoa - 152; Fungi - 442; Plants - 243; Viruses - 0; Other Eukaryotes - 74 (source: NCBI BLink). | chr1:26107120-26110148 REVERSE
AT1G70850	-2.52	NDE	-0.79	NDE	NDE	MLP34: Symbols: MLP34 | MLP34 (MLP-LIKE PROTEIN 34) | chr1:26715183-26716829 REVERSE
AT1G70880	NDE	-2.62	NDE	NDE	-0.72	Bet v I allergen family protein | chr1:26722881-26723778 REVERSE
AT1G70890	-1.79	NDE	NDE	NDE	NDE	MLP43: Symbols: MLP43 | MLP43 (MLP-LIKE PROTEIN 43) | chr1:26725626-26726544 REVERSE
AT1G72070	NDE	NDE	NDE	-1.58	-1.15	DNAJ heat shock N-terminal domain-containing protein | chr1:27118621-27119708 REVERSE
AT1G72500	NDE	NDE	-0.71	0.90	NDE	inter-alpha-trypsin inhibitor heavy chain-related | chr1:27295336-27298556 REVERSE
AT1G72860	NDE	NDE	NDE	-0.79	NDE	disease resistance protein (TIR-NBS-LRR class), putative | chr1:27416228-27420778 REVERSE
AT1G72920	NDE	1.21	NDE	NDE	NDE	disease resistance protein (TIR-NBS class), putative | chr1:27437928-27439091 FORWARD
AT1G72940	NDE	1.88	NDE	NDE	NDE	disease resistance protein (TIR-NBS class), putative | chr1:27442238-27443856 FORWARD
AT1G72950	NDE	NDE	NDE	NDE	-0.70	disease resistance protein (TIR-NBS class), putative | chr1:27444585-27445980 FORWARD
AT1G73260	-3.05	NDE	NDE	-2.09	-2.01	KTI1: Encodes a trypsin inhibitor involved in modulating programmed cell death in plant-pathogen interactions.
AT1G73325	NDE	NDE	NDE	-0.67	-0.67	trypsin and protease inhibitor family protein / Kunitz family protein | chr1:27567518-27568186 REVERSE
AT1G73330	NDE	NDE	-1.17	-0.71	NDE	DR4: encodes a plant-specific protease inhibitor-like protein whose transcript level in root disappears in response to progressive drought stress. The decrease in transcript level is independent from abscisic acid level.
AT1G73620	NDE	NDE	0.80	0.68	0.85	thaumatin-like protein, putative / pathogenesis-related protein, putative | chr1:27681408-27683130 FORWARD
AT1G74310	NDE	NDE	NDE	NDE	-0.64	HSP101: Encodes ClpB1, which belongs to the Casein lytic proteinase/heat shock protein 100 (Clp/Hsp100) family. Involved in refolding of proteins which form aggregates under heat stress. Also known as AtHsp101. AtHsp101 is a cytosolic heat shock protein required for acclimation to high temperature.
AT1G75030	-1.17	NDE	NDE	NDE	NDE	TLP-3: encodes a PR5-like protein

AT1G75800	NDE	NDE	NDE	0.82	NDE	pathogenesis-related thaumatin family protein | chr1:28458782-28460854 FORWARD
AT1G75830	NDE	NDE	NDE	NDE	-0.80	LCR67: Predicted to encode a PR (pathogenesis-related) protein. Belongs to the plant defensin (PDF) family with the following members: At1g75830/PDF1.1, At5g44420/PDF1.2a, At2g26020/PDF1.2b, At5g44430/PDF1.2c, At2g26010/PDF1.3, At1g19610/PDF1.4, At1g55010/PDF1.5, At2g02120/PDF2.1, At2g02100/PDF2.2, At2g02130/PDF2.3, At1g61070/PDF2.4, At5g63660/PDF2.5, At2g02140/PDF2.6, At5g38330/PDF3.1 and At4g30070/PDF3.2.
AT1G76180	NDE	NDE	-1.85	NDE	-1.04	ERD14: Encodes a dehydrin protein whose expression is induced early on in response to dehydration stress. This gene's expression to cold occurs in two waves, with early induction occurring within 1 h and secondary induction occurring 5 h after the beginning of cold stress. Expression is also induced in response to ABA but not in response to 2,4-D, BA, and GA3. ERD14 protein is capable of binding Ca2+, especially when the protein is phosphorylated.
AT1G77260	-2.15	NDE	NDE	NDE	NDE	dehydration-responsive protein-related | chr1:29023772-29026957 REVERSE
AT1G78040	-1.23	NDE	NDE	NDE	NDE	pollen Ole e 1 allergen and extensin family protein | chr1:29345838-29347107 FORWARD
AT2G01520	-2.09	NDE	NDE	NDE	NDE	MLP328: Symbols: MLP328 | MLP328 (MLP-LIKE PROTEIN 328); copper ion binding | chr2:235925-237147 FORWARD
AT2G01530	-1.42	NDE	NDE	NDE	NDE	MLP329: Symbols: MLP329 | MLP329 (MLP-LIKE PROTEIN 329); copper ion binding | chr2:239705-240520 FORWARD
AT2G04030	NDE	NDE	0.96	0.87	NDE	CR88: Encodes a chloroplast-targeted 90-kDa heat shock protein located in the stroma involved in red-light mediated deetiolation response. Mutants are resistant to chlorate, have elongated hypocotyls in light, and affect the expression of NR2, CAB, and RBCS but NOT NR1 and NiR.
AT2G14080	NDE	NDE	NDE	-0.69	NDE	disease resistance protein (TIR-NBS-LRR class), putative | chr2:5925186-5929600 FORWARD
AT2G14610	-4.07	NDE	NDE	NDE	NDE	PR1: PR1 gene expression is induced in response to a variety of pathogens. It is a useful molecular marker for the SAR response. Though the Genbank record for the cDNA associated to this gene is called 'PR-1-like', the sequence actually corresponds to PR1. Expression of this gene is salicylic-acid responsive.
AT2G15970	-1.53	NDE	-1.59	-0.90	-2.21	COR413-PM1: encodes an alpha form of a protein similar to the cold acclimation protein WCOR413 in wheat. Expression is induced by short-term cold-treatment, water deprivation, and abscisic acid treatment.
AT2G19970	NDE	2.36	NDE	0.68	0.86	pathogenesis-related protein, putative | chr2:8623730-8624476 REVERSE
AT2G19990	NDE	NDE	-0.75	-1.21	NDE	PR-1-LIKE: Encodes a PR-1-like protein homolog that is differentially expressed in resistant compared to susceptible cultivars by powdery mildew infection. The deduced amino acid sequence has 24 amino acids comprising the signal peptide and 140 amino acids of the mature peptide (15 kDa). Northern blot analysis showed accumulation of the corresponding mRNA 12 h after inoculation of resistant barley cultivars with Erysiphe graminis. Though the Genbank record for the cDNA associated to this gene model is called 'PR-1', the sequence actually corresponds to PR-1-like. Expression of this gene is not salicylic-acid responsive.
AT2G20142	NDE	NDE	NDE	-1.16	-1.07	transmembrane receptor | chr2:8695374-8696643 FORWARD
AT2G20550	NDE	NDE	NDE	0.71	NDE	DNAJ chaperone C-terminal domain-containing protein | chr2:8845802-8847309 REVERSE
AT2G20560	NDE	NDE	NDE	NDE	-0.98	DNAJ heat shock family protein | chr2:8848130-8850010 REVERSE
AT2G21100	-1.13	NDE	NDE	NDE	NDE	disease resistance-responsive protein-related / dirigent protein-related | chr2:9048131-9049404 REVERSE
AT2G21620	-1.87	NDE	NDE	NDE	NDE	RD2: Encodes gene that is induced in response to dessication; mRNA expression is seen 10 and 24 hrs after start of dessication treatment.
AT2G24040	NDE	NDE	NDE	-0.69	NDE	hydrophobic protein, putative / low temperature and salt responsive protein, putative | chr2:10223945-10224630 FORWARD
AT2G24550	NDE	NDE	-0.83	NDE	-1.18	unknown protein | chr2:10427856-10429222 REVERSE
AT2G26150	NDE	NDE	NDE	NDE	-1.38	HSFA2: member of Heat Stress Transcription Factor (Hsf) family. Involved in response to misfolded protein accumulation in the cytosol. Regulated by alternative splicing and non-sense-mediated decay.
AT2G26660	NDE	NDE	NDE	-0.78	NDE	SPX2: Symbols: ATSPX2, SPX2 | SPX2 (SPX DOMAIN GENE 2) | chr2:11338932-11340950 FORWARD
AT2G27140	NDE	NDE	0.79	1.38	1.27	heat shock family protein | chr2:11598379-11599322 REVERSE
AT2G28790	NDE	NDE	1.06	1.58	1.17	osmotin-like protein, putative | chr2:12354430-12355431 REVERSE
AT2G33790	NDE	-2.26	NDE	NDE	NDE	AGP30: pollen Ole e 1 allergen protein containing 14.6% proline residues, similar to arabinogalactan protein (Daucus carota) GI:11322245, SP:Q03211 Pistil-specific extensin-like protein precursor (PELP) {Nicotiana tabacum}; contains Pfam profile PF01190: Pollen proteins Ole e I family
AT2G35795	NDE	NDE	0.72	NDE	NDE	DNAJ heat shock N-terminal domain-containing protein | chr2:15042175-15043575 FORWARD
AT2G39200	NDE	NDE	-0.88	-0.73	-1.36	MLO12: A member of a large family of seven-transmembrane domain proteins specific to plants, homologs of the barley mildew resistance locus o (MLO) protein. The Arabidopsis genome contains 15 genes encoding MLO proteins, with localization in plasma membrane. Phylogenetic analysis revealed four clades of closely-related AtMLO genes. ATMLO6 belongs to the clade IV, with AtMLO2, AtMLO3 and AtMLO12. The gene is expressed during early seedling growth, in root tips and cotyledon vascular system, in floral organs (anthers and stigma), and in fruit abscission zone, as shown by GUS activity patterns. The expression of several phylogenetically closely-related AtMLO genes showed similar or overlapping tissue specificity and analogous responsiveness to external stimuli, suggesting functional redundancy, co-function, or antagonistic function(s).
AT2G40330	NDE	NDE	NDE	-0.85	-0.74	PYL6: Encodes a member of the PYR (pyrabactin resistance)/PYL(PYR1-like)/RCAR (regulatory components of ABA receptor) family proteins with 14 members. PYR/PYL/RCAR family proteins function as abscisic acid sensors. Mediate ABA-dependent regulation of protein phosphatase 2Cs ABI1 and ABI2.
AT2G41010	NDE	NDE	NDE	-1.18	-0.89	CAMBP25: Encodes a novel calmodulin binding protein whose gene expression is induced by dehydration and ionic (salt) and non-ionic (mannitol) osmotic stress. Lines over-expressing this gene are more sensitive and anti-sense lines are more tolerant to osmotic stress, suggesting this gene may be a negative regulator of response to osmotic stress.
AT2G41430	1.09	NDE	NDE	NDE	NDE	ERD15: Encodes hydrophilic protein lacking Cys residues that is expressed in response to drought stress, light stress and treatment with plant-growth-promoting rhizobacteria (Paenibacillus polymyxa), possibly revealing a connection between responses to biotic and abiotic stress. Also identified as a CTC Interacting Domain (CID) protein in a yeast two hybrid screen using the PAB2 protein as bait. Contains PAM2 like domain which mediates interaction with PABC domain in PAB2.
AT2G42750	NDE	NDE	NDE	NDE	0.65	DNAJ heat shock N-terminal domain-containing protein | chr2:17793307-17795680 FORWARD
AT2G43510	NDE	NDE	0.91	-1.02	NDE	TI1: Member of the defensin-like (DEFL) family. Encodes putative trypsin inhibitor protein which may function in defense against herbivory.
AT2G43530	NDE	2.53	NDE	NDE	NDE	Encodes a defensin-like (DEFL) family protein.
AT2G43590	NDE	NDE	NDE	-1.05	-0.75	chitinase, putative | chr2:18081331-18082767 REVERSE
AT2G43610	NDE	NDE	2.03	0.68	0.63	glycoside hydrolase family 19 protein | chr2:18087840-18089224 REVERSE
AT2G44490	NDE	NDE	NDE	NDE	-1.05	PEN2: Encodes a glycosyl hydrolase that localizes to peroxisomes and acts as a component of an inducible preinvasion resistance mechanism. Required for mlo resistance.
AT2G45130	NDE	NDE	-0.69	-1.04	NDE	SPX3: Symbols: ATSPX3, SPX3 | SPX3 (SPX DOMAIN GENE 3) | chr2:18606410-18607853 FORWARD
AT2G47440	NDE	NDE	NDE	-0.84	NDE	DNAJ heat shock N-terminal domain-containing protein | chr2:19469719-19471886 FORWARD
AT2G47540	NDE	NDE	0.95	NDE	NDE	pollen Ole e 1 allergen and extensin family protein | chr2:19505860-19506598 FORWARD
AT2G47710	NDE	1.14	NDE	0.71	0.71	universal stress protein (USP) family protein | chr2:19554834-19556065 REVERSE
AT2G47770	NDE	NDE	-0.89	-1.46	-1.43	TSPO: Encodes a membrane-bound protein designated AtTSPO (Arabidopsis thaliana TSPO-related). AtTSPO is related to the bacterial outer membrane tryptophan-rich sensory protein (TspO) and the mammalian mitochondrial 18 kDa Translocator Protein (18 kDa TSPO), members of the TspO/MBR domain-containing membrane proteins. Mainly detected in dry seeds, but can be induced in vegetative tissues by osmotic or salt stress or abscisic acid treatment. Located in endoplasmic reticulum and the Golgi stacks.
AT3G01420	NDE	NDE	-1.97	-1.76	-1.77	DOX1: Encodes an alpha-dioxygenase involved in protection against oxidative stress and cell death. Induced in response to Salicylic acid and oxidative stress. Independent of NPR1 in induction by salicylic acid.

AT3G02840	NDE	NDE	NDE	NDE	-0.97	immediate-early fungal elicitor family protein | chr3:618409-619686 FORWARD
AT3G04210	NDE	NDE	NDE	-1.01	NDE	disease resistance protein (TIR-NBS class), putative | chr3:1105963-1108069 REVERSE
AT3G04720	-4.15	NDE	NDE	-1.91	-1.15	PR4: Encodes a protein similar to the antifungal chitin-binding protein hevein from rubber tree latex. mRNA levels increase in response to ethylene and turnip crinkle virus infection.
AT3G05880	-2.12	NDE	-1.32	-1.81	-1.58	RCI2A: Induced by low temperatures, dehydration and salt stress and ABA. Encodes a small (54 amino acids), highly hydrophobic protein that bears two potential transmembrane domains.
AT3G05950	-1.23	NDE	1.22	-1.60	-1.70	germin-like protein, putative | chr3:1781067-1782007 REVERSE
AT3G06340	NDE	NDE	NDE	-0.79	NDE	DNAJ heat shock N-terminal domain-containing protein | chr3:1920207-1923311 REVERSE
AT3G07770	NDE	NDE	0.70	NDE	NDE	Hsp89.1: ATP binding | chr3:2479548-2484194 FORWARD
AT3G09440	NDE	NDE	NDE	0.70	NDE	heat shock cognate 70 kDa protein 3 (HSC70-3) (HSP70-3) | chr3:2903199-2905724 REVERSE
AT3G10985	-1.82	-1.14	-1.59	NDE	-0.99	SAG20: A senescence-associated gene whose expression is induced in response to treatment with Nep1, a fungal protein that causes necrosis.
AT3G11660	NDE	NDE	NDE	-1.17	NDE	NHL1: encodes a protein whose sequence is similar to tobacco hairpin-induced gene (HIN1) and Arabidopsis non-race specific disease resistance gene (NDR1). Expression of this gene is induced by cucumber mosaic virus. Localization of the gene product is similar to that of NHL3 (plasma membrane) but it is yet inconclusive.
AT3G12500	NDE	1.01	NDE	NDE	NDE	HCHIB: encodes a basic chitinase involved in ethylene/jasmonic acid mediated signalling pathway during systemic acquired resistance based on expression analyses.
AT3G13650	NDE	1.64	NDE	NDE	NDE	disease resistance response | chr3:4462892-4463872 FORWARD
AT3G17020	-0.88	NDE	NDE	NDE	NDE	universal stress protein (USP) family protein | chr3:5802536-5804157 REVERSE
AT3G17210	-1.21	NDE	NDE	NDE	NDE	HS1: Encodes a heat stable protein with antimicrobial and antifungal activity.
AT3G19690	NDE	NDE	NDE	-1.59	-0.65	pathogenesis-related protein, putative | chr3:6842220-6842856 REVERSE
AT3G23190	NDE	NDE	NDE	NDE	-0.95	lesion inducing protein-related | chr3:8279289-8280912 FORWARD
AT3G23300	-1.41	NDE	NDE	NDE	0.94	dehydration-responsive protein-related | chr3:8333025-8336146 FORWARD
AT3G28940	-1.92	NDE	NDE	NDE	NDE	avirulence-responsive protein, putative / avirulence induced gene (AIG) protein, putative | chr3:10968100-10969387 REVERSE
AT3G44670	-1.32	NDE	NDE	NDE	NDE	ATP binding / nucleoside-triphosphatase/ nucleotide binding / protein binding / transmembrane receptor | chr3:16216950-16221617 FORWARD
AT3G46230	1.59	NDE	NDE	1.95	NDE	HSP17.4: member of the class I small heat-shock protein (sHSP) family, which accounts for the majority of sHSPs in maturing seeds
AT3G47540	-1.40	-1.02	0.69	1.21	1.07	chitinase, putative | chr3:17521029-17522269 FORWARD
AT3G48450	-2.29	NDE	1.00	1.85	1.18	nitrate-responsive NOI protein, putative | chr3:17944029-17944888 REVERSE
AT3G50460	-0.97	NDE	NDE	NDE	NDE	HR2: Homolog of RPW8
AT3G50970	NDE	NDE	-2.05	-1.88	-2.63	LTI30: Belongs to the dehydrin protein family, which contains highly conserved stretches of 7-17 residues that are repetitively scattered in their sequences, the K-, S-, Y- and lysine rich segments. LTI29 and LTI30 double overexpressors confer freeze tolerance. Located in membranes. mRNA upregulated by water deprivation and abscisic acid.
AT3G52430	NDE	NDE	NDE	NDE	-0.70	PAD4: Encodes a lipase-like gene that is important for salicylic acid signaling and function in resistance (R) gene-mediated and basal plant disease resistance. PAD4 can interact directly with EDS1, another disease resistance signaling protein. Expressed at elevated level in response to green peach aphid (GPA) feeding, and modulates the GPA feeding-induced leaf senescence through a mechanism that doesn't require camalexin synthesis and salicylic acid (SA) signaling. Required for the ssi2-dependent heightened resistance to GPA.
AT3G54420	NDE	NDE	NDE	NDE	-0.90	EP3: encodes an EP3 chitinase that is expressed during somatic embryogenesis in 'nursing' cells surrounding the embryos but not in embryos themselves. The gene is also expressed in mature pollen and growing pollen tubes until they enter the receptive synergid, but not in endosperm and integuments as in carrot. Post-embryonically, expression is found in hydathodes, stipules, root epidermis and emerging root hairs.
AT3G58020	NDE	NDE	NDE	0.68	NDE	heat shock protein binding | chr3:21479480-21481349 FORWARD
AT3G59930	-4.03	NDE	NDE	NDE	NDE	Encodes a defensin-like (DEFL) family protein.
AT3G62020	NDE	-1.10	NDE	NDE	NDE	GLP10: germin-like protein (GLP10)
AT3G62190	-1.39	NDE	NDE	NDE	NDE	DNAJ heat shock N-terminal domain-containing protein | chr3:23021121-23023332 FORWARD
AT3G62730	NDE	1.12	NDE	NDE	NDE	unknown protein | chr3:23207854-23209471 REVERSE
AT4G05420	NDE	NDE	0.76	NDE	NDE	DDB1A: Structurally similar to damaged DNA binding proteins.DDB1a is part of a 350 KDa nuclear localized DET1 protein complex. This complex may physically interact with histone tails and while bound to chromatin- repress transcription of genes involved in photomorphogenesis.
AT4G09940	-1.43	NDE	NDE	NDE	NDE	avirulence-responsive family protein / avirulence induced gene (AIG1) family protein | chr4:6231786-6233397 FORWARD
AT4G10270	3.46	3.59	NDE	NDE	NDE	wound-responsive family protein | chr4:6374734-6375210 FORWARD
AT4G11190	NDE	1.37	NDE	NDE	NDE	disease resistance-responsive family protein / dirigent family protein | chr4:6826677-6827383 FORWARD
AT4G11230	NDE	NDE	NDE	0.70	NDE	respiratory burst oxidase, putative / NADPH oxidase, putative | chr4:6840491-6845591 REVERSE
AT4G11340	NDE	NDE	NDE	-0.77	NDE	Toll-Interleukin-Resistance (TIR) domain-containing protein | chr4:6894208-6899130 REVERSE
AT4G11650	-2.99	NDE	NDE	NDE	NDE	OSM34: osmotin-like protein
AT4G12400	NDE	NDE	NDE	NDE	-0.91	Hop3: stress-inducible protein, putative | chr4:7338659-7341361 REVERSE
AT4G13235	NDE	1.46	0.80	0.89	1.18	EDA21: Encodes a defensin-like (DEFL) family protein.
AT4G13820	NDE	NDE	NDE	0.68	0.63	disease resistance family protein / LRR family protein | chr4:8007989-8010745 REVERSE
AT4G13830	NDE	NDE	NDE	NDE	1.03	J20: DnaJ-like protein (J20); nuclear gene
AT4G14630	NDE	NDE	-1.90	-3.13	-2.87	GLP9: germin-like protein with N-terminal signal sequence that may target it to the vacuole, plasma membrane and/or outside the cell.
AT4G15740	NDE	NDE	NDE	-1.54	-0.88	C2 domain-containing protein | chr4:8964745-8966318 REVERSE
AT4G15910	NDE	NDE	2.39	1.28	NDE	DI21: encodes a gene whose transcript level in root and leaves increases to progressive drought stress. The transcript level is also affected by changes of endogenous or exogenous abscisic acid level. It appears to be a member of plant-specific gene family that includes late embryo-abundant and zinc- IAA-induced proteins in other plants.
AT4G17910	NDE	NDE	NDE	0.84	0.93	transferase, transferring acyl groups | chr4:9951173-9953979 REVERSE
AT4G18030	NDE	NDE	NDE	0.97	NDE	dehydration-responsive family protein | chr4:10012361-10015815 REVERSE
AT4G19810	NDE	NDE	NDE	-1.01	-1.00	ChiC: glycosyl hydrolase family 18 protein | chr4:10763934-10765753 REVERSE
AT4G19920	NDE	-1.16	NDE	NDE	-0.68	disease resistance protein (TIR class), putative | chr4:10797217-10798388 REVERSE
AT4G22214	NDE	-1.27	NDE	NDE	NDE	Encodes a defensin-like (DEFL) family protein.
AT4G22230	NDE	1.34	NDE	NDE	NDE	Encodes a defensin-like (DEFL) family protein.
AT4G23670	NDE	NDE	NDE	0.90	NDE	major latex protein-related / MLP-related | chr4:12332560-12333804 REVERSE
AT4G23680	-1.41	NDE	NDE	NDE	NDE	major latex protein-related / MLP-related | chr4:12336186-12337482 REVERSE
AT4G23690	NDE	1.41	NDE	NDE	NDE	DIR6: disease resistance-responsive family protein / dirigent family protein | chr4:12338871-12339747 REVERSE
AT4G24190	-0.77	NDE	NDE	NDE	NDE	SHD: encodes an ortholog of GRP94, an ER-resident HSP90-like protein and is involved in regulation of meristem size and organization. Single and double mutant analyses suggest that SHD may be required for the correct folding and/or complex formation of CLV proteins. Lines carrying recessive mutations in this locus exhibits expanded shoot meristems, disorganized root meristems, and defective pollen tube elongation. Transcript is detected in all tissues examined and is not induced by heat. Endoplasmin supports the protein secretory pathway and has a role in proliferating tissues.
AT4G25780	NDE	NDE	-0.81	NDE	NDE	pathogenesis-related protein, putative | chr4:13121068-13121955 FORWARD
AT4G25790	NDE	NDE	NDE	0.69	0.99	allergen V5/Tpx-1-related family protein | chr4:13122152-13123347 REVERSE

AT4G27320	NDE	NDE	NDE	-0.87	-0.93	PHOS34: Contains a universal stress protein domain. Protein is phosphorylated in response to Phytophthora infestans zoospores and xylanase.
AT4G28240	-1.19	NDE	-0.68	NDE	NDE	wound-responsive protein-related | chr4:13997292-13998347 REVERSE
AT4G28480	-1.13	NDE	NDE	NDE	NDE	DNAJ heat shock family protein | chr4:14073042-14075271 FORWARD
AT4G29920	NDE	NDE	NDE	0.70	NDE	heat shock protein-related | chr4:14632653-14635885 REVERSE
AT4G30650	-2.85	NDE	NDE	-1.07	-1.23	hydrophobic protein, putative / low temperature and salt responsive protein, putative | chr4:14954320-14954968 FORWARD
AT4G30660	NDE	NDE	NDE	-0.94	-1.14	hydrophobic protein, putative / low temperature and salt responsive protein, putative | chr4:14955373-14956118 FORWARD
AT4G33300	NDE	NDE	-0.74	-0.73	-0.70	ADR1-L1: Symbols: ADR1-L1 | ADR1-L1 (ADR1-like 1); ATP binding / protein binding | chr4:16051051-16056526 REVERSE
AT4G33720	-0.80	NDE	NDE	NDE	NDE	pathogenesis-related protein, putative | chr4:16182752-16183533 FORWARD
AT4G33730	NDE	NDE	NDE	0.69	NDE	pathogenesis-related protein, putative | chr4:16185098-16185616 FORWARD
AT4G36010	NDE	NDE	-0.82	NDE	-0.79	pathogenesis-related thaumatin family protein | chr4:17039192-17041141 REVERSE
AT4G36040	NDE	NDE	-0.68	NDE	NDE	J11: DNAJ heat shock N-terminal domain-containing protein (J11) | chr4:17049171-17050274 REVERSE
AT4G37910	NDE	NDE	1.28	0.93	1.11	mtHsc70-1: Symbols: mtHsc70-1 | mtHsc70-1 (mitochondrial heat shock protein 70-1); ATP binding | chr4:17825080-17828180 REVERSE
AT4G38410	NDE	NDE	NDE	NDE	-1.43	dehydrin, putative | chr4:17980940-17981719 FORWARD
AT4G38660	NDE	NDE	NDE	0.72	0.88	thaumatin, putative | chr4:18066177-18068205 REVERSE
AT4G39640	NDE	NDE	NDE	NDE	-1.01	GGT1: The gene encodes a gamma-glutamyltransferase (AKA gamma-glutamyl transpeptidase, EC 2.3.2.2) that is located in vascular tissues (predominantly phloem) of leaves and is involved in the degradation of glutathione. The encoded enzyme also mitigates oxidative stress by metabolizing GSSG (oxidized form of GSH - glutathione) in the apoplast.
AT5G06320	NDE	NDE	NDE	NDE	-0.90	NHL3: encodes a protein whose sequence is similar to tobacco hairpin-induced gene (HIN1) and Arabidopsis non-race specific disease resistance gene (NDR1). Expression of this gene is induced by cucumber mosaic virus, spermine and Pseudomonas syringae pv. tomato DC3000. The gene product is localized to the plasma membrane.
AT5G07220	NDE	NDE	NDE	NDE	-0.75	BAG3: A member of Arabidopsis BAG (Bcl-2-associated athanogene) proteins, plant homologs of mammalian regulators of apoptosis. Plant BAG proteins are multi-functional and remarkably similar to their animal counterparts, as they regulate apoptotic-like processes ranging from pathogen attack, to abiotic stress, to plant development.
AT5G09590	NDE	NDE	NDE	0.77	NDE	MTHSC70-2: heat shock protein 70 (Hsc70-5); nuclear
AT5G09980	NDE	NDE	1.33	NDE	NDE	PROPEP4: Symbols: PROPEP4 | PROPEP4 (Elicitor peptide 4 precursor) | chr5:3122731-3123964 FORWARD
AT5G17970	1.72	NDE	NDE	NDE	NDE	disease resistance protein (TIR-NBS-LRR class), putative | chr5:5948999-5951619 REVERSE
AT5G18750	NDE	NDE	NDE	0.80	NDE	DNAJ heat shock N-terminal domain-containing protein | chr5:6255177-6257932 FORWARD
AT5G20150	-3.40	NDE	NDE	-1.17	0.75	SPX1: Symbols: ATSPX1, SPX1 | SPX1 (SPX DOMAIN GENE 1) | chr5:6802372-6803625 FORWARD
AT5G20970	NDE	NDE	NDE	1.74	1.35	heat shock family protein | chr5:7122800-7124001 FORWARD
AT5G22690	NDE	NDE	NDE	-0.84	-0.83	disease resistance protein (TIR-NBS-LRR class), putative | chr5:7541271-7545182 FORWARD
AT5G23400	2.14	NDE	NDE	NDE	NDE	disease resistance family protein / LRR family protein | chr5:7880335-7882634 FORWARD
AT5G24620	NDE	NDE	NDE	0.83	1.05	thaumatin-like protein, putative | chr5:8430768-8432414 FORWARD
AT5G33355	NDE	-1.05	NDE	NDE	NDE	Encodes a defensin-like (DEFL) family protein.
AT5G38350	NDE	NDE	NDE	NDE	-1.13	disease resistance protein (NBS-LRR class), putative | chr5:15328659-15331528 FORWARD
AT5G38950	NDE	1.12	NDE	NDE	NDE	germin-like protein-related | chr5:15590693-15591050 FORWARD
AT5G39100	NDE	-1.40	NDE	NDE	NDE	GLP6: germin-like protein (GLP6)
AT5G39150	-2.33	-1.51	NDE	NDE	NDE	germin-like protein, putative | chr5:15669898-15670808 REVERSE
AT5G39180	-1.77	-1.46	NDE	NDE	NDE	germin-like protein, putative | chr5:15683464-15684353 REVERSE
AT5G39730	-1.81	NDE	NDE	NDE	NDE	avirulence-responsive protein-related / avirulence induced gene (AIG) protein-related | chr5:15901679-15902870 FORWARD
AT5G40020	NDE	-1.77	NDE	NDE	NDE	pathogenesis-related thaumatin family protein | chr5:16022753-16024316 REVERSE
AT5G40910	-1.18	NDE	NDE	NDE	NDE	disease resistance protein (TIR-NBS-LRR class), putative | chr5:16395507-16399129 FORWARD
AT5G42020	NDE	NDE	NDE	0.95	NDE	BIP2: Luminal binding protein (BiP2) involved in polar nuclei fusion during proliferation of endosperm nuclei.
AT5G42050	NDE	NDE	NDE	NDE	0.67	FUNCTIONS IN: molecular_function unknown; EXPRESSED IN: 22 plant structures; EXPRESSED DURING: 13 growth stages; CONTAINS InterPro DOMAIN/s: Development and cell death domain (InterPro:IPR013989), Kelch related (InterPro:IPR013089); BEST Arabidopsis thaliana protein match is: unknown protein (TAIR:AT3G27090.1); Has 5084 Blast hits to 2870 proteins in 85 species: Archae - 0; Bacteria - 24; Metazoa - 270; Fungi - 58; Plants - 157; Viruses - 8; Other Eukaryotes - 4567 (source: NCBI BLink). | chr5:16815489-16817301 FORWARD
AT5G45050	-1.59	NDE	NDE	NDE	NDE	TTR1: Encodes a member of WRKY Transcription Factor (Group II-e) that confers resistance to tobacco ringspot nepovirus.
AT5G45490	NDE	NDE	-1.14	-1.30	-1.85	disease resistance protein-related | chr5:18431003-18432397 FORWARD
AT5G47120	NDE	NDE	NDE	NDE	-0.78	BI1: Encodes BI-1, a homolog of mammalian Bax inhibitor 1. Functions as an attenuator of biotic and abiotic types of cell death. Bax-induced cell death can be downregulated by ectopically expressing AtBI in planta.
AT5G47130	NDE	NDE	NDE	NDE	-0.72	Bax inhibitor-1 family / BI-1 family | chr5:19141016-19142011 FORWARD
AT5G48030	NDE	NDE	0.77	NDE	NDE	GFA2: encodes a mitochondrially targeted DNAJ protein involved in female gametophyte development.
AT5G49910	-1.25	NDE	NDE	NDE	NDE	cpHsc70-2: Stromal heat shock protein involved in protein import into chloroplast.
AT5G51440	NDE	NDE	NDE	NDE	-0.92	23.5 kDa mitochondrial small heat shock protein (HSP23.5-M) | chr5:20891163-20892211 FORWARD
AT5G51630	-1.25	NDE	NDE	NDE	NDE	disease resistance protein (TIR-NBS-LRR class), putative | chr5:20970030-20974916 FORWARD
AT5G52060	NDE	NDE	-1.05	NDE	-0.88	BAG1: A member of Arabidopsis BAG (Bcl-2-associated athanogene) proteins, plant homologs of mammalian regulators of apoptosis. Plant BAG proteins are multi-functional and remarkably similar to their animal counterparts, as they regulate apoptotic-like processes ranging from pathogen attack, to abiotic stress, to plant development.
AT5G52640	NDE	NDE	-0.99	NDE	-1.38	HSP90.1: Encodes a cytosolic heat shock protein AtHSP90.1. AtHSP90.1 interacts with disease resistance signaling components SGT1b and RAR1 and is required for RPS2-mediated resistance.
AT5G54855	-1.19	NDE	NDE	NDE	NDE	pollen Ole e 1 allergen and extensin family protein | chr5:22282767-22284387 FORWARD
AT5G59720	2.36	NDE	NDE	NDE	NDE	HSP18.2: encodes a low molecular weight heat shock protein that contains the heat shock element in the promoter region. Expression is induced in response to heat shock.
AT5G60340	NDE	NDE	0.83	NDE	NDE	AAK6: maoC-like dehydratase domain-containing protein | chr5:24273879-24275891 REVERSE
AT5G64890	NDE	NDE	-0.81	-1.99	-1.39	PROPEP2: Symbols: PROPEP2 | PROPEP2 (Elicitor peptide 2 precursor) | chr5:25934568-25935521 FORWARD
AT5G64900	NDE	NDE	NDE	-1.16	-1.09	PROPEP1: Encodes a putative 92-aa protein that is the precursor of AtPep1, a 23-aa peptide which activates transcription of the defensive gene defensin (PDF1.2) and activates the synthesis of H2O2, both being components of the innate immune response.
AT5G64930	-0.81	NDE	NDE	0.89	0.78	CPR5: Regulator of expression of pathogenesis-related (PR) genes. Participates in signal transduction pathways involved in plant defense (systemic acquired resistance -SAR).
AT5G65210	NDE	NDE	-0.79	NDE	NDE	TGA1: Encodes TGA1, a redox-controlled regulator of systemic acquired resistance. TGA1 targets the activation sequence-1 (as-1) element of the promoter region of defense proteins. TGA1 are S-nitrosylated.

RNAseq and Proteomics for Analysing Complex Oomycete Plant Interactions

Dharani D. Burra, Ramesh R. Vetukuri, Svante Resjö, Laura J. Grenville-Briggs and Erik Andreasson*

Resistance Biology, Department of Plant Protection Biology, Swedish University of Agricultural Sciences
*Corresponding author: Erik.Andreasson@slu.se

Abstract

The oomycetes include some of the most devastating plant pathogens. In this review we discuss the latest results from oomycete and plant studies with emphasis on interaction studies. We focus on the outcomes of RNAseq and proteomics studies and some pitfalls of these approaches. Both pathogenic interactions and biological control are discussed. We underline the usefulness of studies at several levels of complexity from studies of one organism, up to two or more and within agricultural fields (managed settings) up to wild ecosystems. Finally we identify areas of future interest such as detailed interactome studies, dual RNAseq studies, peptide modification studies and population/meta omics with or without biological control agents.

Introduction

The recent flurry of technological advancements by next-generation sequencing (NGS) and mass spectrometry (MS) have enabled genome-scale capturing of biological processes at the molecular level. Rather than being limited to a handful of measured compounds, it is now possible to capture thousands of molecules in a single study. However, these techniques have mostly been applied in the laboratory and during controlled growing conditions. The generation of such large datasets—often referred to as "-omics" data—depends partly on new pipelines for experimental set-ups, sampling, data analysis and visualization. The integration of different types of data generated in such experiments and correct biological interpretation of large datasets remain major challenges. Despite these drawbacks, the power of omics based analysis lies in its ability to facilitate a broader understanding of biological systems.

The oomycetes are a lineage of filamentous Eukaryotes that resemble fungi in their hyphal growth and heterotrophic absorptive nutrition, but are in fact more closely related to the heterokont (brown) algae and are therefore members of the Kingdom Stramenophila. Members of this group may be saprotrophic or pathogenic on a wide range of hosts, including plants, algae and animals. The most notorious and well-studied, are the plant pathogens of the genus *Phytophthora,* and include the potato late blight pathogen, *Phytophthora infestans* (responsible for the Irish potato famine in the 1840s and still a major pesticide target) and the agent of sudden oak death, *Phytophthora ramorum* responsible for global losses in forests and in horticulture. The oomycetes exhibit a high level of diversity in the nature of their pathogenic relationships. At one end of the scale, are the biotrophs, such as the downy mildews, which typically only infect a single host plant species and obtain nutrients from living cells. Members of the Phytophthora genus may have broader host ranges, for example *Phytophthora cinnamomi* can infect more than 1000 species of tree, whereas *P. infestans* is restricted to Solanaceae primarily infecting potato, tomato and tobacco plants, and infecting a wide range of organs ranging from leaves, stems to tubers. Most species of Pythium also have a broad host range; some are destructive necrotrophs such as *Pythium ultimum*, which destroy plant tissue to obtain nutrients, whilst others, e.g. *Pythium oligandrum* are mycoparasitic against a wide range of fungi and oomycetes and may have potential to be used as biological control agents (BCAs) against numerous plant diseases. Phytophthora species colonise plants in a hemibiotrophic manner, entering and evading defence responses in a similar manner to biotrophic pathogens, but later shifting to necrotrophic growth at the end of the disease cycle.

The plant immune system is complex and includes functions like specificity (specific targeting of pathogen molecules), memory (memory of prior attacks) and self-tolerance (ability to distinguish self from non-self) similar to the innate immune system of animals (Spoel and Dong, 2012). All these functions are mediated by multiple molecules, therefore rendering this system ideal for omics based investigations. The plant immune system is unique in its bi-layered organization. The first layer termed as PTI (PAMP triggered immunity) involves defense to surface-exposed conserved pathogen associated molecules termed PAMPs (Pathogen associated molecular patterns). This induces rapid responses including a Ca^{2+} burst and MAP kinase activation that in turn regulates transcription factor activity, affecting the expression of genes involved in defense compound production (Bigeard et al., 2015). In response to this, pathogens have evolved a way to circumvent PTI responses via production of specialized molecules called effectors. To counter these effectors, plants produce molecules called resistance proteins (coded by R genes) that can specifically recognize these effectors; this forms the second layer of plant immunity termed effector triggered immunity (ETI). ETI responses generally lead to a specialized form of programmed cell death (PCD) called the hypersensitive response (HR). The observation of a tussle at the molecular level between the plant and the pathogen led to the development of the "Zig-Zag" model of plant immunity (Jones and Dangl, 2006). However, this simplistic bi-layered view is being challenged lately by discoveries that point towards a continuum wherein each layer feeds into the other (Thomma et al., 2011). Observations of the overwhelming effects of other variables (e.g. the presence of multiple pathogens with varying infection strategies) in this interaction has led to the call for additional models to better describe plant-pathogen interactions quantitatively and over correct time scales (Pritchard and Birch, 2014). Recent discoveries have also identified that various pathogens have evolved effectors that hijack a suite of "master-regulators" in the host allowing the pathogen the opportunity to perturb numerous other functions (Weßling et al., 2014). Similarly plants have also evolved immune signaling networks that can efficiently fine tune responses depending on the pathogen, or effector presented (Kim et al., 2014), indicating the highly complex and dynamic interplay between pathogen(s) and host plants.

This interplay is very well exemplified by oomycete-host plant patho-systems. For example, Steinbrenner et al. (2015) show data that indicate that two resistance protein domains from AtRPP1, are involved in direct interaction with *Hyaloperenospora arabidopsidis* effector ATR1; one to associate with the effector, and the other to sensitize the receptor and facilitate activation. This is in contrast to the interaction between P. *infestans* effector Avr2 and host resistance protein R2 wherein recognition of Avr2 by R2 is indirect, mediated via another host protein BSL1 (Saunders et al., 2012). The observed "plasticity" in several oomycete genomes (Raffaele and Kamoun, 2012) that enables them to quickly overcome host resistance mechanisms and further counter adaptation by host plants continuously drives this race for superiority between oomycete pathogens and their hosts.

In recent years, omics based investigations on oomycete-host plant pathosystems have been largely driven by genomics analysis (Dong et al., 2015; Judelson, 2012). A concerted effort by the research community has resulted in the genomes of more than sixteen different oomycetes, with a major focus on plant pathogens that exhibit biotrophic, hemibiotrophic or necrotrophic lifestyles on a variety of host plants. These genomics approaches have led to the identification of many large virulence gene families that have been predicted to encode numerous hydrolytic enzymes for the degradation of host carbohydrates, extracellular toxins (e.g. NLP toxins) and at least three families of effector proteins that enter host cells, the RXLR effectors, CHXC effectors and Crinkler (CRN) proteins (Bozkurt et al., 2012; Fawke et al., 2015).Comparative genomic analysis of pathogens with diverse host ranges, reveal striking differences in pathogenicity mechanisms. E.g. the broad host range nectrophic plant pathogen *Pythium ultimum* contains expanded families of enzymes responsible for degradation of easily accessible plant carbohydrates such as pectins, starch and sucrose and an apparent absence of RXLR effectors (Lévesque et al., 2010). On the contrary, the obligate biotrophic downy mildew *Hyaloperonospora arabidopsidis* contains reduced numbers of genes encoding such hydrolytic enzymes (Baxter et al., 2010), whilst a loss of heterozygosity may have driven the rapid adaptation of *Phytophthora capsici* to new hosts and a swift ability to overcome fungicides (Lamour et al., 2012).

With regards to plants, genome sequences of only two, namely Arabidopsis and Rice were complete and available between the years 2000 and 2006. However in the last 7 years, the number of plant genomes sequenced has rapidly increased (Bolger et al., 2014). The majority of economically important plants (including several "true hosts" of oomycetes) that have been sequenced are polyploid and have large genomes. Earlier sequencing methods were therefore unable to sequence these complex genomes. New developments in sequencing methodologies and reduced costs have enabled sequencing of these genomes (Bolger et al., 2014). However for some plants, such as potato, the heterozygosity and tetraploid nature of the genome still makes it difficult to sequence. Therefore to circumvent this problem in potato a doubled monoploid was sequenced (Consortium, 2011). Further characterization of polyploid genomes is complicated. For example, resistance gene (R gene) analysis in the genomes of tomato and potato revealed that most R genes were present as tandem duplications (Andolfo et al., 2013; Jupe et al., 2012) and many of them seem to be non-functional (Mace et al., 2014) ; Lenman et al., 2015). Therefore just through *in silico* identification of the gene it becomes difficult to predict its functionality, since information regarding, for example, which copy of the R gene is expressed and activated is required, just as it is for pathogen-derived effectors.

Incorporation of RNAseq and proteomic methodologies are facilitating improved analysis of gene functions in economically important plants and both plant pathogenic and mycoparasitic oomycetes. Some progress towards this end has been made by linking proteomics data with gene information in the model plant Arabidopsis (Usadel et al., 2012). We foresee integration of RNAseq and proteomic data sources into existing comparative analysis platforms such as "Phytozome" (http://phytozome.jgi.doe.gov/pz/portal.html) and "PLAZA" (http://plaza.psb.ugent.be/). This will enable seamless knowledge transfer between model and true host plants and will drive future research associated with understanding the role of genome complexity in pathogen responses.

Insights into oomycete effector biology using RNAseq

Sanger sequencing of ESTs and differential expression analysis using custom made Affymetrix GeneChips gave early clues as to the dynamics of pre-infection development and effector gene expression in *P. infestans* (Judelson et al., 2008; Randall et al., 2005). In the post-genomics era of oomycete research, and with possibilities for large-scale deep sequencing of transcriptomes, it is now possible to document the dynamics of *P. infestans* development throughout the disease cycle, from biotrophy to necrotrophy. One such recent study showed that an induction of apoplastic and cytoplasmic effectors and a tight temporal regulation of pathways involved in suppression of host defence responses characterized the biotrophic phase of infection (Zuluaga et al., 2015a). Four effectors that suppress necrosis were identified and such proteins may serve to prolong the biotrophic stage. 818 genes that had not been previously reported in the *P. infestans* genome were also identified. Such "orphan genes" may represent errors in annotation, non-coding RNAs, expressed pseudogenes or alternate splice variants (Zuluaga et al., 2015a), and thereby illustrate the power of deep sequencing approaches using RNAseq technology.

Next-generation sequencing and RNAseq techniques have been, and continue to be, very useful for the study of the infection biology of recalcitrant oomycetes such as the downy mildews, which as obligate biotrophs are hard to work with in a laboratory setting. Large-scale EST sequencing of the downy mildew pathogen *Hyaloperonospora arabidopsidis* during infection of Arabidopsis revealed the presence of isolate-specific effectors (Cabral et al., 2011). Recently, the Jones laboratory has developed a high-throughput mRNA expression profiling method (Expression Profiling through Random Sheared cDNA tag Sequencing [EXPRSS]) which detects the differential expression of more genes with a higher sensitivity than microarray or traditional RNAseq methods (Rallapalli et al., 2014). Expression analysis of both host and pathogen using this method has identified that *H. arabidopsidis* suppresses responsiveness to salicylic acid (SA) in host cells containing pathogen haustoria through which host-translocated effectors are delivered. An *H. arabidopsidis* RXLR effector, HaRxL62, which enhances host susceptibility was highly expressed during this assay and was shown to suppress host responsiveness to SA (Asai et al., 2014).

Massive parallel transcriptome sequencing of lettuce infected with the downy mildew *Bremia lactucae* revealed expression of approximately 78 effectors with RXLR-like features, as well as elicitins, necrosis inducing peptides, glucanase inhibitors and lectins and an enrichment of cysteine-rich proteins (Stassen et al., 2012). RXLR and CRN-type effectors were also identified in a sunflower (*Helianthus annuus)* - downy mildew (*Plasmopara halstedii)* interaction trancriptome sequenced by 454 FLX pyrosequencing (As-Sadi et al., 2011). An extensive mRNAseq analysis of expression of pathogen genes over a time course of infection of cucumber (*cucumis stivus)* by the downy mildew *Pseudoperonospora cubensis* identified expression of RXLR-like effectors and other pathogenicity genes. Interestingly, comparison of these expression profiles with expression data from a timecourse of *P. infestans* infection of potato (*Solanum tuberosum)* revealed a substantial overlap in molecular events and gene expression in virulence of the biotrophic *Ps. cubensis* and the hemibiotrophic *P. infestans* (Savory et al., 2012). Similarly, Kunjeti et al. (2012) found that many of the transcripts identified from *Phytophthora phaseoli* infection of lima bean by RNAseq were homologous to known or putative virulence genes in *P. infestans.* Surprisingly, although RXLR and elicitin type effectors were upregulated during host plant infection, CRN effectors were not. *Phytophthora capsici* is a soilborne necrotroph with a broad host range. RNAseq of pre-infection spore development in this pathogen identified major differences between the expressed gene content of the mycelium, zoospores and germinating cysts, and a total of 98 predicted effector genes were identified, including members of the RXLR and CRN classes. Nineteen of these effector genes were shown to also be expressed during infection and were able to suppress host cell death (Chen et al., 2013). RXLR and CRN effectors are therefore expressed by a wide range of oomycete pathogens during colonization and infection of host plants, and may be ancient and conserved pathogenicity mechanisms in the oomycetes. However, pyrosequencing of the transcriptome of the interaction between the brown alga *Ectocarpus siliculosus* by the obligate biotroph and basal oomycete *Eurychasma dicksonii* failed to positively identify putative RXLR and CRN effector transcripts, suggesting either that they are only present in low abundance in this interaction or that they arose later on in the oomycete lineage after the oomycetes left their marine environment (Grenville-Briggs et al., 2011).

Small RNA (sRNA) studies on oomycete pathogens have demonstrated that some effectors are regulated by sRNAs. This regulation allows for the development of variation in pathogenicity and virulence toward plant resistance genes (Qutob et al., 2013; Vetukuri et al., 2012). Silencing of the *Phytophthora sojae*, effector *Avr3a*, is reported to be controlled by 25 nt *cis*-acting sRNAs, and the silencing signal appears to be heritable and confers virulence on plants carrying the Avr3a-recognizing *R* gene *Rps3a* (Qutob et al., 2013). Studies in both fungi and oomycetes have suggested the existence of bidirectional transport of sRNAs. For instance 21-22nt *Botrytis cinerea* sRNAs induced during the infection process target host genes in *Arabidopsis* and tomato by hijacking host sRNA machinery (Weiberg et al., 2013), whereas sRNAs from the potato targeting oomycete *P.infestans* are reported from a host induced gene silencing sRNA screen (Jahan et al., 2015), suggesting that RNA has a multifaceted role in host-pathogen interactions.

RNAseq for improved understanding of plant-oomycete interactions

Over the past few years, RNAseq technology has fostered the move from studying model plant patho-systems to true host-pathogen interactions. RNAseq has been rapidly outcompeting traditional microarrays because unlike microarrays, which are limited to the availability of sequenced genome/prior information, RNAseq is independent of the same (Wang et al., 2009). Since the procedure is not limited to sequenced genomes, the possibilities of identifying new components regulating plant-pathogen interactions are dramatically improved. Recently Zuluaga et al. (2015b) used 454 pyrosequencing to identify 94 000 different tomato transcripts that altered in response to *Phytophthora infestans* inoculation of leaves. Microarray analysis using the tomato genome (Jupe et al., 2013b) would not have detected the novel components such as alternative spliced transcripts or fusion transcripts that constitute the large number of transcripts found by Zuluaga et al. (2015b). Oomycetes like *P. infestans* cause disease in different tissues or organs; a strategy that uses specific RNAseq analyses can be helpful in elucidating novel components involved in tissue or organ specific oomycete interactions. Similarly, RNAseq has also enabled researchers to perform detailed exploration of alternative splicing/miRNA processes involved in plant defense (Wang et al., 2010). For instance, RNAseq has been used to detect novel miRNAs involved in plant-oomycete interactions. Zhao et al. (2015) assayed miRNA profiles of 9 near isogenic lines (NILs) of soybean, each carrying a unique resistance gene (Rps; resistance to *Phytophthora sojae*) in a susceptible background, in response to *Phytophthora sojae* using RNAseq. The study identified 369 putative miRNAs to be differentially expressed in the NILs and the background, 63 of these could not be classified into known miRNA families and hence were regarded as unique to this patho-system. The study also found that miRNAs were repressed in the resistant NILs. RNAseq enabled transcriptomic analysis of the same material was also performed (Lin et al., 2014). Based on target identification analysis of miRNA and corresponding RNA data, the authors were able to show that general down regulation of miRNAs after pathogen inoculation results in the up regulation of corresponding defense genes leading to observed resistance. Alternative splicing is another facet of RNA mediated regulation of immunity that can be uncovered using RNAseq (Mandadi and Scholthof, 2015), although there has been no report on the use of RNAseq to specifically uncover host alternative splicing events in response to oomycetes, RNAseq has revealed evidence of alternative splicing in effector coding transcripts in the oomycete cucurbit pathogen *Pseudoperonospora cubensis* (Burkhardt et al., 2015).

Plant disease resistance is mediated by highly specific resistance genes (R genes); currently genome dependent map based cloning methods that are laborious and time consuming are used to identify these R genes. RNAseq has been used to circumvent these traditional approaches. For instance, Strauß et al. (2012) used an innovative RNAseq based approach wherein transcription activator-like effector (TALE) protein AvrBS4 was used as a probe to identify the corresponding *Bs4C* resistance gene in pepper. Analysis of oomycete genomes has revealed the presence of vast repertoires of effectors with probably different pathogenicity strategies, however there has been slow progress in identifying their targets in host. Therefore RNAseq combined with strategies such as effector screens (Du et al., 2015), that facilitate isolated response of novel oomycete effectors, could be used to identify R genes or effector targets involved in plant-oomycete interactions. With regards to R gene mapping, advancements to traditional methods through the use of high throughput genomic data are already underway, for example with the use of RenSeq it is possible to specifically enrich for R genes from genomic libraries as shown by Jupe et al. (2013a) in potato. Using this method, the authors were able to identify many more NB-LRR R genes in comparison to the published potato genome sequence. Although RenSeq is based on genomic libraries, the possibility to apply this technique to long-read based PAC-bio sequencing that reduces assembly issues as well as the use of expression datasets or exomes could simplify the identification of R genes involved in oomycete resistance.

Use of dual RNAseq methodology that allows for simultaneous quantification of gene expression in the host and the pathogen during the infection process can enable the discovery of multiple effectors produced by an oomycete during infection. Furthermore it may assist in the identification of candidates for corresponding host targets as negative correlation between repressed transcripts and effector transcript up regulation is expected. Simple microarrays from which much information about general defense responses in arabidopsis, potato and tomato to various biotic interactions have been gained (Ali et al., 2014; Lodha and Basak, 2012) are not particularly amenable for studies aimed towards effector target identification (Westermann et al., 2012). Some progress however has been made with the use of composite microarrays. In a recent study, custom-made composite microarrays were employed to study the *Phytophthora capsici* -tomato patho-system (Jupe et al., 2013b). Although it is relatively easy without a trained bioinformatician to analyze data obtained from microarrays, the use of microarrays in fact is not ideal if the focus is on identification of lowly expressed effector targets or elucidation of various other factors that enable transcriptional plasticity (e.g. miRNAs). Hayden et al. (2014) used a dual RNAseq approach to investigate the *Phytophthora ramorum- Notholithocarpus densiflorus* patho-system, and although they did find induction of effector proteins 5 days after inoculation (dai), they were unable to detect pathogen associated transcripts 1 dai, and the authors cite poor sequencing depth as a reason for this observation. Higher sequencing depth is an even more important pre-requisite in this regard as unlike other host-pathogen patho-systems, oomycete infections might be characterized by the low relative abundance of pathogen associated transcripts (Derevnina and Michelmore, 2015). Inclusion of crossing populations with differing resistance in the experimental set up and the use of dual RNAseq could help in elucidating novel facets of effector interactions with host targets in plant-oomycete interactions.

Although RNAseq offers the benefit of generating large amounts of data for analysis, the challenge has been to extract and generate knowledge from the data produced (Sboner et al., 2011). New developments like use of RNAseq in bulked segregant analysis (BSA) that facilitates fine mapping of genes is being done (Martin et al., 2013). Much needed focus has also been put onto improving experimental designs that can yield targeted results (Yang and Wei, 2015). Technical aspects that need redress are issues related to inherent biases in library construction and data analyses methods/storage. These have been reviewed widely (Kaufmann and Busch, 2013; Martin et al., 2013; McGettigan, 2013). There is also a need for simultaneous improvements in statistical tools, since the current tools for the analysis of RNAseq datasets have been shown to have lower power in identifying differentially regulated transcripts in comparison to statistical tools that analyze microarrays (Kaufmann and Busch, 2013). In the recent past, improvements have been made to traditional probe dependent methodologies, and some of these could assist in the development of standardized data analysis methods for RNAseq data processing, analysis, integration, visualization and storage.

Proteomics of Oomycetes

Proteomics can be defined as the study of the whole protein complement of an organism, tissue, cell or subcellular compartment. The intracellular levels of mRNA and protein correlate, but with a broad range of variation (Schwanhausser et al., 2013; Schwanhäusser et al., 2011). In addition, proteins in an intracellular compartment may be synthesized in advance and transported there (e.g. secreted proteins). This makes it difficult to predict the level of a protein from mRNA expression data alone. Proteomics offers the opportunity to obtain direct information about protein abundance, as well as information about post translational modifications.

Early proteomic studies of oomycetes were conducted by two-dimensional gel electrophoresis (2DGE). Krämer et al. (1997) showed for the first time, that the asexual life stages and infection structures of *P. infestans* could be produced *in vitro* in a synchronised fashion, and that the switch between spore types and germination was accompanied by the synthesis of stage-specific proteins. Ebstrup et al. (2005) built upon this work, by further identifying differentially expressed proteins during cyst germination and appressorium formation. A major limitation of the early studies was that very few sequences were available. For this reason, many of the early research articles are 2D gel based studies comparing life stages or organisms, but not identifying protein spots that change in abundance. Another limitation was that 2D-gel based studies are restricted to the subset of a proteome that can be resolved by 2DGE such as non-basic, non-membrane and average sized proteins. However, with extensive new sequence data, new instrumentation for HPLC-MSMS based proteomics as well as new methods for quantitative analysis, both label free and chemical label based methods, opens up new possibilities for a coming-of-age of proteomics in oomycete and plant research.

A number of authors have used proteomics to compare different oomycete life stages. In a study of *Phytophthora palmivora* (the causal agent of blackpod of cocoa), 1 % of proteins resolved by 2DGE were specific for each of the pre-infection stages of the asexual lifecycle and 30% co-migrated with *P. infestans* proteins (Shepherd et al., 2003), giving a glimpse of the comparative power to come from such studies. Some of the first evidence for the presence and secretion of extracellular effectors from *P. infestans* was provided by a combined *in silico* EST data mining approach and 2DGE of secreted proteins (Torto et al., 2003). Grenville-Briggs et al. (2005) used a combination of 2DGE and Suppressive Subtractive Hybridisation (SSH) to assess the nutritional requirements of *P. infestans* asexual development, revealing the importance of amino acid biosynthesis in *P. infestans* appressorium production. Savidor et al. (2008) were the first to use large scale quantitative HPLC-MSMS proteomics methods in oomycete research when they compared mycelia and germinating cysts of *P. sojae* and *P. ramorum*. In this project they identified 3897 *P. ramorum* and 2970 *P. sojae* proteins. The quantitative analysis yielded candidates for early infection, vegetative growth and other proteins important for determining host specificity.Grenville-Briggs et al. (2010) isolated cell wall proteins from *P. infestans* vegetative mycelium, sporulating mycelium, and germinating cysts with appressoria. 31 cell wall associated proteins were identified. All proteins identified in germinating cysts with appressoria were either putative effector or pathogen-associated molecular pattern (PAMP) molecules, indicating that the cell wall is an important reservoir for this type of molecules. Hosseini et al. (2015) compared hyphae and germinating cysts of *P. sojae* and the recently discovered species *P. pisi*. Using HPLC-MSMS and quantitative analyses, 58 candidates for infection and 23 candidates for vegetative growth were identified. The secretome of *P. infestans* was recently mapped by Meijer et al. (2014) wherein over 200 secreted proteins from *P. infestans* were identified.

Resjö et al. (2014) performed the first phosphoproteomics study of any oomycete. Six life stages of *P. infestans* were compared and more than 2000 phosphosites were identified. The Crinkler (CRN) group of effectors was extensively phosphorylated and a number of phosphorylations characteristic for appressoria, such as transport and signal transduction proteins were demonstrated. In the future, the study of other post-translational modifications, such as methylation and acelylation might prove useful in order to increase our understanding of the mechanisms behind host resistance and immunity as has been shown in other plant-pathogen systems.

Proteomics of plant-oomycete interactions

The first proteomics study that investigated the effect of *P. sojae* infection on soybean root cell wall proteins was performed by Mithöfer et al. (2002). The proteins were analysed using one-dimensional SDS-PAGE, and protein bands of interest were excised and analysed using ESI-MS/MS analysis. Like many early proteomics studies, this one was hampered by a lack of sequence data. Since the soybean genome had not been sequenced at that time, the authors had to rely on the presence of proteins with sufficiently high homology in other databases in order to identify the analysed protein. For this reason, only four out of eight excised bands could be identified.

Subsequent studies have frequently used a two-dimensional SDS-PAGE based approach. This technique has the advantage that the quantitative analysis is independent of protein identification. Hence, different conditions (such as infected/healthy) can be compared and the number of (detectable) proteins that are differently abundant in the different conditions can be determined without any protein identification. Proteins can then be identified using techniques and sequence data are available. In the cases where the genome of the organism in question has not been sequenced, the most common approaches are to use ESTs or sequences from species with a high degree of sequence similarity. When 2D-gel based experiments designed to study the effect of oomycete infection are compared, some common themes emerge. Typically, 20-80 proteins that display significant changes in abundance in response to infection are identified and these proteins are often metabolic enzymes, pathogenesis proteins (PR) proteins and proteins involved in the production of reactive oxygen species during the oxidative burst.

The work of the group of Frank Colditz, on the *Medicago truncatula - Aphanomyces euteiches* pathosystem provides a number of interesting examples of different experimental approaches. In one of their earlier studies, a healthy *M. truncatula* was compared with plants infected with *A. euteiches* (Colditz et al., 2004). Twelve proteins with differential abundance were identified. Among these were heat shock proteins and proteins belonging to the PR 10 family. In a subsequent experiment, the PR10-1 gene of *M. truncatula* was silenced and the effect of *A. euteiches* infection of the transgenic plants was compared with the effect of infection of wild type plants (Colditz et al., 2007). The transgenic plants were shown to be more resistant to *A. euteiches.* The proteomic analyses demonstrated increased abundance of other families of PR proteins in the PR10-1-silenced line. In another paper published in 2009, *M. truncatula* cells grown in suspension culture were used to compare the effect of infection with *A. euteiches* zoospores with elicitation using filtered medium from *A. euteiches* culture (Trapphoff et al., 2009). The filtered medium contained a mixture of secreted *A. euteiches* proteins and cell wall proteins released from dead cells. The purpose was to compare the effect of infection with the earliest reaction of the *M. truncatula* cells, triggered by PAMPs from *A. euteiches*. 44 proteins were identified. Infection resulted in increased abundance of 38 proteins, while elicitation resulted in the increased abundance of 16 proteins. Of these, 10 where induced by both treatments. The proteins induced by infection were involved in production of ROS, redox reactions, signal transduction, cell wall restructuring, PR proteins, carbohydrate metabolism and amino acid biosynthesis. The proteins induced by elicitation were mainly involved in the production of ROS and redox reactions. Interestingly, PR proteins and proteins involved in carbohydrate metabolism were not induced by elicitation. Another paper published later included proteomics to compare the effects on *M. truncatula* infection by *A. euteiches* with co inoculation with two symbiotic soil microorganisms (Schenkluhn et al., 2010). The proteins induced by pathogenic infection were similar to those described previously, whilst co-infection of pathogenic and symbiotic microorganisms did not induce any PR-proteins and induced less of other pathogen-induced protein patterns.

The first proteomics study of oomycete-plant infection that used a technique other than 2D SDS-PAGE was performed by Palmieri et al. (2012). This study was also the first to benefit from a complete genome sequence. They used gel-free proteomics to study the mechanism by which the beneficial microorganism *Trichoderma harzianum* T39 reduces infection of *Plasmopara viticola* in grapevine. 800 proteins were identified and quantified, 246 of which displayed changes in abundance between some of the different conditions. The combination of gel free methods and a more comprehensive database resulted in a dramatical increase in the number of identified and quantified proteins, compared to 2D-gel based approaches. This is a general trend and one of the major reasons that gel free methods have become much more commonly used in recent years. As a rule, gel-free methods (i.e. HPLC-MSMS based) result in hundreds to thousands of identified proteins. While identification is straightforward, quantification can be more challenging. Palmieri et al. (2012) used iTRAQ, a quantification method based on chemical labeling of the samples, but various other label free methods also exist.

Apoplastic effectors will often degrade secreted proteins from the infected plant. Ali et al. (2014) used this fact in a study where they combined RNA measurements and proteomics in order to get a more global understanding of these interactions. They isolated apoplastic molecules from one susceptible and two resistant potato cultivars during *P. infestans* infection. A total of 1639 proteins were included in the quantitative data analysis. Of these, 1075 were found to vary in at least one comparison between the different plants and stages of infection. To identify potential apoplastic effector targets, proteins with reduced abundance and increased RNA during infection were selected. The idea behind this strategy was that the plant will increase transcription of defense related proteins in response to infection (which will lead to increased mRNA), but the pathogen will use apoplastic effectors to hydrolyze the corresponding protein (which will lead to decreased abundance of the protein). Another strategy was to search for proteins that displayed decreased abundance in the susceptible cultivar but not in the resistant cultivars. Among the effector target candidates identified by that analysis were the protease C14, which has previously been identified as an effector target. Ali et al. (2014) also demonstrated the usefulness of combining protein sequences generated de novo from RNAseq data with published databases. Using this strategy, they were able to increase the number of identified proteins by 17 %, as compared to using only the published protein database.

It has so far been very difficult to identify pathogen proteins in samples from infected plants. The reason for this is probably that the plant proteins are so abundant compared to the proteins of the pathogen that it becomes impossible to detect them. True interactome studies are an important goal for the future, but will probably require the use of techniques such as laser microdissection for specific isolation of infected tissue. However, even so it may be problematic to design proper controls. Another option is to use targeted proteomic methods such as selected reaction monitoring (SRM). In this type of HPLC-MSMS analysis, pre-selected peptides are monitored throughout an entire HPLC gradient (Picotti et al., 2009). This makes it possible to detect the peptides of interest with very high sensitivity, even in a complex mixture. By monitoring several peptides from a single protein, pre-specified proteins can be detected with high sensitivity. Hopefully, these methods, combined with the ever-increasing sensitivity of new mass spectrometers will make such studies possible in the future. Another technique that may be useful for future analysis is SWATH acquisition. In conventional MSMS analysis, only selected peptides are subjected to fragmentation and identification in the mass spectrometer. In SWATH acquisition, all peptides are fragmented, and the resulting masses are assigned to their respective parents based on elution time and other properties. The result is a dataset that can be re-analysed retrospectively, to look for certain proteins or post-translational modifications (Gillet et al., 2012).

Oomycete biological control and omics

One promising method for increasing disease resistance in crops and controlling oomycete pathogens is by addition of microbial antagonists that suppress pathogens also called biological control agents (BCA). There are a number of microorganisms that belong to bacterial, fungal and oomycetes lineages that mycoparatize oomycete pathogens such as *Burkholderia* spp.(Heungens and Parke, 2000), *Bacillus* spp.(Baker et al., 1983), *Streptomyces* spp. (Xiao et al., 2002), *Candida* spp. (Saligkarias et al., 2002), *Trichoderma* spp.(Anandaraj and Sarma, 1995; Sawant et al., 1995) and *Pythium* spp. (Gerbore et al., 2014; Horner et al., 2012). Despite the presence of BCAs in the market for over two decades, the usage of BCAs compared to conventional pesticides is very limited due to low success rate in controlling plant diseases (Alabouvette et al., 2006). The commercialization of BCA has been constrained by lack of understanding of the complex interactions between BCA- host plant - surrounding microbial flora, and pathogens in the field. Genomics, transcriptomics and proteomics can aid in successful commercialization of BCAs by identification of key virulence genes, selection of virulent microbial isolates, identification of genes involved in the production and secretion of antimicrobial compounds and fitness, as well as other characteristics that affect cultural variability and stability of a BCA. More specifically, laboratory conditions which affect sporulation and germination of spores should be carefully addressed since these factors affect maintenance, preservation and the selection of isolates for commercialization.

Recent genome and transcriptome studies on BCAs identifed virulence factors, other important genes and pathways involved in mycoparasitism (Baroncelli et al., 2015; Gao et al., 2011; Horner et al., 2012; Velasquez et al., 2014). Interestingly, in a recent pilot transcriptome sequencing study of the mycoparasitic oomycete, *Pythium oligandrum*, several transcripts with similarity to effectors expressed by plant pathogenic oomycetes were identified, such as glucanases, proteases, protease inhibitors, Crinkler effectors, and elicitors (Horner et al., 2012). This suggests that oomycetes with different hosts and different modes of growth have nevertheless evolved similar infection strategies (Grenville-Briggs et al., 2013; Horner et al., 2012). We foresee more detailed comparative analysis of effector biology of mycoparasitic and plant pathogenic oomycetes, also including the host in the future (Figure 1).

As mentioned earlier a proteomic study on priming of defenses in grapevine against the oomycete *Plasmopara viticola,* the causative agent of downy mildew with biocontrol agent *Trichoderma harzianum* T39, demonstrated accumulation of proteins involved in signal transduction activating different defense pathways, reactive oxygen species and callose formation at infection sites (Palmieri et al., 2012). Studies on interactions of BCA *Metschnikowia fructicola*, and its target fungal pathogen *Penicillium digitatum*, and plant host red grapefruit have revealed the biology and mechanism associated with bio-control activity (Hershkovitz et al., 2013). In this study over 250 genes have shown differential expression and may potentially affect the biocontrol activity of *M. fructicola*. Presently, there is no study focusing on BCA-oomycete pathogen or host multipart interactions in actual field conditions which are of utmost importance for effective control of plant diseases.

Field-omics and crop systems biology

The recent omics advances enable genome-scale capturing of complex biological processes at the molecular level in crop fields. This opens up new possibilities for understanding the relationships between genotype, environment and management interventions such as fungicide treatments. However, combining different types of data obtained from the field and understanding the biological relevance of large datasets remain challenging. We have earlier pointed out the need to create a cross-disciplinary platform for experimental design, sampling and subsequent analysis of large-scale molecular data from field trials by introducing the term "Field-omics" (Alex-andersson et al., 2014). Field-omics strives to couple information from genomes, transcriptomes, proteomes, metabolomes and metagenomes to the long-established practice in crop science of conducting field trials as well as to adapt current strategies for recording and analysing field data to facilitate integration with '-omics' data." This concept creates a basis for crop systems biology, which models complex crop traits by combining functional genomics with crop physiology and biochemistry (Yin and Struik, 2010). Future exciting models might be built on for example how plants develop over the whole season, based on several different inputs including long term weather forecasts and molecular information that may predict the severity of Phytophthora outbreaks and could be a future tool for crop selection relating to a certain field in a specific year.

Multistress conditions affects pathogen interactions, and therefore field studies are necessary, particularly since in such conditions non-dominating resistance genes may be in play. Natural light conditions also influence the infection success of pathogens (Kangasjärvi et al., 2012), which is important to consider since plant-pathogen interactions in the laboratory are mostly conducted in relatively low and constant light regimes. Field resistance, a phenomena only observed in a field situation and not in the laboratory, has been observed to pathogens such as *Magnaporthe oryzae* (Miah et al., 2013), and *Phytophthora infestans* (Rietman et al., 2012). This phenonema is assumed to be controlled by more than a single genetic locus and dependent on environmental and management cues. For example, in the late blight-resistant potato cultivar, Sarpo Mira, a discrepancy between resistance in laboratory tests and

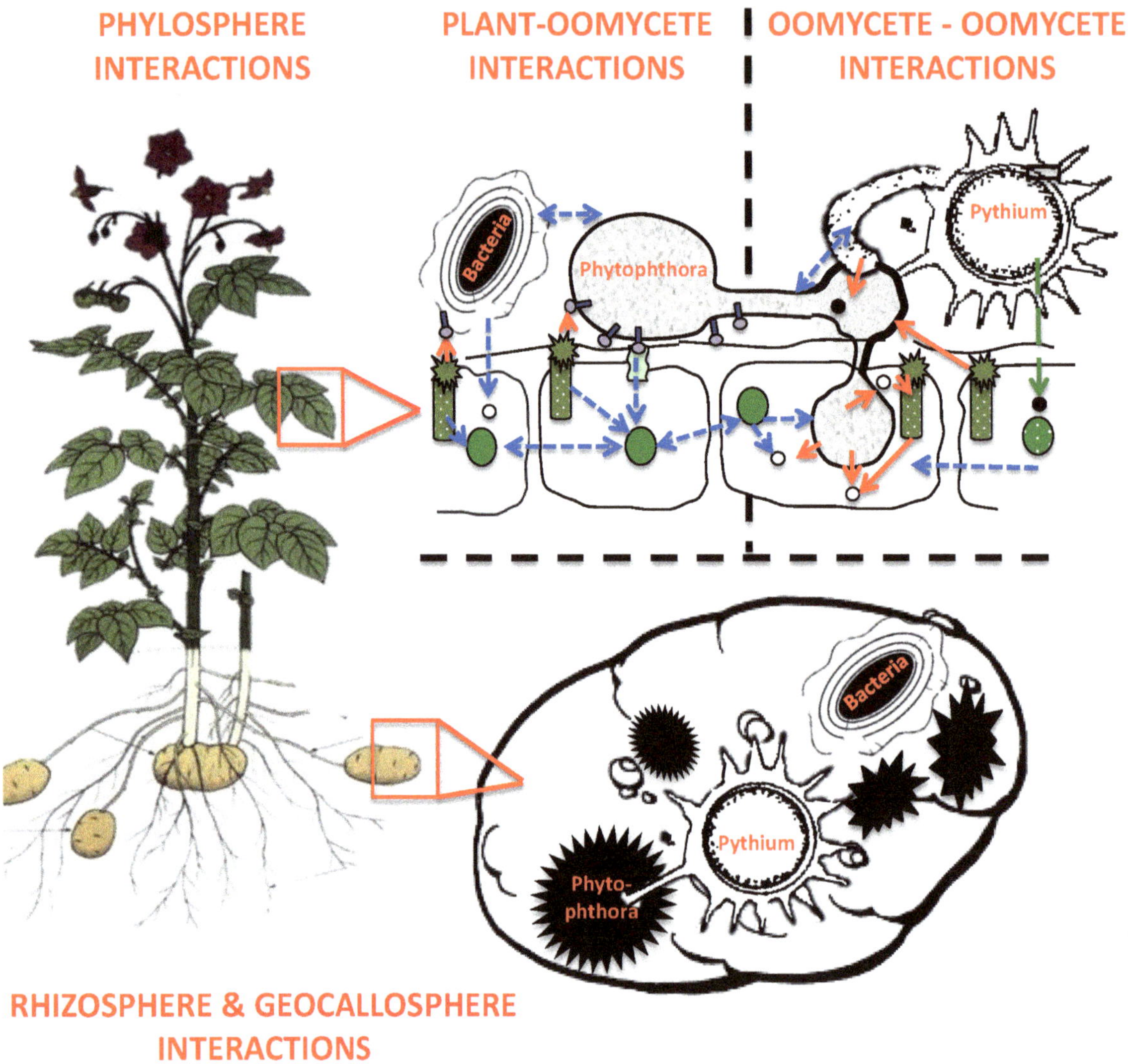

Figure1. Schematic representation of potential plant-oomycete and oomycete-oomycete interactions under field conditions as exemplified by *Phytophthora infestans* interactions with potato and with the mycoparasitic oomycete *Pythium oligandrum* and Bacteria in the phyllosphere, rhizosphere and geocallosphere.

field trials was discovered and later linked to an additional resistant factor conferring field resistance (Rietman et al., 2012). It is also clear that resistant oomycete-plant interactions can influence other biotic interactions such as insect oviposition (Abreha et al., 2015).

RNA samples from different vineyards were used in a genome-wide expression study and several transcripts for pathogenesis-related (PR) proteins were found to be independent of external stress cues (Dal Santo et al., 2013). In cultivated and wild potato, large scale secretome sampling revealed surprisingly few samples with high levels of PR proteins (less than 30% in average) even with a high disease pressure (Ali et al., 2014; Åsa Lankinen personal communcation). These proteins have been used as classical biomarkers for stress-induction in the laboratory, but might be unsuitable as biomarkers in the field where they seem to be absent or present constitutively.

Obtaining global proteome samples from the field would be labor-intensive and require complicated fractionation steps impeding large-scale analysis. Consequently, targeted approaches are currently necessary in order to obtain biologically relevant results. The secretome fraction containing apoplastic proteins constitutes the interface between intracellular processes and the environment, and is of great interest for plant pathogen interaction studies, since it includes apoplastic effectors and many signaling and enzymatic components. It is fast and relatively easy to isolate from other cellular compartments, even in the field, with a low level of contamination (Alexandersson et al., 2013); Andreasson et al., 2015) but it needs processing before freezing the samples. Another way to overcome the complexity associated with protein quantification and identification by MS is to analyse a few peptides by selected reaction monitoring (SRM). As mentioned above, SRM has a high quantitative accuracy and can discriminate between protein isoforms in contrast to shotgun proteomics (Addona et al., 2009), but requires detailed knowledge about the system. Another way of getting interesting fractions is to isolate for example the phosphoproteome, or methylated proteins, but this requires more samples as the technical variation increases. These studies will be crucial to set a baseline for molecular field studies and to learn about the differences in global molecular mechanisms compared to a laboratory setting.

To explore the general effect of spatial variance or management methods, isolation of RNA is still preferable since it is relatively easy to sample if the specimen can be flash frozen and gives a global molecular view. Proteomics is "closer to the phenotype" than RNA but more dependent on stability of different fractions or specific molecules (see above). Metabolites might currently be more suitable to study variance in plant-tissue composition for desired properties, especially if the genes affecting the metabolites of key-interest are poorly characterized.

Samples from the field also include different micro-organisms. Despite improvements in microbial cultivation techniques, culturable organisms so far identified in complex environmental samples are likely to represent only a minor fraction of the organisms present (Sørensen et al., 2009). Meta-transcriptomics, data from a community of several different organsims, sequenced directly from the field can facilitate our understanding of the significance of plant-microbe interactions in an ecological context. Coupling metagenomics and meta-transcriptomics to agricultural practices is likely to have a large impact on the understanding of biocontrol agents, plant defense induction, pathogen spread, spraying-regimes and integrated pest management. Bernard et al. (2012) detected shifts in soil microbial community structure using fatty acid methyl ester (FAME) profile analysis of soils from organic and conventional potato production systems. These results demonstrate that cultural practices such as organic amendments, rotations, and the use of biological control agents can significantly impact local microbial communities in both organic and conventional potato farming systems. The success of an oomycete as a pathogen may be influenced by the community of microbes occupying the same ecological niche, and the oomycetes themselves may also influence the colonisation of plants by other microbes. For example, the presence of downy mildew *(Bremia lactucae)* on romine lettuce promotes subsequent colonisation by human enteric pathogens *E. coli* 0157:H7 and *Salmonella enterica* (Simko et al., 2015). Previously, oomycetes in soil have been targeted with limited success (Coince et al., 2013), however the metatranscriptomics approach used by Geisen et al. (2015) revealed the lineage to be ubiquitous, abundant and active in many soil types. Espindola et al. (2015) have recently developed a new approach to utilise such data by analysis of metagenomics data with electronic-probe diagnostic nucleic acid analysis (EDNA) for rapid and sensitive detection of oomycete sequences in metasequence data.These data suggest oomycetes might play important roles in structuring natural plant communities. Their ubiquitous nature may also contribute to the rapid development of oomycete diseases in several agricultural or natural systems. One of the complicated issues in rhizosphere metatranscriptome studies is the number of different compartments to sample, such as: 1) soil close to roots, 2) root surface microbes (rhizosphere epiphytic community), 3) root colonising microbes (endophytes), 4) tuber/stolen associated community (geocaulosphere microbes) 5) leaf material (pathogens and endophytes), and 6) vascular tissues. Lundberg et al. (2012) surveyed the root microbiome of the model plant Arabidopsis using metatranscriptomic approaches and were able to identify distinct groups of microbes that were highly abundant in individual compartments. Peiffer et al. (2013) surveyed the metatranscriptome of the maize root microbiome under field conditions and identified substantial diversity in bacterial species present in the rhizosphere in relation to bulk soil analysis. Such results should facilitate expanded studies to identify robust plant-microbe interactions at the level of individual polymorphisms by genome wide association, so potential for that plant-microbiome interactions can ultimately be incorporated into plant breeding. However, the majority of species identified in similar studies are prokaryotic in nature and studies incorporating analysis of oomycetes or other Eukaryotes from other agriculturally relevant crop species are largely lacking.

Multi-stress experiments require also that spatial and temporal variance is taken into account and recorded. Micro-climate and soil composition may vary not only between fields but also over single plots. Sampling is also dependent on plant developmental stage and variance of biotic factors such as disease or symbiont load as well as management. This has profound implications on study-design and sampling strategies for field trials (Alexandersson et al., 2014). Phenotyping and remote sensing of crops in the field is a rapidly developing area. One of the main challenges is still the ability to handle, analyze and visualize large-scale data in a way that it really advances biological understanding. Interestingly, compared to ecosystems and clinical data, the agricultural field as an entity is a simpler and a more controlled structure and thus has the power to connect between the ecological and greenhouse system levels, and could serve as a model for ecosystems studies in the wild.

Conclusions

Omics based technologies have and are revolutionizing the approach to study complex and agriculturally important systems like host plant-oomycete patho-systems (Figure 2). Genome sequencing data is shedding light on the blue-print that causes complexities observed in plants and oomycete pathogens. Transcriptomics and proteomics are now adding layers to this by unraveling functionality that arises from this blue-print. Much of our understanding of plant-oomycete patho-systems has been driven by studying both the host and the oomycete separately. However, with improvements in transcriptomic and proteomic methodologies, we foresee an increased in

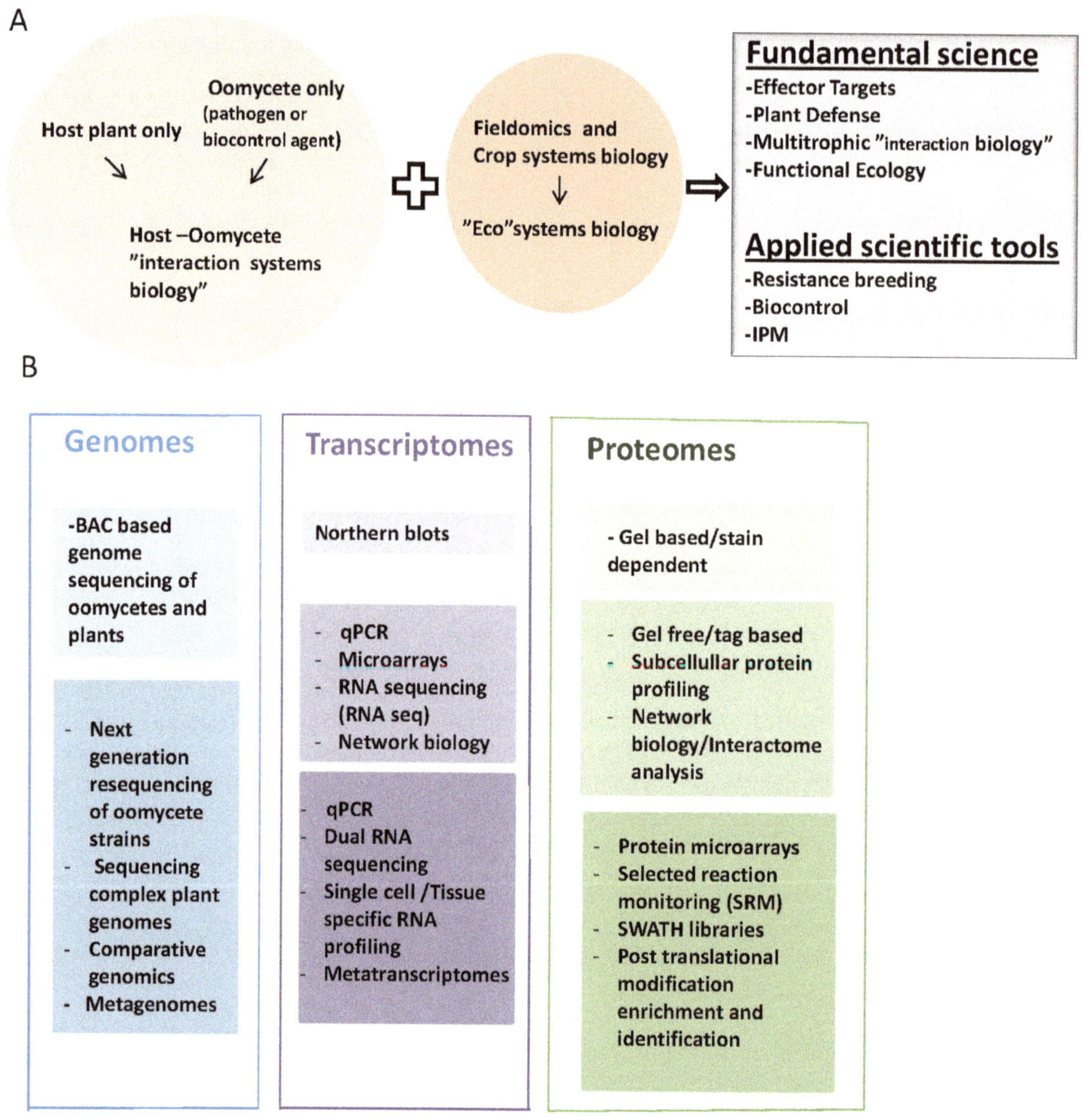

Figure 2. A. Suggested strategy and goals for the study of host-oomycete patho-systems. B. Various techniques discussed in the review that aid in deepened understanding of host-oomycete pathosystems. Techniques listed in boxes shaded in lighter gradient of color indicate older techniques, while techniques listed in boxes with increasing color gradient refer to current and future technologies respectively.

studies of these individual systems together as an "interaction system" in specific tissues. These interactions can also include biocontrol agents due to the rising prospects of their use as effective control strategies in an integrated pest management system. Another level of understanding of these multiple interaction systems in the context of crops and their ecosystems can be gained through fieldomics where the samples are taken in the field in order to grasp the multi-stress environment, currently most probably lacking mechanistic proteomic and post transcriptional analyses. Together, these studies will create a basis for crop systems biology and new understanding at the ecosystem level. Crop losses incurred due to oomycete infections are immense and current control strategies through fungicides and traditional R gene based resistance breeding programs will be strengthened, since new methods can be developed with data obtained from Omics driven research. At the same time, these studies will also improve fundamental biological and biochemical understanding of this "interaction-system".

Acknowledgements
We would like to thank all the members of our resistance biology group for valuable discussions as well as the Swedish foundation for strategic research and the Swedish foundation for environmental strategic research for financial support.

References

Abreha, K.B., Alexandersson, E., Vossen, J.H., Anderson, P., and Andreasson, E. (2015). Inoculation of Transgenic Resistant Potato by Phytophthora infestans Affects Host

Plant Choice of a Generalist Moth. PLoS ONE *10*, e0129815.

Addona, T.A., Abbatiello, S.E., Schilling, B., Skates, S.J., Mani, D., Bunk, D.M., Spiegelman, C.H., Zimmerman, L.J., Ham, A.-J.L., and Keshishian, H. (2009). Multi-site assessment of the precision and reproducibility of multiple reaction monitoring-based measurements of proteins in plasma. Nature biotechnology *27*, 633-641.

Alabouvette, C., Olivain, C., and Steinberg, C. (2006). Biological control of plant diseases: the European situation. European Journal of Plant Pathology *114*, 329-341.

Alexandersson, E., Ali, A., Resjö, S., and Andreasson, E. (2013). Plant secretome proteomics. Frontiers in plant science *4*, 9.

Alexandersson, E., Jacobson, D., Vivier, M.A., Weckwerth, W., and Andreasson, E. (2014). Field-omics—understanding large-scale molecular data from field crops. Frontiers in plant science *5*, 286.

Ali, A., Alexandersson, E., Sandin, M., Resjö, S., Lenman, M., Hedley, P., Levander, F., and Andreasson, E. (2014). Quantitative proteomics and transcriptomics of potato in response to Phytophthora infestans in compatible and incompatible interactions. BMC genomics *15*, 497.

Anandaraj, M., and Sarma, Y. (1995). Diseases of black pepper (Piper nigrum L.) and their management. Journal of Spices and Aromatic Crops, *4* , 17-23.

Andolfo, G., Sanseverino, W., Rombauts, S., Van de Peer, Y., Bradeen, J.M., Carputo, D., Frusciante, L., and Ercolano, M.R. (2013). Overview of tomato (Solanum lycopersicum) candidate pathogen recognition genes reveals important Solanum R locus dynamics. New Phytologist *197*, 223-237.

Andreasson, E., Abreha, K.B., Resjö, S. (2015). The isolation of plant organelles and structures: isolation of apoplast. Meth. Mol. Biol. In press.

As-Sadi, F., Carrere, S., Gascuel, Q., Hourlier, T., Rengel, D., Le Paslier, M.-C., Bordat, A., Boniface, M.-C., Brunel, D., and Gouzy, J. (2011). Transcriptomic analysis of the interaction between Helianthus annuus and its obligate parasite Plasmopara halstedii shows single nucleotide polymorphisms in CRN sequences. BMC genomics *12*, 498.

Asai, S., Rallapalli, G., Piquerez, S.J., Caillaud, M.-C., Furzer, O.J., Ishaque, N., Wirthmueller, L., Fabro, G., Shirasu, K., and Jones, J.D. (2014). Expression profiling during Arabidopsis/downy mildew interaction reveals a highly-expressed effector that attenuates responses to salicylic acid. PLoS Pathog *10,* e1004443.

Baker, C.J., Stavely, J., Thomas, C., Sasser, M., and MacFall, J. (1983). Inhibitory effect of Bacillus subtilis on Uromyces phaseoli and on development of rust pustules on bean leaves. Phytopathology *73*, 1148-1152.

Baroncelli, R., Piaggeschi, G., Fiorini, L., Bertolini, E., Zapparata, A., Pè, M.E., Sarrocco, S., and Vannacci, G. (2015). Draft whole-genome sequence of the biocontrol agent Trichoderma harzianum T6776. Genome announcements *3*, e00647-00615.

Baxter, L., Tripathy, S., Ishaque, N., Boot, N., Cabral, A., Kemen, E., Thines, M., Ah-Fong, A., Anderson, R., and Badejoko, W. (2010). Signatures of adaptation to obligate biotrophy in the Hyaloperonospora arabidopsidis genome. science *330*, 1549-1551.

Bernard, E., Larkin, R.P., Tavantzis, S., Erich, M.S., Alyokhin, A., Sewell, G., Lannan, A., and Gross, S.D. (2012). Compost, rapeseed rotation, and biocontrol agents significantly impact soil microbial communities in organic and conventional potato production systems. Applied Soil Ecology *52*, 29-41.

Bigeard, J., Colcombet, J., and Hirt, H. (2015). Signaling Mechanisms in Pattern-Triggered Immunity (PTI). Molecular Plant, *8*, 521-539.

Bolger, M.E., Weisshaar, B., Scholz, U., Stein, N., Usadel, B., and Mayer, K.F.X. (2014). Plant genome sequencing — applications for crop improvement. Current Opinion in Biotechnology *26*, 31-37.

Bozkurt, T.O., Schornack, S., Banfield, M.J., and Kamoun, S. (2012). Oomycetes, effectors, and all that jazz. Current opinion in plant biology *15*, 483-492.

Burkhardt, A., Buchanan, A., Cumbie, J.S., Savory, E.A., Chang, J.H., and Day, B. (2015). Alternative Splicing in the Obligate Biotrophic Oomycete Pathogen Pseudoperonospora cubensis. Molecular Plant-Microbe Interactions *28*, 298-309.

Cabral, A., Stassen, J., Seidl, M.F., Bautor, J., Parker, J.E., and Van den Ackerveken, G. (2011). Identification of Hyaloperonospora arabidopsidis transcript sequences expressed during infection reveals isolate-specific effectors. PLoS One *6*, e19328.

Chen, X.-R., Xing, Y.-P., Li, Y.-P., Tong, Y.-H., and Xu, J.-Y. (2013). RNA-Seq reveals infection-related gene expression changes in Phytophthora capsici. PloS one *8*, e74588.

Coince, A., Caël, O., Bach, C., Lengellé, J., Cruaud, C., Gavory, F., Morin, E., Murat, C., Marcais, B., and Buee, M. (2013). Below-ground fine-scale distribution and soil versus fine root detection of fungal and soil oomycete communities in a French beech forest. Fungal Ecology *6*, 223-235.

Colditz, F., Niehaus, K., and Krajinski, F. (2007). Silencing of PR-10-like proteins in Medicago truncatula results in an antagonistic induction of other PR proteins and in an increased tolerance upon infection with the oomycete Aphanomyces euteiches. Planta *226*, 57-71.

Colditz, F., Nyamsuren, O., Niehaus, K., Eubel, H., Braun, H.-P., and Krajinski, F. (2004). Proteomic approach: identification of Medicago truncatula proteins induced in roots after infection with the pathogenic oomycete Aphanomyces euteiches. Plant Molecular Biology *55*, 109-120.

Consortium, P.G.S. (2011). Genome sequence and analysis of the tuber crop potato. Nature *475*, 189-195.

Dal Santo, S., Tornielli, G.B., Zenoni, S., Fasoli, M., Farina, L., Anesi, A., Guzzo, F., Delledonne, M., and Pezzotti, M. (2013). The plasticity of the grapevine berry transcriptome. Genome Biol *14*, r54.

Derevnina, L., and Michelmore, R.W. (2015). Wheat rusts never sleep but neither do sequencers: will pathogenomics transform the way plant diseases are managed? Genome biology *16*, 44.

Dong, S., Raffaele, S., and Kamoun, S. (2015). The two-speed genomes of filamentous pathogens: waltz with plants.

Du, J., Verzaux, E., Chaparro-Garcia, A., Bijsterbosch, G., Keizer, L.C.P., Zhou, J., Liebrand, T.W.H., Xie, C., Govers, F., Robatzek, S., *et al.* (2015). Elicitin recognition confers enhanced resistance to Phytophthora infestans in potato. Nature Plants *1*.

Ebstrup, T., Saalbach, G., and Egsgaard, H. (2005). A proteomics study of *in vitro* cyst germination and appressoria formation in Phytophthora infestans. Proteomics *5*, 2839-2848.

Espindola, A., Schneider, W., Hoyt, P.R., Marek, S.M., and Garzon, C. (2015). A new approach for detecting fungal and oomycete plant pathogens in next generation sequencing metagenome data utilising electronic probes. International Journal of Data Mining and Bioinformatics *12*, 115-128.

Fawke, S., Doumane, M., and Schornack, S. (2015). Oomycete Interactions with Plants: Infection Strategies and Resistance Principles. Microbiology and Molecular Biology Reviews *79*, 263-280.

Gao, Q., Jin, K., Ying, S.-H., Zhang, Y., Xiao, G., Shang, Y., Duan, Z., Hu, X., Xie, X.-Q., and Zhou, G. (2011). Genome sequencing and comparative transcriptomics of the model entomopathogenic fungi Metarhizium anisopliae and M. acridum. PLoS Genet *7*, e1001264.

Geisen, S., Tveit, A.T., Clark, I.M., Richter, A., Svenning, M.M., Bonkowski, M., and Urich, T. (2015). Metatranscriptomic census of active protists in soils. The ISME journal.

Gerbore, J., Benhamou, N., Vallance, J., Le Floch, G., Grizard, D., Regnault-Roger, C., and Rey, P. (2014). Biological control of plant pathogens: advantages and limitations seen through the case study of Pythium oligandrum. Environmental Science and Pollution Research *21*, 4847-4860.

Gillet, L.C., Navarro, P., Tate, S., Röst, H., Selevsek, N., Reiter, L., Bonner, R., and Aebersold, R. (2012). Targeted data extraction of the MS/MS spectra generated by data-independent acquisition: a new concept for consistent and accurate proteome analysis. Molecular and Cellular Proteomics *11*, O111. 016717.

Grenville-Briggs, L., Gachon, C.M., Strittmatter, M., Sterck, L., Küpper, F.C., and van West, P. (2011). A molecular insight into algal-oomycete warfare: cDNA analysis of Ectocarpus siliculosus infected with the basal oomycete Eurychasma dicksonii. PloS one *6*, e24500.

Grenville-Briggs, L.J., Avrova, A.O., Bruce, C.R., Williams, A., Whisson, S.C., Birch, P.R., and van West, P. (2005). Elevated amino acid biosynthesis in Phytophthora infestans during appressorium formation and potato infection. Fungal Genetics and Biology *42*, 244-256.

Grenville-Briggs, L.J., Avrova, A.O., Hay, R.J., Bruce, C.R., Whisson, S.C., and Van West, P. (2010). Identification of appressorial and mycelial cell wall proteins and a survey of the membrane proteome of Phytophthora infestans. Fungal biology *114*, 702-723.

Grenville-Briggs, L.J., Horner, N.R., Phillips, A.J., Beakes, G.W., and Van West, P. (2013). A family of small tyrosine rich proteins is essential for oogonial and oospore cell wall development of the mycoparasitic oomycete Pythium oligandrum. Fungal biology *117*, 163-172.

Hayden, K.J., Garbelotto, M., Knaus, B.J., Cronn, R.C., Rai, H., and Wright, J.W. (2014). Dual RNA-seq of the plant pathogen Phytophthora ramorum and its tanoak host. Tree Genetics and Genomes *10*, 489-502.

Hershkovitz, V., Sela, N., Taha-Salaime, L., Liu, J., Rafael, G., Kessler, C., Aly, R., Levy, M., Wisniewski, M., and Droby, S. (2013). De-novo assembly and characterization of the transcriptome of Metschnikowia fructicola reveals differences in gene expression following interaction with Penicillium digitatum and grapefruit peel. BMC genomics *14*, 168.

Heungens, K., and Parke, J. (2000). Zoospore homing and infection events: effects of the biocontrol bacterium Burkholderia cepacia AMMDR1 on two oomycete pathogens of pea (Pisum sativum L.). Applied and environmental microbiology *66*, 5192-5200.

Horner, N.R., Grenville-Briggs, L.J., and Van West, P. (2012). The oomycete Pythium oligandrum expresses putative effectors during mycoparasitism of Phytophthora infestans and is amenable to transformation. Fungal biology *116*, 24-41.

Hosseini, S., Resjö, S., Liu, Y., Durling, M., Heyman, F., Levander, F., Liu, Y., Elfstrand, M., Jensen, D.F., Andreasson, E., and Karlsson, M. (2015). Comparative proteomic analysis of hyphae and germinating cysts of Phytophthora pisi and Phytophthora sojae. Journal of proteomics *117*, 24-40.

Jahan, S.N., Åsman, A.K., Corcoran, P., Fogelqvist, J., Vetukuri, R.R., and Dixelius, C. (2015). Plant-mediated gene silencing restricts growth of the potato late blight pathogen Phytophthora infestans. Journal of experimental botany, *66*, 2785-2794.

Jones, J.D.G., and Dangl, J.L. (2006). The plant immune system. Nature *444*, 323-329.

Judelson, H.S. (2012). Dynamics and innovations within oomycete genomes: insights into biology, pathology, and evolution. Eukaryotic cell *11*, 1304-1312.

Judelson, H.S., Ah-Fong, A.M.V., Aux, G., Avrova, A.O., Bruce, C., Cakir, C., da Cunha, L., Grenville-Briggs, L., Latijnhouwers, M., Ligterink, W., *et al.* (2008). Gene Expression Profiling During Asexual Development of the Late Blight Pathogen Phytophthora infestans Reveals a Highly Dynamic Transcriptome. Molecular Plant-Microbe Interactions *21*, 433-447.

Jupe, F., Pritchard, L., Etherington, G.J., MacKenzie, K., Cock, P.J., Wright, F., Sharma, S.K., Bolser, D., Bryan, G.J., and Jones, J.D. (2012). Identification and localisation of the NB-LRR gene family within the potato genome. BMC genomics *13*, 75.

Jupe, F., Witek, K., Verweij, W., Śliwka, J., Pritchard, L., Etherington, G.J., Maclean, D., Cock, P.J., Leggett, R.M., and Bryan, G.J. (2013a). Resistance gene enrichment sequencing (RenSeq) enables reannotation of the NB-LRR gene family from sequenced plant genomes and rapid mapping of resistance loci in segregating populations. The Plant Journal *76*, 530-544.

Jupe, J., Stam, R., Howden, A.J., Morris, J.A., Zhang, R., Hedley, P.E., and Huitema, E. (2013b). Phytophthora capsici-tomato interaction features dramatic shifts in gene expression associated with a hemi-biotrophic lifestyle. Genome Biol *14*, R63.

Kangasjärvi, S., Neukermans, J., Li, S., Aro, E.-M., and Noctor, G. (2012). Photosynthesis, photorespiration, and light signalling in defence responses. Journal of Experimental Botany *63*, 1619-1636.

Kaufmann, K., and Busch, W. (2013). Plant genomics: from weed to wheat. Genome biology *14*, 308.

Kim, Y., Tsuda, K., Igarashi, D., Hillmer, R.A., Sakakibara, H., Myers, C.L., and Katagiri, F. (2014). Mechanisms underlying robustness and tunability in a plant immune signaling network. Cell host and microbe *15*, 84-94.

Krämer, R., Freytag, S., and Schmelzer, E. (1997). *In vitro* formation of infection structures of Phytophthora infestans is associated with synthesis of stage specific polypeptides. European Journal of Plant Pathology *103*, 43-53.

Kunjeti, S.G., Evans, T.A., Marsh, A.G., Gregory, N.F., Kunjeti, S., Meyers, B.C., Kalavacharla, V.S., and Donofrio, N.M. (2012). RNA-Seq reveals infection-related global gene changes in Phytophthora phaseoli, the causal agent of lima bean downy mildew. Molecular plant pathology *13*, 454-466.

Lamour, K.H., Mudge, J., Gobena, D., Hurtado-Gonzales, O.P., Schmutz, J., Kuo, A., Miller, N.A., Rice, B.J., Raffaele, S., and Cano, L.M. (2012). Genome sequencing and mapping reveal loss of heterozygosity as a mechanism for rapid adaptation in the vegetable pathogen Phytophthora capsici. Molecular Plant-Microbe Interactions *25*, 1350-1360.

Lenmam, L., Ali, A., Mühlenbock, P., Carlson-Nilsson, U., Liljeroth, E., Vleeshouwer, W., Andreasson, E. (2015). Effector-driven marker development and cloning of resistance genes against *Phytophthora infestans* in potato breeding clone SW93-1015. Theor. Appl. Gen. In press.

Lévesque, C.A., Brouwer, H., Cano, L., Hamilton, J.P., Holt, C., Huitema, E., Raffaele, S., Robideau, G.P., Thines, M., and Win, J. (2010). Genome sequence of the necrotrophic plant pathogen Pythium ultimum reveals original pathogenicity mechanisms and effector repertoire. Genome biology *11*, R73.

Lin, F., Zhao, M., Baumann, D.D., Ping, J., Sun, L., Liu, Y., Zhang, B., Tang, Z., Hughes, E., and Doerge, R.W. (2014). Molecular response to the pathogen Phytophthora sojae among ten soybean near isogenic lines revealed by comparative transcriptomics. BMC genomics *15*, 18.

Lodha, T., and Basak, J. (2012). Plant-pathogen interactions: what microarray tells about it? Molecular biotechnology *50*, 87-97.

Lundberg, D.S., Lebeis, S.L., Paredes, S.H., Yourstone, S., Gehring, J., Malfatti, S., Tremblay, J., Engelbrektson, A., Kunin, V., and Del Rio, T.G. (2012). Defining the core Arabidopsis thaliana root microbiome. Nature *488*, 86-90.

Mace, E., Tai, S., Innes, D., Godwin, I., Hu, W., Campbell, B., Gilding, E., Cruickshank, A., Prentis, P., and Wang, J. (2014). The plasticity of NBS resistance genes in sorghum is driven by multiple evolutionary processes. BMC plant biology *14*, 253.

Mandadi, K.K., and Scholthof, K.-B.G. (2015). Genome-wide analysis of alternative splicing landscapes modulated during plant-virus interactions in Brachypodium distachyon. The Plant Cell *27*, 71-85.

Martin, L.B., Fei, Z., Giovannoni, J.J., and Rose, J.K. (2013). Catalyzing plant science research with RNA-seq. Frontiers in plant science *4*, 66.

McGettigan, P.A. (2013). Transcriptomics in the RNA-seq era. Current opinion in chemical biology *17*, 4-11.

Meijer, H.J., Mancuso, F.M., Espadas, G., Seidl, M.F., Chiva, C., Govers, F., and Sabidó, E. (2014). Profiling the secretome and extracellular proteome of the potato late blight pathogen Phytophthora infestans. Molecular and Cellular Proteomics *13*, 2101-2113.

Miah, G., Rafii, M., Ismail, M., Puteh, A., Rahim, H., Asfaliza, R., and Latif, M. (2013). Blast resistance in rice: a review of conventional breeding to molecular approaches. Molecular biology reports *40*, 2369-2388.

Mithöfer, A., Müller, B., Wanner, G., and Eichacker, L.A. (2002). Identification of defence-related cell wall proteins in Phytophthora sojae-infected soybean roots by ESI-MS/MS. Molecular Plant Pathology *3*, 163-166.

Palmieri, M.C., Perazzolli, M., Matafora, V., Moretto, M., Bachi, A., and Pertot, I. (2012). Proteomic analysis of grapevine resistance induced by Trichoderma harzianum T39 reveals specific defence pathways activated against downy mildew. Journal of experimental botany *63*, 6237-6251.

Peiffer, J.A., Spor, A., Koren, O., Jin, Z., Tringe, S.G., Dangl, J.L., Buckler, E.S., and Ley, R.E. (2013). Diversity and heritability of the maize rhizosphere microbiome under field conditions. Proceedings of the National Academy of Sciences *110*, 6548-6553.

Picotti, P., Bodenmiller, B., Mueller, L.N., Domon, B., and Aebersold, R. (2009). Full dynamic range proteome analysis of S. cerevisiae by targeted proteomics. Cell *138*, 795-806.

Pritchard, L., and Birch, P.R. (2014). The zigzag model of plant-microbe interactions: is it time to move on? Molecular plant pathology *15*, 865-870.

Qutob, D., Chapman, B.P., and Gijzen, M. (2013). Transgenerational gene silencing causes gain of virulence in a plant pathogen. Nature communications *4*, 1349.

Raffaele, S., and Kamoun, S. (2012). Genome evolution in filamentous plant pathogens: why bigger can be better. Nat Rev Micro *10*, 417-430.

Rallapalli, G., Kemen, E.M., Robert-Seilaniantz, A., Segonzac, C., Etherington, G.J., Sohn, K.H., MacLean, D., and Jones, J.D. (2014). EXPRSS: an Illumina based high-throughput expression-profiling method to reveal transcriptional dynamics. BMC genomics *15*, 341.

Randall, T.A., Dwyer, R.A., Huitema, E., Beyer, K., Cvitanich, C., Kelkar, H., Ah Fong, A.M., Gates, K., Roberts, S., and Yatzkan, E. (2005). Large-scale gene discovery in the oomycete Phytophthora infestans reveals likely components of phytopathogenicity shared with true fungi. Molecular Plant-Microbe Interactions *18*, 229-243.

Resjö, S., Ali, A., Meijer, H.J., Seidl, M.F., Snel, B., Sandin, M., Levander, F., Govers, F., and Andreasson, E. (2014). Quantitative label-free phosphoproteomics of six different life stages of the late blight pathogen Phytophthora infestans reveals abundant phosphorylation of members of the CRN effector family. Journal of proteome research *13*, 1848-1859.

Rietman, H., Bijsterbosch, G., Cano, L.M., Lee, H.-R., Vossen, J.H., Jacobsen, E., Visser, R.G., Kamoun, S., and Vleeshouwers, V.G. (2012). Qualitative and quantitative late blight resistance in the potato cultivar Sarpo Mira is determined by the perception of five distinct RXLR effectors. Molecular Plant-Microbe Interactions *25*, 910-919.

Saligkarias, I., Gravanis, F., and Epton, H.A. (2002). Biological control of Botrytis cinerea on tomato plants by the use of epiphytic yeasts Candida guilliermondii strains 101 and US 7 and Candida oleophila strain I-182: II. A study on mode of action. Biological Control *25*, 151-161.

Saunders, D.G., Breen, S., Win, J., Schornack, S., Hein, I., Bozkurt, T.O., Champouret, N., Vleeshouwers, V.G., Birch, P.R., and Gilroy, E.M. (2012). Host protein BSL1 associates with Phytophthora infestans RXLR effector AVR2 and the Solanum demissum immune receptor R2 to mediate disease resistance. The Plant Cell *24*, 3420-3434.

Sawant, I.S., Sawant, S., and Nanaya, K. (1995). Biological control of Phytophthora root-rot of coorg mandarin (Citrus reticulata) by Trichoderma species grown on coffee waste. Indian Journal of Agricultural Science *65*, 842-846.

Savidor, A., Donahoo, R.S., Hurtado-Gonzales, O., Land, M.L., Shah, M.B., Lamour, K.H., and McDonald, W.H. (2008). Cross-species global proteomics reveals conserved and unique processes in Phytophthora sojae and Phytophthora ramorum. Molecular and Cellular Proteomics *7*, 1501-1516.

Savory, E.A., Adhikari, B.N., Hamilton, J.P., Vaillancourt, B., Buell, C.R., and Day, B. (2012). mRNA-Seq analysis of the Pseudoperonospora cubensis transcriptome during cucumber (Cucumis sativus L.) infection. PLoS One *7*,4.

Sboner, A., Mu, X.J., Greenbaum, D., Auerbach, R.K., and Gerstein, M.B. (2011). The real cost of sequencing: higher than you think! Genome biology *12*, 125.

Schenkluhn, L., Hohnjec, N., Niehaus, K., Schmitz, U., and Colditz, F. (2010). Differential gel electrophoresis (DIGE) to quantitatively monitor early symbiosis-and pathogenesis-induced changes of the Medicago truncatula root proteome. Journal of proteomics *73*, 753-768.

Schwanhausser, B., Busse, D., Li, N., Dittmar, G., Schuchhardt, J., Wolf, J., Chen, W., and Selbach, M. (2013). Corrigendum: Global quantification of mammalian gene expression control. Nature *495*, 126-127.

Schwanhäusser, B., Busse, D., Li, N., Dittmar, G., Schuchhardt, J., Wolf, J., Chen, W., and Selbach, M. (2011). Global quantification of mammalian gene expression control. Nature *473*, 337-342.

Shepherd, S.J., Van West, P., and Gow, N.A. (2003). Proteomic analysis of asexual development of Phytophthora palmivora. Mycological research *107*, 395-400.

Simko, I., Zhou, Y., and Brandl, M.T. (2015). Downy mildew disease promotes the colonization of romaine lettuce by Escherichia coli O157: H7 and Salmonella enterica. BMC microbiology *15*, 19.

Spoel, S.H., and Dong, X. (2012). How do plants achieve immunity? Defence without specialized immune cells. Nat Rev Immunol *12*, 89-100.

Stassen, J.H., Seidl, M.F., Vergeer, P.W., NIJMAN, I.J., Snel, B., Cuppen, E., and Van den Ackerveken, G. (2012). Effector identification in the lettuce downy mildew Bremia lactucae by massively parallel transcriptome sequencing. Molecular plant pathology *13*, 719-731.

Steinbrenner, A.D., Goritschnig, S., and Staskawicz, B.J. (2015). Recognition and Activation Domains Contribute to Allele-Specific Responses of an Arabidopsis NLR Receptor to an Oomycete Effector Protein. PLoS pathogens *11*, e1004665.

Strauß, T., van Poecke, R.M.P., Strauß, A., Römer, P., Minsavage, G.V., Singh, S., Wolf, C., Strauß, A., Kim, S., Lee, H.-A., *et al.* (2012). RNA-seq pinpoints a Xanthomonas TAL-effector activated resistance gene in a large-crop genome. Proceedings of the National Academy of Sciences *109*, 19480-19485.

Sørensen, J., Haubjerg Nicolaisen, M., Ron, E., and Simonet, P. (2009). Molecular tools in rhizosphere microbiology—from single-cell to whole-community analysis. Plant Soil *321*, 483-512.

Thomma, B.P., Nürnberger, T., and Joosten, M.H. (2011). Of PAMPs and effectors: the blurred PTI-ETI dichotomy. The Plant Cell Online *23*, 4-15.

Torto, T.A., Li, S., Styer, A., Huitema, E., Testa, A., Gow, N.A., van West, P., and Kamoun, S. (2003). EST mining and functional expression assays identify extracellular effector proteins from the plant pathogen Phytophthora. Genome research *13*, 1675-1685.

Trapphoff, T., Beutner, C., Niehaus, K., and Colditz, F. (2009). Induction of distinct defense-associated protein patterns in Aphanomyces euteiches (Oomycota)-elicited and-inoculated Medicago truncatula cell-suspension cultures: a proteome and phosphoproteome approach. Molecular plant-microbe interactions *22*, 421-436.

Usadel, B., Schwacke, R., Nagel, A., and Kersten, B. (2012). GabiPD - The GABI Primary Database integrates plant proteomic data with gene-centric information. Frontiers in Plant Science *3*, 154.

Wang, L., Li, P., and Brutnell, T.P. (2010). Exploring plant transcriptomes using ultra high-throughput sequencing. Briefings in Functional Genomics *9*, 118-128.

Wang, Z., Gerstein, M., and Snyder, M. (2009). RNA-Seq: a revolutionary tool for transcriptomics. Nature reviews. Genetics *10*, 57-63.

Weiberg, A., Wang, M., Lin, F.-M., Zhao, H., Zhang, Z., Kaloshian, I., Huang, H.-D., and Jin, H. (2013). Fungal small RNAs suppress plant immunity by hijacking host RNA interference pathways. Science *342*, 118-123.

Velasquez, P.F.Q., Abiff, S.K., Fins, K.C., Conway, Q.B., Salazar, N.C., Delgado, A.P., Dawes, J.K., Douma, L.G., and Tartar, A. (2014). Transcriptome analysis of the entomopathogenic oomycete Lagenidium giganteum reveals putative virulence factors. Applied and environmental microbiology *80*, 6427-6436.

Weßling, R., Epple, P., Altmann, S., He, Y., Yang, L., Henz, S.R., McDonald, N., Wiley, K., Bader, K.C., and Gläßer, C. (2014). Convergent targeting of a common host protein-network by pathogen effectors from three kingdoms of life. Cell host and microbe *16*, 364-375.

Westermann, A.J., Gorski, S.A., and Vogel, J. (2012). Dual RNA-seq of pathogen and host. Nat Rev Micro *10*, 618-630.

Vetukuri, R.R., Åsman, A.K., Tellgren-Roth, C., Jahan, S.N., Reimegård, J., Fogelqvist, J., Savenkov, E., Söderbom, F., Avrova, A.O., and Whisson, S.C. (2012). Evidence for small RNAs homologous to effector-encoding genes and transposable elements in the oomycete Phytophthora infestans.

Xiao, K., Kinkel, L.L., and Samac, D.A. (2002). Biological control of Phytophthora root rots on alfalfa and soybean with Streptomyces. Biological Control *23*, 285-295.

Yang, C., and Wei, H. (2015). Designing microarray and RNA-seq experiments for greater systems biology discovery in modern plant genomics. Molecular plant *8*, 196-206.

Yin, X., and Struik, P.C. (2010). Modelling the crop: from system dynamics to systems biology. Journal of Experimental Botany *61*, 2171-2183.

Zhao, M., Cai, C., Zhai, J., Lin, F., Li, L., Shreve, J., Thimmapuram, J., Hughes, T.J., Meyers, B.C., and Ma, J. (2015). Coordination of microRNAs, phasiRNAs, and NB-LRR genes in response to a plant pathogen: Insights from analyses of a set of soybean Rps gene near-isogenic lines. The Plant Genome *8*, 1.

Zuluaga, A.P., Vega-Arreguín, J.C., Fei, Z., Ponnala, L., Lee, S.J., Matas, A.J., Patev, S., Fry, W.E., and Rose, J.K.C. (2015a). Transcriptional dynamics of Phytophthora infestans during sequential stages of hemibiotrophic infection of tomato. Molecular Plant Pathology,

Zuluaga, A.P., Vega–Arreguín, J.C., Fei, Z., Matas, A.J., Patev, S., Fry, W.E., and Rose, J.K. (2015b). Analysis of the tomato leaf transcriptome during successive hemibiotrophic stages of a compatible interaction with the oomycete pathogen Phytophthora infestans. Molecular plant pathology.

Other Publications of Interest

Xu, J.-P. (2014). Next-generation Sequencing: Current Technologies and Applications. Caister Academic Press, U.K. ISBN: 978-1-908230-33-1

He, Z. (2014). Microarrays: Current Technology, Innovations and Applications. Caister Academic Press, U.K. ISBN: 978-1-908230-49-2

Poptsova, M.S. (2014). Genome Analysis: Current Procedures and Applications. Caister Academic Press, U.K. ISBN: 978-1-908230-29-4

Fuentes, M. and LaBaer, J. (2014). Proteomics: Targeted Technology, Innovations and Applications. Caister Academic Press, U.K. ISBN: 978-1-908230-46-1

Murillo, J., Vinatzer, B.A., Jackson, R.W. and Arnold, D.L. (2015). Bacteria-Plant Interactions: Advanced Research and Future Trends. Caister Academic Press, U.K. ISBN: 978-1-908230-58-4

Caranta, C., Aranda, M.A., Tepfer, M. and Lopez-Moya, J.J. (2011). Recent Advances in Plant Virology. Caister Academic Press, U.K. ISBN: 978-1-904455-75-2

Nannipieri, P., Pietramellara, G. and Renella, G. (2014). Omics in Soil Science. Caister Academic Press, U.K. ISBN: 978-1-908230-32-4

Herold, K.E. and Rasooly, A. (2009). Lab-on-a-Chip Technology: Fabrication and Microfluidics. Caister Academic Press, U.K. ISBN: 978-1-904455-46-2

Omics Approaches for the Engineering of Pathogen Resistant Plants

Diego F. Gomez-Casati[1]*, María A. Pagani[1], María V. Busi[1] and Vijai Bhadauria[2]

[1]Centro de Estudios Fotosintéticos y Bioquímicos (CEFOBI-CONICET), Universidad Nacional de Rosario, Suipacha 531, 2000, Rosario, Argentina
[2]Crop Development Centre / Department of Plant Sciences, University of Saskatchewan, Saskatoon, SK S7N 5A8 Canada
*Corresponding author.
Email: gomezcasati@cefobi-conicet.gov.ar

Abstract
The attack of different pathogens, such as bacteria, fungi and viruses has a negative impact on crop production. In counter such attacks, plants have developed different strategies involving the modification of gene expression, activation of several metabolic pathways and post-translational modification of proteins, which culminate into the accumulation of primary and secondary metabolites implicated in plant defense responses. The recent advancement in omics techniques allows the increase coverage of plants transcriptomes, proteomes and metabolomes during pathogen attack, and the modulation of the response after the infection. Omics techniques also allow us to learn more about the biological cycle of the pathogens in addition to the identification of novel virulence factors in pathogens and their host targets. Both approaches become important to decipher the mechanism underlying pathogen attacks and to develop strategies for improving disease-resistant plants.

In this review, we summarize some of the contribution of genomics, transcriptomics, proteomics, metabolomics and metallomics in devising the strategies to obtain plants with increased resistance to pathogens. These approaches constitute important research tools in the development of new technologies for the protection against diseases and increase plant production.

Introduction
Plants are always exposed to various pathogens attack, affecting their growt and productivity, causing economically important diseases. In order to ensure sustainable food supply, it will be necessary to engineering plants with a broad spectrum disease resistance. One of the major goals of plant research is the better characterization of the plant immune system and elucidate how this is manipulated by various pathogens (Piquerez et al., 2014).

After pathogen attack, plants undergo metabolic adaptations, however, these microorganisms are specialized in manipulate the host metabolism in order to reduce or abolish the defensive response (Balmer et al., 2013). The control of pathogen proliferation has been achieved by conducting breeding programs for either qualitative or quantitative resistance to particular pathogens, or through the use of pesticides. The first approach has been moderately effective to generate disease resistant plants. In addition, the use of naturally occurring resistance genes (*R* genes) leads to the obtention of disease resistance in many plants of economic significance. In contrast, the use of pesticides to control pathogens has detrimental effect on beneficial micro fauna and flora besides the emergence of pesticide-resistant strains (Wally and Punja, 2010).

The relatively simple approach to generate plant disease resistant plants is the introduction of single genes coding for rate-limiting enzymes or genes from other organisms to develop new cellular metabolic pathways. Moreover, the introduction of known antimicrobial peptides also constitutes a promising approach and it is discussed further in this review (Oksman-Caldentey and Saito, 2005). However, sometimes it is highly difficult to achieve success with the strategy of expression of a single gene and it becomes essential to understand the entire metabolic networks between genes, transcripts, proteins and metabolites in biological systems (Oksman-Caldentey and Saito, 2005). In this sense, it becomes relevant to carry out a comprehensive analysis using functional genomics technologies such as transcriptomics, proteomics, and metabolomics for the characterization of the plant-pathogen interactions to better understand the genetic and metabolic adaptations of a particular plant species to the infection. This becomes highly important to design better plants with a sustained response to the attack of specific pathogenic organisms. In this work, we review the current and future biotechnological approaches integrating the various omics technologies for the production of plant disease resistant plants (Figure 1).

Transcriptomic approaches to increase pathogen resistance
There are various applications of transcriptomics in plants. The transcriptomic analysis constitutes a dynamic link between the genome, proteome and the cellular phenotype. While the final products of gene expression are proteins, the analysis and quantification of levels of mRNAs is a valuable molecular tool (Aharoni and Vorst, 2001). Using high throughput technologies, such as microarrays or RNA-Seq techniques, it is possible to obtain gene expression profiles from the hosts (plants) or the gene expression changes of host-associated pathogen. This information could be useful to have an integrative view of a plant's response to a treatment (i.e. pathogen attack) and to provide new insights into biological processes. In

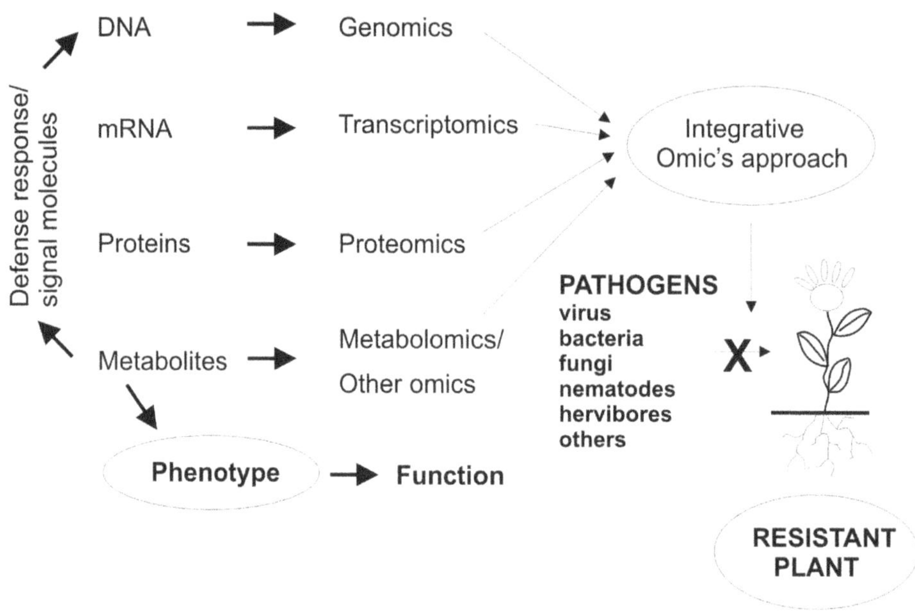

Figure 1. Omics strategies for the obtention of plants resistant to pathogen infection. An integrative Omics approach is essential to obtain plants with a wide spectrum of disease resistance.

addition, it is possible to investigate the function of novel and/or several uncharacterized genes from both organisms.

In order to understand the molecular components underlying plant pathogen, gene expression profiling was first applied to *Arabidopsis thaliana* (Schenk et al., 2000). In this study, changes in the expression patterns of more than 2300 selected genes were examined simultaneously by cDNA microarray analysis in *Arabidopsis* after inoculation with the fungal pathogen *Alternaria brassicicola*, and the treatment with signaling molecules involved in plant defense such as salicylic acid, methyl jasmonate or ethylene. Results showed the existence of a substantial network of regulatory interactions and coordination events occurring during plant defense among the different defense signaling pathways including interactions between the salicylate and jasmonate pathways which were previously thought to act in an antagonistic fashion.

Plants can detect the presence of disease-causing microorganisms, such as bacteria, viruses or fungi by recognizing specific molecules delivered by the pathogens during the infection process (Martin et al., 2003). In this sense, plants have developed an immune system to defend themselves against pathogen attack. The pathogenic organisms target different proteins into plant cells and interfere with this immune response, which, in turn, lead to disease development. However, many plants express different resistance proteins that could detect the presence of specific effectors and triggers a defense mechanism. To understand the molecular basis of this response, recently Pombo and colleagues (2014) have identified genes involved in specific immune responses to *Pseudomonas syringae* in tomato plants using RNA-Seq technology (Pombo et al., 2014). These studies are highly relevant due to the economic relevance of tomato as it is one of the most important cultivated plant in the world, and its production is affected by several pathogens, including viruses, bacteria, fungi and nematodes. In this study, several genes including one encoding a serine/threonine tyrosine protein kinase (Epk1) had been identified. Epk1 acts upstream of the effector triggered immunity associated MAP kinase signaling cascade, and its deficiency leads to the development of the disease triggered by *P. syringae*. In addition, other proteins involved in defense against pathogens were described such as nucleotide binding-leucine rich repeat receptors (NB-LRR) (Bernoux et al., 2011). Whereas several attempts were carried out to engineering disease resistance in economically important crop plants, many of these have failed. However, it has been shown that the overexpression of a serine/threonine kinase (Pto) in tomato induces gene expression changes leading to increase the immune response against *P. syringae* conferring protection to disease (Mysore et al., 2003). In addition, the overexpression of NPR1, a gene involved in systemic acquired resistance in plants, results in increased disease resistance in *Malus x domestica* (Malnoy et al., 2007). This gene had no detrimental effects on plant growth and development, and it has been postulated that could be used for genetic engineering of non-specific resistance in plants (Cao et al., 1998). There still many open questions about the mechanisms and the participation of these disease resistance proteins in the immune response, however, in addition to conferring greater resistance to diseases, these genes can be used to develop specific reporter genes useful to study the response of different plants to pathogens.

Using transcriptomic approaches, it has been also discovered several genes that respond to virus attack. Moreover, it appeared that RNA constitutes another sequence-specific plant defense mechanism against virus invasion. It was reported that the replication of DNA and RNA viruses is associated with the accumulation of small RNAs that are involved in the specific cleavage of viral transcripts, however this could be suppressed by virus proteins that inhibit the host defense response (Czosnek et al., 2013). This mechanism known as virus induced gene silencing (VIGS) has been routinely probed using the *Tobacco rattle virus* (TRV) in *Nicotiana benthamiana* or *A. thaliana* and also in some *Solanum* species to assess functions of candidate genes and as a way to discover new genes required for diverse pathways (Brigneti et al., 2004).

One of the most studied tomato pathogens is a geminivirus infecting tomato (tomato yellow leaf curl virus [TYLCV]). It has been described the production of genetically engineered plants to resist infection by the TYLCV by the expression of viral proteins, or gene silencing strategies but to date, breeding remains a method of choice to obtain plants resistant to TYLCV (Czosnek et al., 2013; Shepherd et al., 2009).

However, recently, using comparative transcriptomic and metabolomic analyses, Sade and collaborators had reported the alteration of several genes and metabolites of tomato cultivars in response to the TYLCV. Some of the metabolites with altered levels were many amino acids, polyamines, phenolic and indolic metabolites, all leading to the synthesis of defense compounds (Sade et al., 2015). In addition, they reported the induction of a hexose transporter gene (*LeHT1*) after TYLCV infection (Sade et al., 2013; Sade et al., 2015). When this virus infects a tomato cultivar, there are several changes in sugar metabolism such as a decrease in photosynthesis, an increase in invertase expression and the release of hexoses that lead to the activation of the defense response. When this mechanism fails promotes the virus replication and the establishment of the disease. It could be possible that the enhanced amounts of internal hexoses could activate phytohormone-mediate responses, osmoregulates the cell homeostasis and efficiently activates plant defense responses (Sade et al., 2013). Thus, the overexpression of *LeHT1* could be a promissory strategy for obtaining tomato plants resistant to virus.

Proteomic analyses to increase disease resistance

Recent efforts in sequencing pathogen genomes it has allowed us to know new insights regarding the attack mechanisms employed by different microorganisms (Grant et al., 2013). In addition, functional genomics approaches are making an important impact on the knowledge on the use of microbes as biological agents to control diseases (Lorito et al., 2010). In conjunction with genomics, proteomics has contributed substantially to the large-scale functional assignment of candidate proteins and, by using this approach, several antimicrobial proteins expressed during phytopathogenic interactions have also been identified mainly by high resolution of two-dimensional polyacrylamide gel electrophoresis coupled with mass spectrometry (Mehta et al., 2008).

Other omics approaches to generate plant-pathogen resistant plants involve the identification and the expression of peptides with antimicrobial properties such as antibacterial or antifungal genes (Ceasar and Ignacimuthu, 2012). Many of these antimicrobial peptides are listed in the PhytAMP database (www.phytamp.pfba-lab-tun.org) (Hammami et al., 2009) and other databases such as CAMP (Collection of Anti-Microbial Peptides, www.camp.bicnirrh.res.in) (Thomas et al., 2010). The compilation of such information in databases would therefore facilitate the study of the potential of several peptides as alternatives in response to increasing antibiotic resistance or for increasing plant resistance to pathogens by genetic engineering.

On the other hand, many filamentous fungi, such as *Trichoderma*, the most widely applied biocontrol fungi, have been extensively studied using genomics, transcriptomics, proteomics, metabolomics and secretomics. These organisms to possess anti-microbial peptides and several genes have been identified and transferred to plants for improving tolerance to biotic and abiotic stress, as well as for bioremediation and herbicide tolerance (Lorito et al., 2010; Nicolas et al., 2014). It was reported that the introduction of a gene coding for an endochitinase from the fungus *Trichoderma harzianum* in tobacco and potato plants. The transgenic lines showed a increased resistance to the infection with the several pathogens such as *Alternaria alternata*, *Alternaria solani*, *Botrytis cinerea*, and *Rhizoctonia solani* (Lorito et al., 1998). Similar findings were obtained using an endochitinase or exochitinase gene in apple (Bolar et al., 2001), demonstrating that these genes can be used to control diseases in plants (Lorito et al., 1998).

The expression of other antimicrobial peptides such as thanatin(S) confers a broad spectrum of antimicrobial activity. In order to investigate the effect of thanatin, *Arabidopsis* plants were transformed with this gene and analyzed against pathogen attack. Results show that transgenic plants have increased antifungal activity against *Botrytis cinerea* and powdery mildew, as well as antibacterial activity against *Pseudomonas syringae* pv. Tomato (Wu et al., 2013). Thus, it was postulated that thanatin(S) could be an ideal candidate for the construction of transgenic crops possessing a broad-spectrum resistance to phytopathogens (Wu et al., 2013).

Other group of economically important plants is the group of ornamental plants, where a high level of disease resistance is required due to the low tolerance of blemishes. There are several pathogens that affect many species such as rose cultivars. The most important fungal diseases of this species are black spot and powdery mildew, caused by *Diplocarpon rosae* and *Podosphaera pannosa*, respectively, as well as botrytis and downy mildew (Marchant et al., 1998). In addition, the rose rosette constitutes a lethal viral pathogen, which is having a considerable importance at the present. It was reported previously that the introduction of a chitinase gene into blackspot-susceptible rose reduces the severity of the disease in correlation with the levels of chitinase expression (Marchant et al., 1998). However, other efforts to obtain transgenic rose lines expressing fungal resistance genes were not highly effective in counteract the effects after the infection by pathogens, and RNAi technology is emerging as one of the most promising solutions to this problem (Debener and Byrne, 2014).

Finally, several proteins located and/or involved in strengthening the cell wall of plants constitute another level of plant disease protection. In this sense, proteomic approaches of the apoplast fluid have been done to better understand plant-pathogen crosstalk. The apoplastic space is the first physiological compartment that reflects the effects of the pathogen-host interaction (Delaunois et al., 2014). Thus, the identification of the proteins present in this compartment becomes essential to elucidate the mechanism of pathogen attack and plant defense. It was reported the apoplastic proteome changes after infection with *Verticillium longisporum* in *A. thaliana* and *Brassica napus* (Floerl et al., 2012); *Pseudomonas syringae* in *A. thaliana* and *Medicago sativa* (De-la-Pena et al., 2008); *Agrobacterium tumefaciens* in *N. benthamiana* (Goulet et al., 2010) and *Magnaporthe oryzae* in *Oryza sativa* (Shenton et al., 2012).

Thus, it is important to build a comprehensive inventory of the experimentally identified plant-pathogen secretome to predict secreted proteins more accurately, and then to address the question of their biological role. Apoplastic proteome analyses of plant-pathogen interactions have provided a better understanding of plant defense responses. However, the lack of published studies using quantitative and *in vivo* proteomic techniques is still striking. Considering the apoplast proteomic studies conducted so far, there still many open questions. A systematic integration of the proteomic approaches will provide useful information that will allow us to better predict and manipulate plant responses to pathogens. One of the main challenges in the future will be to characterize the roles of individual secreted proteins involved in plant-pathogen interactions (Delaunois et al., 2014). However, further studies, such as the engineering of deficient and overexpressing plants of these apoplastic proteins are needed in order to assign functional roles to each one in plant-pathogen interactions.

Metabolomics approaches

Is widely known that many metabolites contribute to the phenotypic properties of plant tissues, such as the color, aroma and taste; in addition as previously mentioned they are involved also in stress and pathogen responses. Thus, the simultaneous identification and quantification of metabolites is necessary to understand the dynamics of the metabolome, analyze fluxes in metabolic pathways and decipher the role of each metabolite following various stimuli (Gomez-Casati et al., 2013). The challenge of this omics applied to pathogen attack and plant defense is to find changes in biochemical pathways, and metabolic networks that might correlate with the physiological and developmental phenotype of a cell, tissue, or organism (Gomez-Casati et al., 2013). Thus, we can identify changes in metabolite levels that are induced after the infection and thus, develop different strategies to obtain transgenic plants with increased levels of these metabolites that could potentially confer greater resistance to diseases.

Most plants reconfigure their metabolism in order to increase the concentration of defense compounds, which act against to pathogen attack. However, many plant pathogens manipulate the host metabolism to offset the defense responses, inducing favorable nutritional conditions. Advances in metabolomics techniques leads to the generation of large metabolic profiles that it was shown to be specific of each plant tissue during pathogen infection.

One group of plants that are economically important and are under continuous attack by different pathogens, such as viruses, bacteria and fungi, are cereals like maize, rice, wheat, sorghum and barley. Until 2000, most cereal metabolomic studies were based on the evaluation of different compounds such as vitamins, sterols, phenolic and volatile compounds and few metabolites related to biotic or abiotic stresses (Khakimov et al., 2014). Recent advances in integrated transcriptomics and metabolomics technologies lead to the screening of diverse plant cereals in regard to pathogen-resistant genotypes as well as biochemical phenotypes (Langridge and Fleury, 2011).

Bollina et al. (Bollina et al., 2010) had been reported the presence of near 500 metabolites in barley cultivars resistant to *Fusarium* head blight (FHB), one of the most important diseases in cereals such as maize, barley and wheat. Interestingly, most of the metabolites belong from the phenylpropanoid, flavonoid, fatty acid and terpenoid metabolic pathways. Other studies carried out in maize showed the involvement of benzoxazinones (BX) in the resistance against the fungus *Setophaeria turcica (Ahmad et al., 2011)*. In addition plants also produce several secondary metabolites in response to nematodes and pest herbivores, such as benzoxazinoids (Niemeyer, 2009); flavonoids, such as C-glycosyl flavones, which had been identified as an effective protectant against corn earworm (Lee et al., 1998) and also against *Pratylenchus* and *Heterodera*, two nematodes of cereals (Soriano et al., 2004); and alkaloids, such as hordenine and gramine, which protects plants against grasshoppers attack (Hinks and Olfert, 1992).

Considering the recent advances in cereal metabolomics in disease resistance, targeting metabolic pathways becomes a promissory strategy for the obtention of transgenic cereals with increased resistance to pathogen infection. Recently, it was reported the production of rice lines that accumulates high levels of momilactone, whose effectiveness in protecting against *Magnaporthe grisea* and *Xanthomonas oryzae* has been proved (Kurusu et al., 2010; Sawada et al., 2004). Thus, using integrated metabolomics and transcriptomics data, the manipulation of several biosynthetic pathways could be possible to design different strategies for crop improvement.

On the other hand, a novel approach is the modulation of the synthesis of compounds, such as phytoalexins to protect plants against infections. Phytoalexins are low molecular weight metabolites with antimicrobial properties, which exhibit toxicity against prokaryotic and eukaryotic organisms, and are *de novo* synthesized during biotic stress (Ahuja et al., 2012). Genetic engineering of plants to increase phytoalexin compounds for disease resistance requires the manipulation of a single or a few genes directly involved in their biosynthetic pathways or involved in their signaling/regulatory pathways (Jeandet et al., 2013). The most abundant examples are related to the production of resveratrol, one of the most abundant phytoalexins in plants. It is synthesized by stilbene synthase (STS) through the phenylpropanoid acid pathway with phenylalanine or tyrosine as precursors. The first approach to increase resveratrol levels in plants was the introduction of two STS genes from grapevine in tobacco, conferring resistance to *Botrytis cinerea* infection (Hain et al., 1990; Hain et al., 1993). Likewise, it has been reported that the protection to *Phoma medicaginis* in *Medicago sativa* transformed with a STS gene (Hipskind and Paiva, 2000). Similar approaches using the introduction of different STS genes conferred resistance to pathogen attack to several crops such as rice (*Pyricularia oryzae* resistance), barley and wheat (*Botrytis cinerea* resistance) (Jeandet et al., 2013).

Metallomics approaches

Metallomics is the study of metalloproteins or any other metal-containing biomolecule as well as the entirety of metal and metalloid species within a cell or tissue type. Therefore, metallomics can be considered a branch of metabolomics, even though the metals are not typically considered as metabolites. But because of the interactions and functional connections of metal ions and their species with genes and proteins, metallomics is related to genomics and proteomics, resulting in a trans-disciplinary research area (Mounicou et al., 2009). Metals are essential for the majority of living organisms but become toxic when present in excess. In the context of plant-pathogen interactions, availability and toxicity of metals can define the outcome of the disease. Metals have profound effects at different levels, such as plant health, plant defense signaling and alterations to the environment that the pathogen experiences *in planta*. Mineral deficiency is generally related to an increased plant disease. Moreover, the pathogen is also affected by them in terms of mineral nutrition, regulation of virulence gene expression and toxicity. There are several examples that illustrate each of these levels of metal-organism interactions in the framework of plant-pathogen attack, and where we might intervene in favor of disease resistance. Reactive oxygen species (ROS) have key role in plant defense mechanisms as potent antimicrobials generated *in situ* in response to pathogenic proteins (Lamb and Dixon, 1997) and also as signaling molecules that induce systemic responses (Reczek and Chandel, 2015). Redox active metals, like iron and copper among others, are important actors in ROS generation as they can participate in Fenton reactions by which hydroxyl radicals may be generated from H_2O_2 (Pierre and Fontecave, 1999). In addition, metals can affect plant-pathogen interactions via their inclusion in metalloenzymes, such as superoxide dismutase (three isoforms with Fe, Mn or Cu/Zn metallic cofactors) and catalase (containing four heme groups) that are responsible of recovering or normalizing ROS levels to avoid further damage in plant cells after neutralizing the pathogen attack (Mittler, 2002). There is plenty of literature referring that availability of inorganic fertilizers can bring the demise of many diseases through improved plant resistance (Datnoff et al., 2006; Engelhard, 1989; Graham, 1983; Huber and Haneklaus, 2007). Since plants obtain their nutrients almost exclusively from the soil, in summary, a good starting point to decrease the incidence of plant diseases is to apply a good fertilization strategy, carefully considering micronutrient levels. In this way, plant growth is improved, while health is potentiated through the generation of ROS -acting both as antimicrobial and as signals for further defense responses- and its subsequent control.

As metals are toxic when present in high concentrations, this property can be employed in disease control and resistance. In fact, copper has long been used as fungicide and anti-bacterial agent in agriculture (Floyd, 1991), being widely used in organic farming, in a diversity of crops, and particularly in viticulture and orchard settings (Van Zweiten et al., 2007). However, recently a number a problems derived from the use of copper as plant protectant have been reported. Copper is toxic to both pathogens and vegetables, and as a result of the repeated applications, bioaccumulation in the soil might occur with the consequence of phytotoxicity (Rusjan et al., 2007). Furthermore, another issue is the development of pathogen resistance to metals, as an outcome of prolonged exposure, because of the horizontal transfer of copper resistance determimants (Schenk and Pscheidt, 1998). Plants also employ this defense strategy by overloading the pathogen with an excess of metals, with the objective of disrupting the microbial homeostatic and tolerance mechanisms (Fones et al., 2010). Metal hyper-accumulating plants are an extreme example of this tactics. They can accumulate exceptionally high levels of metals in the aboveground parts, reaching concentrations that would be toxic to most other plant species (Baker and Brooks, 1989). While it is not clear how this trait has evolved independently multiple times in the plant kingdom, there is substantial evidence showing that these plants can use high concentrations of accumulated metals to defend themselves against attack by pathogenic microorganisms and herbivores (Poschenrieder et al., 2006). The "elemental defence hypothesis" postulates that the metals act directly to deter (herbivores) or kill pests, although they can also act in other indirect ways. Hyper-accumulator plants have received considerable attention due to the possibility of exploiting their accumulation traits for practical applications, such as plant-pathogen interactions. An interesting breakthrough that has emerged from comparative physiological and transcriptomic analyses of hyper-accumulators and related non-hyper-accumulators is that most key steps of hyper-accumulation rely on different regulation and expression of genes found in both kinds of plants (constitutive overexpression of genes encoding transmembrane metal transporters, such as members of ZIP, HMA, MATE, YSL and MTP families) (Rascioa and Navari-Izzo, 2011). This might imply that by only increasing gene expression of a few endogenous candidate genes it would be possible to augment metal accumulation in economically important plants, and thus strengthen their disease resistance. However, in spite of important progress made in recent years, the complexity of hyper-accumulation is far from being understood and several aspects of this astonishing feature still await explanation, making the extrapolation of their unusual trait to other crops distant from practical application.

Finally, in contrast to the previous strategy, plants many times employ the opposite tactic: withholding of metals. This is an unquestionable plan of attack considering that microorganisms and hosts are in constant competition for metal ions (Hammer and Skaar, 2012). Among them, iron is a key target considering its low availability in aerobic environments. The host strategy of restraining iron to limit pathogen growth by sequestration with chelating proteins such as transferrin and lactoferrin is well known in mammalian systems (Ong et al., 2006). Plant ferredoxin-like proteins, heterologously overexpressed in tobacco protected the plant from the soft rot infection by Erwinia carotovora (Huang et al., 2006). Moreover, *Arabidopsis* upregulates ferredoxin in response to the detection of iron-siderophores from pathogens (Dellagi et al., 2005), so the iron chelating strategy to protect plant from pathogen attack might be functional in vegetables likewise. Iron is not the only metal which may be restrained as a form of defence. Zinc can also be withheld, and there is a nice example involving rough lemon protection from its pathogen *Altemaria altemata* (Nishimura et al., 2013). The disease determinant of the fungus is ACR-toxin, and zinc is needed in its biosynthetic pathway. The presence of the citrus metallothionein RlemMT1 -a protein with zinc binding properties- in the culture medium caused suppression of ACR toxin production by the fungus and a significant decrease in symptoms on rough lemon leaves. Since metallothioneins are frequently overexpressed by hosts during biotic stresses (Figueira et al., 2012), they likely can act as plant defense factor by controlling metal availability. As a result, they are potential targets, together with ferredoxins and other metal binding proteins, of transgenic approaches to improve plant disease resistance. Nevertheless, care should be taken, as alteration in cell metal levels can have profound effects in plant physiology given the many roles in which they are involved.

Conclusions

Omics approaches to generate plant disease resistant plants have emerged as a promissory alternative to classical resistance introgression strategies, such as molecular breeding. These approaches combine molecular, physiological and metabolic aspects of plant-pathogen interaction that are being characterized by functional genomics, transcriptomics, proteomics, metabolomics and other new omics techniques, such as metallomics. Despite technological advances, the characterization of the plant response after pathogen infection are far from over. The expression of one or more genes leading to the increase of specific proteins and/or metabolites involved in pathogen defense responses is one of the most successful widely used strategy, however, it is extremely necessary to evaluate if such genetic manipulation has undesirable effects on the growth and development of the resistant plants.

Acknowledgements

This work was supported by grants from ANPCyT (PICT 0512, 0982, 2188 and 2184). DGC, MAP and MVB are research members from CONICET.

References

Aharoni, A., and Vorst, O. (2001). DNA microarrays for functional plant genomics. Plant Mol Biol *48*, 99-118.

Ahmad, S., Veyrat, N., Gordon-Weeks, R., Zhang, Y., Martin, J., Smart, L., Glauser, G., Erb, M., Flors, V., Frey, M., *et al.* (2011). Benzoxazinoid metabolites regulate innate immunity against aphids and fungi in maize. Plant Physiol *157*, 317-327.

Ahuja, I., Kissen, R., and Bones, A.M. (2012). Phytoalexins in defense against pathogens. Trends Plant Sci *17*, 73-90.

Baker, A., and Brooks, R. (1989). Terrestrial higher plants which hyperaccumulate metallic elements. A review of their distribution, ecology and phytochemistry. Biorecovery *1*.

Balmer, D., Flors, V., Glauser, G., and Mauch-Mani, B. (2013). Metabolomics of cereals under biotic stress: current knowledge and techniques. Front Plant Sci *4*, 82.

Bernoux, M., Ellis, J.G., and Dodds, P.N. (2011). New insights in plant immunity signaling activation. Curr Opin Plant Biol *14*, 512-518.

Bolar, J.P., Norelli, J.L., Harman, G.E., Brown, S.K., and Aldwinckle, H.S. (2001). Synergistic activity of endochitinase and exochitinase from Trichoderma atroviride (T. harzianum) against the pathogenic fungus (Venturia inaequalis) in transgenic apple plants. Transgenic Res *10*, 533-543.

Bollina, V., Kumaraswamy, G.K., Kushalappa, A.C., Choo, T.M., Dion, Y., Rioux, S., Faubert, D., and Hamzehzarghani, H. (2010). Mass spectrometry-based metabolomics application to identify quantitative resistance-related metabolites in barley against Fusarium head blight. Mol Plant Pathol *11*, 769-782.

Brigneti, G., Martin-Hernandez, A.M., Jin, H., Chen, J., Baulcombe, D.C., Baker, B., and Jones, J.D. (2004). Virus-induced gene silencing in Solanum species. Plant J *39*, 264-272.

Cao, H., Li, X., and Dong, X. (1998). Generation of broad-spectrum disease resistance by overexpression of an essential regulatory gene in systemic acquired resistance. Proc Natl Acad Sci U S A *95*, 6531-6536.

Ceasar, S.A., and Ignacimuthu, S. (2012). Genetic engineering of crop plants for fungal resistance: role of antifungal genes. Biotechnol Lett *34*, 995-1002.

Czosnek, H., Eybishtz, A., Sade, D., Gorovits, R., Sobol, I., Bejarano, E., Rosas-Diaz, T., and Lozano-Duran, R. (2013). Discovering host genes involved in the infection by the Tomato Yellow Leaf Curl Virus complex and in the establishment of resistance to the virus using Tobacco

Rattle Virus-based post transcriptional gene silencing. Viruses *5*, 998-1022.
Datnoff, L.E., Elmer, W., and Huber, D.M., eds. (2006). Mineral Nutrition and Plant Disease. APS Press, St. Paul, MN. (St. Paul, MN, APS Press).
De-la-Pena, C., Lei, Z., Watson, B.S., Sumner, L.W., and Vivanco, J.M. (2008). Root-microbe communication through protein secretion. J Biol Chem *283*, 25247-25255.
Debener, T., and Byrne, D.H. (2014). Disease resistance breeding in rose: current status and potential of biotechnological tools. Plant Sci *228*, 107-117.
Delaunois, B., Jeandet, P., Clement, C., Baillieul, F., Dorey, S., and Cordelier, S. (2014). Uncovering plant-pathogen crosstalk through apoplastic proteomic studies. Front Plant Sci *5*, 249.
Dellagi, A., Rigault, M., Segond, D., Roux, C., Kraepiel, Y., Cellier, F., Briat, J.F., Gaymard, F., and Expert, D. (2005). Siderophore-mediated upregulation of Arabidopsis ferritin expression in response to Erwinia chrysanthemi infection. Plant J *43*, 262-272.
Engelhard, A.W., ed. (1989). Soilborne Plant Pathogens: Management of Diseases with Macro and Microelements (St. Paul, MN, APS Press).
Figueira, E., Branco, D., Antunes, S.C., Goncalves, F., and Freitas, R. (2012). Are metallothioneins equally good biomarkers of metal and oxidative stress? Ecotoxicol Environ Saf *84*, 185-190.
Floerl, S., Majcherczyk, A., Possienke, M., Feussner, K., Tappe, H., Gatz, C., Feussner, I., Kues, U., and Polle, A. (2012). Verticillium longisporum infection affects the leaf apoplastic proteome, metabolome, and cell wall properties in Arabidopsis thaliana. PLoS One *7*, e31435.
Floyd, R. (1991). Bordeaux mixture and similar copper fungicides. Farmnote *78/1991*.
Fones, H., Davis, C.A., Rico, A., Fang, F., Smith, J.A., and Preston, G.M. (2010). Metal hyperaccumulation armors plants against disease. PLoS Pathog *6*, e1001093.
Gomez-Casati, D.F., Zanor, M.I., and Busi, M.V. (2013). Metabolomics in plants and humans: applications in the prevention and diagnosis of diseases. Biomed Res Int *2013*, 792527.
Goulet, C., Goulet, M.C., and Michaud, D. (2010). 2-DE proteome maps for the leaf apoplast of Nicotiana benthamiana. Proteomics *10*, 2536-2544.
Graham, R.D. (1983). Effect of nutrient stress on susceptibility of plants to disease with particular reference to the trace elements. Adv Bot Res *10*, 221-276.
Grant, M.R., Kazan, K., and Manners, J.M. (2013). Exploiting pathogens' tricks of the trade for engineering of plant disease resistance: challenges and opportunities. Microb Biotechnol *6*, 212-222.
Hain, R., Bieseler, B., Kindl, H., Schroder, G., and Stocker, R. (1990). Expression of a stilbene synthase gene in Nicotiana tabacum results in synthesis of the phytoalexin resveratrol. Plant Mol Biol *15*, 325-335.
Hain, R., Reif, H.J., Krause, E., Langebartels, R., Kindl, H., Vornam, B., Wiese, W., Schmelzer, E., Schreier, P.H., Stocker, R.H., *et al.* (1993). Disease resistance results from foreign phytoalexin expression in a novel plant. Nature *361*, 153-156.
Hammami, R., Ben Hamida, J., Vergoten, G., and Fliss, I. (2009). PhytAMP: a database dedicated to antimicrobial plant peptides. Nucleic Acids Res *37*, D963-968.
Hammer, N.D., and Skaar, E.P. (2012). The impact of metal sequestration on Staphylococcus aureus metabolism. Curr Opin Microbiol *15*, 10-14.
Hinks, C.F., and Olfert, O. (1992). Cultivar resistance to grasshoppers in temperate cereal crops and grasses: A review. J Orthop Res *1*, 1-9.
Hipskind, J.D., and Paiva, N.L. (2000). Constitutive accumulation of a resveratrol-glucoside in transgenic alfalfa increases resistance to Phoma medicaginis. Mol Plant Microbe Interact *13*, 551-562.
Huang, H.-E., Ger, M.-J., Chen, C.-Y., Yip, M.-K., Chung, M.-C., and Feng, T.-T. (2006). Plant ferredoxin-like protein (PFLP) exhibits an anti-microbial ability against soft-rot pathogen Erwinia carotovora subsp. carotovora *in vitro* and *in vivo*. Plant Sci *171*, 17-23.
Huber, D.M., and Haneklaus, S. (2007). Managing nutrition to control plant disease. Landbauforschung Volkenrode *57*, 313-322.
Jeandet, P., Clement, C., Courot, E., and Cordelier, S. (2013). Modulation of phytoalexin biosynthesis in engineered plants for disease resistance. Int J Mol Sci *14*, 14136-14170.
Khakimov, B., Bak, S., and Balling Engelsen, S. (2014). High-throughput cereal metabolomics: Current analytical technologies, challenges and perspectives. J Cereal Sci *59*, 393-418.
Kurusu, T., Hamada, J., Nokajima, H., Kitagawa, Y., Kiyoduka, M., Takahashi, A., Hanamata, S., Ohno, R., Hayashi, T., Okada, K., *et al.* (2010). Regulation of microbe-associated molecular pattern-induced hypersensitive cell death, phytoalexin production, and defense gene expression by calcineurin B-like protein-interacting protein kinases, OsCIPK14/15, in rice cultured cells. Plant Physiol *153*, 678-692.
Lamb, C., and Dixon, R.A. (1997). The Oxidative Burst in Plant Disease Resistance. Annu Rev Plant Physiol Plant Mol Biol *48*, 251-275.
Langridge, P., and Fleury, D. (2011). Making the most of 'omics' for crop breeding. Trends Biotechnol *29*, 33-40.
Lee, E.A., Byrne, P.F., McMullen, M.D., Snook, M.E., Wiseman, B.R., Widstrom, N.W., and Coe, E.H. (1998). Genetic mechanisms underlying apimaysin and maysin synthesis and corn earworm antibiosis in maize (Zea mays L.). Genetics *149*, 1997-2006.
Lorito, M., Woo, S.L., Garcia, I., Colucci, G., Harman, G.E., Pintor-Toro, J.A., Filippone, E., Muccifora, S., Lawrence, C.B., Zoina, A., *et al.* (1998). Genes from mycoparasitic fungi as a source for improving plant resistance to fungal pathogens. Proc Natl Acad Sci U S A *95*, 7860-7865.
Lorito, M., Woo, S.L., Harman, G.E., and Monte, E. (2010). Translational research on Trichoderma: from 'omics to the field. Annu Rev Phytopathol *48*, 395-417.
Malnoy, M., Jin, Q., Borejsza-Wysocka, E.E., He, S.Y., and Aldwinckle, H.S. (2007). Overexpression of the apple MpNPR1 gene confers increased disease resistance in Malus x domestica. Mol Plant Microbe Interact *20*, 1568-1580.
Marchant, R., Davey, M.R., Lucas, J.A., Lamb, C.J., Dixon, R.A., and Brian Power, J. (1998). Expression of a

chitinase transgene in rose (Rosa hybrida L.) reduces development of blackspot disease (Diplocarpon rosae Wolf) Mol Breeding *4*, 187-194.

Martin, G.B., Bogdanove, A.J., and Sessa, G. (2003). Understanding the functions of plant disease resistance proteins. Annu Rev Plant Biol *54*, 23-61.

Mehta, A., Brasileiro, A.C., Souza, D.S., Romano, E., Campos, M.A., Grossi-de-Sa, M.F., Silva, M.S., Franco, O.L., Fragoso, R.R., Bevitori, R., *et al.* (2008). Plant-pathogen interactions: what is proteomics telling us? FEBS J *275*, 3731-3746.

Mittler, R. (2002). Oxidative stress, antioxidants and stress tolerance. Trends Plant Sci *7*, 405-410.

Mounicou, S., Szpunar, J., and Lobinski, R. (2009). Metallomics: the concept and methodology. Chem Soc Rev *38*, 1119-1138.

Mysore, K.S., D'Ascenzo, M.D., He, X., and Martin, G.B. (2003). Overexpression of the disease resistance gene Pto in tomato induces gene expression changes similar to immune responses in human and fruitfly. Plant Physiol *132*, 1901-1912.

Nicolas, C., Hermosa, R., Rubio, B., Mukherjee, P.K., and Monte, E. (2014). Trichoderma genes in plants for stress tolerance- status and prospects. Plant Sci *228*, 71-78.

Niemeyer, H.M. (2009). Hydroxamic acids derived from 2-hydroxy-2H-1,4-benzoxazin-3(4H)-one: key defense chemicals of cereals. J Agric Food Chem *57*, 1677-1696.

Nishimura, S., Tatano, S., Miyamoto, Y., Ohtani, K., Fukumoto, T., Gomi, K., Tada, Y., Ichimura, K., and Akimitsu, K. (2013). A zinc-binding citrus protein metallothionein can act as a plant defense factor by controlling host-selective ACR-toxin production. Plant Mol Biol *81*, 1-11.

Oksman-Caldentey, K.M., and Saito, K. (2005). Integrating genomics and metabolomics for engineering plant metabolic pathways. Curr Opin Biotechnol *16*, 174-179.

Ong, S.T., Ho, J.Z., Ho, B., and Ding, J.L. (2006). Iron-withholding strategy in innate immunity. Immunobiology *211*, 295-314.

Pierre, J.L., and Fontecave, M. (1999). Iron and activated oxygen species in biology: the basic chemistry. Biometals *12*, 195-199.

Piquerez, S.J., Harvey, S.E., Beynon, J.L., and Ntoukakis, V. (2014). Improving crop disease resistance: lessons from research on Arabidopsis and tomato. Front Plant Sci *5*, 671.

Pombo, M.A., Zheng, Y., Fernandez-Pozo, N., Dunham, D.M., Fei, Z., and Martin, G.B. (2014). Transcriptomic analysis reveals tomato genes whose expression is induced specifically during effector-triggered immunity and identifies the Epk1 protein kinase which is required for the host response to three bacterial effector proteins. Genome Biol *15*, 492.

Poschenrieder, C., Tolra, R., and Barcelo, J. (2006). Can metals defend plants against biotic stress? Trends Plant Sci *11*, 288-295.

Rascioa, N., and Navari-Izzo, F. (2011). Heavy metal hyperaccumulating plants: How and why do they do it? And what makes them so interesting? Plant Sci *180*, 169-171.

Reczek, C.R., and Chandel, N.S. (2015). ROS-dependent signal transduction. Curr Opin Cell Biol *33*, 8-13.

Rusjan, D., Strlic, M., Pucko, D., and Korosec-Koruza, Z. (2007). Copper accumulation regarding the soil characteristics in Sub-Mediterranean vineyards of Slovenia. Geoderma *141*, 111-118.

Sade, D., Brotman, Y., Eybishtz, A., Cuadros-Inostroza, A., Fernie, A.R., Willmitzer, L., and Czosnek, H. (2013). Involvement of the hexose transporter gene LeHT1 and of sugars in resistance of tomato to tomato yellow leaf curl virus. Mol Plant *6*, 1707-1710.

Sade, D., Shriki, O., Cuadros-Inostroza, A., Tohge, T., Semel, Y., Haviv, Y., Willmitzer, L., Fernie, A.R., Czosnek, H., and Brotman, Y. (2015). Comparative metabolomics and transcriptomics of plant response to Tomato yellow leaf curl virus infection in resistant and susceptible tomato cultivars. Metabolomics *11*, 81-97.

Sawada, K., Hasegawa, M., Tokuda, L., Kameyama, J., Kodama, O., Kohchi, T., Yoshida, K., and Shinmyo, A. (2004). Enhanced resistance to blast fungus and bacterial blight in transgenic rice constitutively expressing OsSBP, a rice homologue of mammalian selenium-binding proteins. Biosci Biotechnol Biochem *68*, 873-880.

Schenk, H.J., and Pscheidt, J.W. (1998). Effect of copper bacteriocides on copper-resistant and -sensitive strains of Pseudomonas syringae pv. syringae. Plant Dis *82*, 397-406.

Schenk, P.M., Kazan, K., Wilson, I., Anderson, J.P., Richmond, T., Somerville, S.C., and Manners, J.M. (2000). Coordinated plant defense responses in Arabidopsis revealed by microarray analysis. Proc Natl Acad Sci U S A *97*, 11655-11660.

Shenton, M.R., Berberich, T., Kamo, M., Yamashita, T., Taira, H., and Terauchi, R. (2012). Use of intercellular washing fluid to investigate the secreted proteome of the rice-Magnaporthe interaction. J Plant Res *125*, 311-316.

Shepherd, D.N., Martin, D.P., and Thomson, J.A. (2009). Transgenic strategies for developing crops resistant to geminiviruses. Plant Sci *176*, 1-11.

Soriano, I.R., Asenstorfer, R.E., Schmidt, O., and Riley, I.T. (2004). Inducible Flavone in Oats (Avena sativa) Is a Novel Defense Against Plant-Parasitic Nematodes. Phytopathology *94*, 1207-1214.

Thomas, S., Karnik, S., Barai, R.S., Jayaraman, V.K., and Idicula-Thomas, S. (2010). CAMP: a useful resource for research on antimicrobial peptides. Nucleic Acids Res *38*, D774-780.

Van Zweiten, M., Stovold, G., and Van Zweiten, L. (2007). Alternatives to Copper for Disease Control in the Australian Organic Industry. (Kingston, Australia).

Wally, O., and Punja, Z.K. (2010). Genetic engineering for increasing fungal and bacterial disease resistance in crop plants. GM Crops *1*, 199-206.

Wu, T., Tang, D., Chen, W., Huang, H., Wang, R., and Chen, Y. (2013). Expression of antimicrobial peptides thanatin(S) in transgenic Arabidopsis enhanced resistance to phytopathogenic fungi and bacteria. Gene *527*, 235-242.

Other Publications of Interest

Xu, J.-P. (2014). Next-generation Sequencing: Current Technologies and Applications. Caister Academic Press, U.K. ISBN: 978-1-908230-33-1

He, Z. (2014). Microarrays: Current Technology, Innovations and Applications. Caister Academic Press, U.K. ISBN: 978-1-908230-49-2

Poptsova, M.S. (2014). Genome Analysis: Current Procedures and Applications. Caister Academic Press, U.K. ISBN: 978-1-908230-29-4

Fuentes, M. and LaBaer, J. (2014). Proteomics: Targeted Technology, Innovations and Applications. Caister Academic Press, U.K. ISBN: 978-1-908230-46-1

Murillo, J., Vinatzer, B.A., Jackson, R.W. and Arnold, D.L. (2015). Bacteria-Plant Interactions: Advanced Research and Future Trends. Caister Academic Press, U.K. ISBN: 978-1-908230-58-4

Caranta, C., Aranda, M.A., Tepfer, M. and Lopez-Moya, J.J. (2011). Recent Advances in Plant Virology. Caister Academic Press, U.K. ISBN: 978-1-904455-75-2

Nannipieri, P., Pietramellara, G. and Renella, G. (2014). Omics in Soil Science. Caister Academic Press, U.K. ISBN: 978-1-908230-32-4

Herold, K.E. and Rasooly, A. (2009). Lab-on-a-Chip Technology: Fabrication and Microfluidics. Caister Academic Press, U.K. ISBN: 978-1-904455-46-2

Oscillating Transcriptome during Rice-*Magnaporthe* Interaction

T.R. Sharma[1]*, Alok Das[2], Shallu Thakur[1], B.N. Devanna[1], Pankaj Kumar Singh[1], Priyanka Jain[1], Joshitha Vijayan[1] and Shrawan Kumar[1]

[1]ICAR-National Research Centre on Plant Biotechnology, PUSA Campus, New Delhi 110 012, India
[2]Division of Plant Biotechnology, ICAR-Indian Institute of Pulses Research, Kanpur 208 024, India
*Corresponding author. Email:trsharma1965@gmail.com
Phone 91-11-25848783, 25841787 Ext 260

Abstract

Rice blast disease caused by the fungus, *Magnaporthe oryzae*, is one of the most devastating diseases of rice. Deciphering molecular mechanism of host-pathogen interactions is of great importance in devising disease management strategies. Transcription being the first step for gene regulation in eukaryotes, basic understanding of the transcriptome is *sine qua non* for devising effective management strategy. The availability of genome sequences of rice and *M. oryzae* has facilitated the process to a large extent. The current review summarizes recent understanding of rice-blast pathosystem, application of transcriptomics approaches to understand the interactions employing different platforms, major determinants in the interaction and possibility of using certain candidate for conditioning enhanced disease resistance (Effector Triggered Immunity and PAMP Triggered Immunity) and downstream signalling in rice. A better understanding of the interaction elements and effective strategies hold potential to reduce yield losses in rice caused by *M. oryzae*.

Introduction

Rice blast caused by *Magnaporthe oryzae* is one of the most serious diseases of rice, resulting severe yield losses (20-100%) across the globe (Khush and Jena, 2009; Sharma et al., 2012). The disease is widely distributed across 85 countries of the world and can be very devastating, when environmental conditions are conducive for disease development. It is extremely difficult to control rice blast disease; therefore, it poses a significant economic and humanitarian problem (Dean et al., 2005). The genome of the fungus *M. oryzae* is rich in repetitive segments and retro-transposons, which enable the fungus to alter pathogenicity or escape from host recognition by altering the effector molecules (Dean et al., 2005). It requires use of novel management strategies to be designed to manage this pathogen. One of the potential alternatives is employing new sources of host disease resistance against continuously evolving and geographically diverse pathogenic races. Pathogens purportedly evade this employed defence by releasing special effector molecules, aptly explained in *zig-zag* model of disease resistance (Jones and Dangl, 2006; Jacob et al., 2013). The blast fungus (*hemibiotropic*) affects all plant parts and appears at all stages of development, leading to death of plants. Major resistance genes (*R* genes) identified to play essential role fall into eight different classes of proteins, with additional subclasses defined on the basis of domain structures as well as their membrane topology (Sharma et al., 2014). The nucleotide binding site and leucine rich repeats (NBS-LRR) type form the largest class of R proteins that have either a coiled coil (CC) or a TOLL/interleukin 1 receptor (TIR) domain at the N-terminus. Each of the identified domains have structural and functional significance; the LRR domain that contains highly conserved segments (HCS) as well as variable segments (VS) is known to be involved in protein-protein or protein-ligand interactions (Matsushima and Miyashita, 2012). More than 700 NBS-LRR genes have been identified in the rice genome (Monosi et al., 2004; Singh et al., 2015) and current genome wide association studies indicate that majority of *R* genes are located on chromosome 11 of rice genome (Wang et al., 2014). The *R* genes are being widely used in breeding programs for protection against various diseases.

Rice-blast pathosystem in past few years has emerged as a model system to study host-pathogen interactions largely because of the availability of genomic sequences of both host as well as the pathogen. (IRGSP, 2005; Dean et al., 2005). A comprehensive and integrated database on rice blast: Genomic Resources of *Magnaporthe oryzae* (GROMO) has been developed for the research community (Thakur et al., 2009). Various studies on the molecular mechanisms of infection of rice with blast fungus, implicate many genes involved in plant defense and pathogen attack (Skamnioti and Gurr, 2009; Chen and Ronald, 2011; Valent and Khang, 2010; Wilson and Talbot, 2009). The emerging *omics* strategies like genomics, transcriptomics, proteomics and metabolomics have played a major role in understanding the interaction, better, like never before. Of these emerging technologies, transcriptomics plays a major role in understanding host- pathogen interaction, since these are the first evidences of expression profile in both host and pathogen.

Transcriptome refers to total set of transcripts in a given organism, or to the specific subset of transcripts present in a particular cell type at specified time frame. It is highly dynamic and changes continually. Transcripts of an organism can be measured based on the hybridization (microarray) and sequencing (RNASeq and HT/RL SAGE) technologies. Emerging high throughput sequencing techniques enable rapid acquisition of huge amounts of transcriptomic sequence data at relatively low costs (Wang et al., 2009). To date, microarray techniques have been predominantly used for gene expression studies employing high quality annotation data. Sequencing technologies offer wider dynamic range, higher technical reproducibility, and better estimate of relative expression levels, compared to microarray technique (Fu et al., 2009; Marioni et al., 2008). The expression profiles of unannotated transcripts cannot be analyzed in microarray because the probes on chips are designed based on annotated data. HT SuperSAGE also requires gene structure information to convert tag counts to corresponding transcript expression levels. However, RNA-Seq can measure the expression levels of all transcripts without prior knowledge, and hence currently being extensively used for transcriptome profiling. Mixed transcriptome technique for simultaneous elucidation of the systems might further provide better insights. Transcriptomic studies thus provide glimpses to understand the interaction between rice and *M. oryzae* at whole genome level (Kawahara et al., 2012; Wang et al., 2014). Due to co-evolution processes, rice have also developed molecular mechanisms to suppress successful infection by pathogens, such as recognition of pathogens through receptors, generation of reactive oxygen species (ROS), expression of pathogenesis-related (PR) proteins, and accumulation of anti-microbial secondary defense compounds (Jwa et al., 2006). The transcriptome analysis thus, provide clues for understanding the rice immune response against blast fungus attack as well as how *M. oryzae* fungus counters host defence.

This article deals with the current understandings and advancement in interaction between rice-*M. oryzae* pathosystems with particular emphasis on the dynamics of transcriptomic profile in rice and *M. oryzae*. Several advancements related to the evolution of resistance genes in rice and avirulence genes in pathogenic races to understand selective forces acting on those loci have also been discussed.

Rice blast: the biggest challenge

The blast fungus was first described by Soong Ying-Shin in the 17th century (Ou, 1985). The rice blast fungus can be found in the literature under several names. *Pyricularia oryzae* was used to refer to the asexual stage of rice blast fungus, as it was found in the field. The rice pathogen was morphologically indistinguishable from pathogens of other hosts, and the entire group was defined under the name *Pyricularia (or Piricularia) grisea* (Rossman et al., 1990). The sexual stage of this fungus was named as *Magnaporthe grisea* until it was shown by phylogenetic analysis and inter-strain fertility tests that *Magnaporthe* isolates should be separated into species that infect *Digitaria* spp. (crab grass) *(M. grisea)*, whereas *M. oryzae* collectively refers to the other characterized isolates, including the rice pathogen (Couch et al., 2005).The fungus *M. oryzae* (Couch and Kohn, 2002) is a hemi-biotrophic filamentous *Ascomycete* devastating rice blast pathogen (Kawasaki, 2004). Hemi-biotrophy is defined on the basis of the pathogen lifestyle that is initiated through biotrophic infection and later switching to necrotrophic growth. In biotrophic infection fungus invades living plant cells followed by necrotrophic infection where the pathogen kills the host cells, ahead of its infection and then invades them. Blast fungus has been long known to infect all the above ground parts of the plant. The most fatal form of the disease occurs when it infects the neck/panicle, which will fail to set seed (Ou, 1985). Interestingly, this fungus was shown to infect roots like a classical root pathogen, forming hyphapodia (Sesma and Osbourn, 2004). The sources of inoculum can be spores, crop residue or secondary hosts (Teng, 1994; Greer and Webster, 2001).

The blast fungus initiates disease cycle when the pyriformed, three-celled, asexual spore lands on a compatible plant host. The spore germinates to form a germ tube under optimal humidity. Appressorial formation begins when the tips of germ tubes swell and bend to leaf surface, known as 'hooking' which indicates recognition of the host (Howard and Valent, 1996). As the appressorium develops the cell wall becomes thinner at the host interface. This wall less region, called the appressorium pore, is surrounded by an 'O ring', which seals the appressorium to the host surface tightly. Melanin formation is critical for generation of enormous turgor pressure required to mechanically breach the plant cell wall. The turgor pressure in these appressoria which is around 80 times atmospheric pressure, is the highest known turgor pressure reported in a living organism. The fungus develops narrow primary invasive hyphae (IH) which subsequently develops into bulbous secondary IH. The invasive hyphae are enclosed in the extra-invasive hyphal membrane (EIHM) produced by plant cells. The hyphae move across the cells *via*. plasmodesmata and engulfs the surrounding cell. The fungus finally colonizes the host to form *eye-shaped* sporulating lesions, completing its life cycle (Figure 1).

Molecular basis of the rice blast disease development

Understanding the molecular basis of the biology of the biotrophic hyphae is an important step towards disease management. Effector proteins are usually synthesized in the invasive hyphae and secreted. Cytoplasmic effectors are secreted into specialized structure of biotrophic interfacial complex (BIC) before being translocated into plant cell cytoplasm. Apoplastic effectors are secreted into the space between the fungal cell wall and extra-invasive hyphal membrane. Few genes that impact biotrophic growth of the blast fungus have been identified, because extensive mutational analyses have mainly identified genes with a role in appressorium structure and function (Talbot, 2003). However, the exciting breakthrough in the field has been the identification of core effector proteins. Effector proteins are secreted by the pathogen into the host cell to manipulate the plant regulatory pathways (Slot and Knogge, 2002).

In cases where the plant has a resistance gene that recognizes the effector, the recognition triggers defense responses and arrests the pathogen growth. This is called the *gene-for-gene* resistance model (Flor, 1971). The rice blast system follows this model where the interaction of the host resistance (*R*) gene with the corresponding avirulence (*AVR*) gene confers resistance to the host. This interaction is very specific to a given *R*- gene. The absence of the corresponding *R*-gene or the *AVR*-gene renders the fungus

Figure 1. Typical phenotypic symptoms of rice blast disease on rice seedlings and leaf under field infection.

virulent (Jia et al., 2000). So far, 15 *M. oryzae* effector proteins have been characterized including nine Avr effectors (PWL1, PWL2, AvrPi-ta, AvrPiz-t, Avr-Pia, AvrPii, Avr-Pik/km/kp, Avr1-CO39, and ACE1) and six newly identified effectors (BAS1, BAS2, BAS3, BAS4, Slp1, and MC69) (Table 1). Identification of the fungal effectors has been a big challenge; since it has been difficult to enrich the biomass of the biotrophic hyphae in the conventional infection methods like spray inoculations and drop inoculations. Recently, a total of 851 *in planta* expressed genes encoding predicted effector proteins have been identified using a genome-wide transcriptome analysis of blast-infected rice leaves (Chen et al., 2013). Several other studies have focused on identifying infection-specific genes in the rice blast system using cDNA libraries and more recently whole genome microarray experiments. In the search for infection specific genes in *M. oryzae*, Takano et al. (2003) developed two cDNA libraries from mycelia grown in liquid cultures and conidia germinated for 8 h on an appressorium-inducing surface. Other factors like efficient nitrogen metabolism in fungi are considered pivotal to allow the fungus to adapt in different environmental conditions including infection of the host plants. Experiments conducted for identification of pathogen proteins responsible for the infection lead to either appressorial specific or *in planta* genes at later stages of infection. Jantasuriyarat et al. (2005) used an Expressed Sequence Tags (EST) sequencing approach to identify the molecular basis for defence responses. They studied resistant, partially resistant and susceptible interactions at 6 hpi and 24 hpi, representing the early stages of infection when the germ tubes were formed and when the appressorial penetration occur on the leaves, respectively. A total of 68,920 EST sequences were identified, of which only 4 sequences showed similarity to blast fungus sequences. Further, EST analysis of infected rice leaves at 74 and 120 hpi when the blast symptoms were visible lead to identification of 24.6% fungal sequences (Kim et al., 2001).

Understanding Rice-*M. oryzae* interaction

Understanding the innate mechanism underlying host-pathogen interactions helps devising strategies to manage diseases. *M. oryzae* is highly tractable and hence serves as a seminal model for plant pathological studies (Bhadauria et al., 2007). Molecular techniques like cDNA-AFLP using doubled haploid (DH) based bulked segregant analysis (BSA) led to the identification of genes that control rice blast (*Magnaporthe oryzae*). (Zheng et al., 2004). A transcript-derived fragment (TDF) was reported to be upregulated upon *M. oryzae* inoculation. Using modified SAGE technique (SuperSAGE), gene expression profiles of both the rice and blast fungus implicate *hydrophobin* gene to be most actively transcribed fungal gene in blast infected rice leaves (Matsumura et al., 2003). Expression profiles of two isogenic lines (G205 and G71) of rice in response to *M. oryzae* infection using cDNA microarray indicate expression of 2,200 expressed sequence tags (ESTs) and 998 genes were identified. Three annotated genes (casein kinase II alpha subunit gene, retrotransposon TOS17 insertion element gene and gene with unknown function) were found significantly induced in G205 only and other 35 genes exhibited differential expression in defence reactions, signal transduction, stress response, photosynthesis and sugar metabolism (Rao et al., 2002).

Table 1. Characterized effectors of *Magnaporthe oryzae*

Effector gene	*M. oryzae* isolate	Encoding protein	Gene function	Reference
MPG1	Guy11	Hydrophobin-like Protein, secreted protein	Role in conidial development and cell surface hydrophobicity	Talbot et al., 1993
PWL1	WGG-FA40	Glycine-rich, hydrophilic protein, secreted protein	Involves in resistance of weeping love grass	Kang et al., 1995; Sweigard et al., 1995
PWL2	Guy11	Glycine-rich, hydrophilic protein, secreted protein	Involves in resistance of weeping love grass	Kang et al., 1995; Sweigard et al., 1995
AvrPita	O-137	Secreted Zn-metallopeptidase protein	Interacts with *Pi-ta*	Orbach et al., 2000
ACE1	Guy11	Polyketide synthase / peptide synthetase	Interacts with *Pi33*	Bohnert et al., 2004
EMP1	70-15	Extracellular matrix protein 1, secreted pritein	Essential in appressorium formation and pathogenicity	Ahn et al., 2004
MHP1	70-15	Class II hydrophobin protein, secreted protein	Required for surface hydrophobicity	Kim et al., 2005
MSP1	70-15	Snodprot1 homolog protein, secreted protein	Essential in appressorium formation and pathogenicity	Jeong et al., 2007
Avr1-CO39	K76-79	Secreted protein	Interacts with *Pi-CO39*	Leong, 2008
AvrPiz-t	81278ZB15	Secreted protein	Interacts with *Piz-t*	Li et al., 2009
AvrPia	Ina168	Secreted protein	Interacts with *Pia*	Miki et al., 2009; Yoshida et al., 2009
AvrPii	Ina168	Secreted protein	Interacts with *Pii*	Yoshida et al., 2009
AvrPik/km/kp	Ina168	Secreted protein	Interacts with *Pik/km/kp*	Yoshida et al., 2009
BAS1	KV1	Secreted protein	Translocated into infected rice cells	Mosquera et al., 2009
BAS2	KV1	Secreted protein	Translocated into infected rice cells	Mosquera et al., 2009
BAS3	KV1	Secreted protein	Confined near to cell wall	Mosquera et al., 2009
BAS4	KV1	Secreted protein	Bounded with growing invasive hyphae	Mosquera et al., 2009
Slp1	Guy11	Secreted protein	Supresses rice basal resistance mechanism	Mentlak et al., 2012
MC69	Ina72	Secreted protein	Required for *M. oryzae* infection	Saitoh et al., 2012
MoCDIP1	Che86061 and KJ201	Secreted protein	Bcl-x1–mediated cell death suppression	Chen et al., 2013a
MoCDIP2	Che86061 and KJ201	Secreted protein	Bcl-x1–mediated cell death suppression	Chen et al., 2013a
MoCDIP3	Che86061 and KJ201	Secreted protein	Bcl-x1–mediated cell death suppression	Chen et al., 2013a
MoCDIP4	Che86061 and KJ201	Secreted protein	Bcl-x1–mediated cell death suppression	Chen et al., 2013a
MoCDIP5	Che86061 and KJ201	Secreted protein	Bcl-x1–mediated cell death suppression	Chen et al., 2013a
Iug6	98-06	Secreted protein	Supresses rice defence mechanism	Dong et al., 2015
Iug9	98-06	Secreted protein	Supresses rice defence mechanism	Dong et al., 2015

Till date, five pairs of *Avr* and *R* genes (*AvrPita* versus *Pita*, *AvrPik* versus *Pik*, *AvrPiz-t* versus *Piz-t*, *Avr-Pia* versus *Pia*, and *Avr1-CO39* versus *Pi-CO39*) have been intensively studied. Both direct and indirect interactions occur between these rice R proteins and blast Avr effectors. *AvrPita* and *Pita*, and *Avr-Pik* and *Pik* interact directly, while *AvrPiz-t* and *Piz-t* interact indirectly. The rice *R*-gene, *Pi-ta* and its corresponding *AVR-Pita* gene from the fungus were cloned and their interaction characterized (Bryan et al., 2000; Orbach et al., 2000). AVR-Pita protein appears to interact directly with Pi-ta protein and transient expression of this avirulence/effector protein in the cytoplasm of rice cells with Pi-ta triggers hypersensitive resistance (Jia et al., 2000). This suggests that the fungus delivers AVR-Pita protein into the cytoplasm of the rice cell. Although AVR-Pita appears to function as a protease inside rice cells, its role in the invasion process is not yet understood. To understand how AVR-Pita and other blast effectors function in promoting rice blast disease, it is first necessary to understand how the fungus co-opts normal plant cell processes for successful colonization of the host tissue.

Cloned blast resistance R-genes

Resistance mechanism in rice-*M. oryzae* pathosystem is governed by direct interaction as enunciated in the gene-for-gene concept (Flor, 1971). Accordingly, for each gene that confers avirulence to the pathogen there is a corresponding gene that confers resistance to the host, and vice-versa. Plant resistance (R) proteins recognize pathogen avirulence (Avr) determinants which in turn trigger signal transduction cascades that can lead to development of resistance by induction of defense responses resulting in the hypersensitive reaction at the site of infection. Although *R*-genes have been used in resistance breeding programs for a long time, they suffer from the disadvantage of being defeated by the co-evolving and variable nature of the pathogen. Hence, development of durably resistant cultivars for rice blast is a continuous and challenging process that demands a better understanding of the disease, especially when we consider the poor durability of many blast-resistant cultivars of rice, which have a typical field life of only 2-3 growing seasons before disease resistance is overcome. Till date, more than 100 *R*-genes has been mapped on rice genome, but only twenty five genes have been cloned and characterized (Sharma et al., 2012). The cloned genes include, *Pib* (Wang et al., 1999), *Pita* (Bryan et al., 2000), *Pi54* (Sharma et al., 2005; Sharma et al., 2010), *Pi-9* (Qu et al., 2006), *Pid2* (Chen et al., 2006), *Pi2* and *Piz-t* (Zhou et al., 2006), *Pi-36* (Liu et al., 2007), *Pi-37* (Lin et al., 2007), *Pikm* (Ashikawa et al., 2008), *Pi5* (Lee et al., 2009), *Pid3* (Shang et al., 2009), *pi21* (Fukuoka et al., 2009), *Pit* (Hayashi et al., 2009), *Pb1* (Hayashi et al., 2010), *Pish* (Takahasi et al., 2010), *Pi-k* (Zhai et al., 2011), *Pik-p* (Yuan et al., 2011), *Pia* (Okuyama et al., 2011), *Pi54rh* (Das et al., 2012), *Pi25* (Chen et al.,2011), *Pi1*(Hua et al., 2012), *NLS-1*(Tang et al., 2011), *Pi54of* (Devanna et al., 2014) and *Pi35* (Fukuoka et al., 2014). Considering the requirement of broad spectrum resistance against highly variable and virulent strains of *M. oryzae,* new sources of resistance are required to develop blast resistant rice varieties. The availability of complete genome sequences of rice in public domain (rgp.dna.affrc.go.jp; www.genomics.org.cn) has been an important resource for the analysis of the genomes of different rice cultivars to develop improved rice varieties. This genome sequence information of rice will also be useful to facilitate rapid and efficient polymerase chain reaction (PCR) based allele mining approach. PCR based primer walking strategy has been used to isolate the useful alleles of cloned and functionally validated rice genes from a wide range of rice cultivars and wild species. The available sequence data of rice will also assist allele mining in other cereals like wheat which have conserved synteny at the genome level (Singh et al., 2007). Mining for allele of major *R-* genes will be important for giving rice breeders direct access to key alleles conferring resistance to biotic stresses. Allele mining for blast resistance has been reported from wild and cultivated species of rice. Readers can refer other exhaustive review on this topic (Sharma et al., 2012). Two major blast resistance genes *viz. Pi-ta* and *Pi54* have been studied extensively. *Pi-ta* alleles have been mined from various land races and different *Oryza* species. *Pi-ta* was conserved before the divergence of these *Oryza* species and there is dimorphic pattern of nucleotide polymorphism and low nucleotide diversity at the LRD region in both resistant and susceptible accessions (Wang et al., 2008; Yoshida and Miyashita, 2009; Lee et al., 2009). The allelic diversity of the *Pi-ta* in 58 US weedy rice accessions and Indian landraces was also reported (Lee et al., 2011; Thakur et al., 2013a). Similarly extensive allele mining has been reported for *Pi54*, *Pi9, Piz(t)* and *Pid3* from different rice germplasm sources (Shang et al., 2009; Liu et al., 2011; Kumari et al., 2013; Thakur et al., 2013b; Thakur et al., 2015).

Rice-*Magnaporthe oryzae* pathosystem

In recent past, considerable effort has been made to understand molecular mechanisms of *M. oryzae* infection towards rice plants. Based on those studies, a number of genes involved in rice blast disease has been cloned and characterized (Talbot, 1995; Hamer and Talbot, 1998; Xu, 2000; Balhadere and Talbot, 2001; Idnurm and Howlett, 2001). Attempts have also been made to explore the expression profiles of rice and *Magnaporthe oryzae* during both compatible and incompatible phases of interaction. These studies, though not yet comprehensive, have given a broad understanding about the probable molecular mechanisms during rice-*M. oryzae* interaction. However, available information on transcriptome analysis of rice and *M. oryzae* during infection process is present in scattered form and no efforts have been made to compile these findings together and draw specific conclusions.

Pre-colonization of rice blast fungus

Appressorium formation is a key process during *M. oryzae* infection on rice plants that leads to invasive growth of hyphae in host tissues. Naturally, conidium of *M. oryzae* is known to form appressorium on hydrophobic surface and several reports on gene expression of *M. oryzae* on hydrophobic surface have been reported (Beckerman and Ebbole, 1996; Fang and Dean, 2000; Kamakura et al., 2002). However, Lee and Dean (1993) have also shown that cAMP could induce appressorium formation even on hydrophilic membranes, if the required concentration (50 mM) of this compound is given to the sporulation. In a related study, Irie et al. (2003) reported many cAMP regulated genes in *M. oryzae.* Among the cAMP-induced, genes those related to sugar metabolism, nucleic acid metabolism, transcription factors and amino acid metabolisms have been annotated. As expected, many of them were also categorised in melanin synthesis genes that presumably involved in appressorium melanisation. High concentration of melanin in appressorium is very important in maintaining the required turgor pressure during rice plant infection. However, the appressorium formation in *M. oryzae* has a tight association with cell cycle regulation. Molecular mechanism of cell cycle mediated regulation of appressorium morphogenesis, controlled by a temperature-sensitive *MgNIMA* gene, has shown that this gene which encodes a protein kinase is essential for mitotic entry (Osmani et al., 1988). Moreover, autophagy is another process to indirectly regulate appressorium formation by conidial cell death, a programmed cell death in order to re-cycle the contents of the fungal spore which is governed by *MgATG8* gene that encodes a protein essential for autophagy process prior to rice plant infection (Veneault-Fourrey et al., 2006). For generation of very high turgor pressure inside appressorium, autophagy process and transportation of lipid bodies and glycogen to the developing appressorium can be readily observed (Thines et al., 2000). Subsequently, fatty acid metabolism in the appressorium is a complex process which requirs an orchestrated action of many triacylglycerol lipases necessary for penetration event during plant infection (Wang et al., 2007). Appressorium function is mainly regulated by two genes *MFP1* and *PEX6*. The *MFP1* gene, encoding a β-oxidation enzyme substantially affect virulence level, while *PEX6*, a peroxisomal biogenesis gene completely controls appressorium function and rice blast disease (Wang et al., 2007; Ramos-Pamplona et al., 2006). *PEX6* gene also helps in generation of Woronin bodies that is essential for proper development and functioning of appressoria by sealing the septal pores damaged in the process of autophagy. By this, a high cellular turgor is created and maintained inside the cell that is necessary for plant infection (Soundararajan et al., 2004). In addition, *Pth2* gene encodes a carnitine acetyl transferase that involves in lipid mobilization across the peroxisomal membrane for further utilization during appressorium maturation and is necessary for penetration-hypha formation (Bhambra et al., 2006).

Highly conserved signaling networks that transfer signals from the environment to the nucleus play a crucial role in regulating host-pathogen interactions. Mitogen-activated protein kinase (MAPK) mediated signalling pathways have now been directly implicated in regulating infection-related development in various phytopathogenic fungi and highlighting the conservation of MAPK signalling as a regulatory component of fungal pathogenicity. In *M. oryzae*, three distinct MAPK mediated signalling pathways have been identified like *PMK1* for pathogenicity MAP kinase, *MPS1* for MAP kinase for penetration and sporulation and *OSM1* for osmo-regulated MAP kinase (Dixon et al., 1999; Xu et al., 1998; Xu, 2000). For normal appressorium formation, the activation of *Pmk1* gene is controlled by *Mst7* (MAPK kinase, MAPKK) and *Mst11* (MAPKK kinase, MAPKKK) genes and that was confirmed by gene replacement method using their mutated form in *M. oryzae*. *Mst50* a SAM containing protein that directly interacts with *Mst11* MAPKKK gene and mutant of this gene abolished appressorium formation and pathogenicity of *M. oryzae* (Park et al., 2006).

Post-colonization of rice blast fungus

In the rice- *M. oryzae* interaction study, several significant findings have been revealed at the stages of appressorium formation, conidial germination and other events like adhesion of conidia to the host surface, germ tube formation, turgor building inside the conidial cell, *etc.* (Howard et al., 1991; Sweigard et al., 1992; Talbot, 2003). All these events are essential for proper appressorium formation and subsequent development of blast disease in rice. Thus, it is necessary to explore the pathways and its regulation mechanisms during colonization of the fungus for actual understanding about the rice- *M. oryzae* pathosystem. Association of the effector molecules in case of *Pita-AvrPita*, *Piz-t-AvrPiz-t*, with the defense system of host plant has been well characterized in this system. The avirulence (*Avr*) protein, encoded by *AVR* gene is recognized by corresponding resistance (R) gene protein, leading to the race-specific recognition (de Wit, 2009). Several recently characterized effectors in rice blast fungus continue to follow the similar theme that most effectors are small novel secreted proteins and generally lack homology with known proteins. The mutant of *Avr* protein escapes detection by its interacting counterpart, *R* gene product and proceeds to a disease development process. In the host-pathogen interaction study of rice- *M. oryzae*, over 40 *Avr* genes have been identified (Kawahara et al., 2012) and among them nine *Avr* genes have been cloned. Among the cloned genes, all of the AVR genes except *ACE1* encode secreted proteins expressed in the invasive hyphae inside the host plant (Kawahara et al., 2012). *ACE1* is specifically expressed in the fungal appressoria and encodes PKS-NRPS protein (Bohnert et al., 2004). *Avr-Pita* and *Avr-Piz-t* are well characterized *Avr* genes of rice blast fungus. *Avr-Pita*, a family of genes (*Avr-Pita1*, *Avr-Pita2* and *Avr-Pita3*) encodes a putative neutral zinc metalloprotease (Jia et al., 2000; Khang et al., 2008). *Avr-Piz-t* acts to suppress pathogen-associated molecular pattern (PAMP)-triggered immunity (PTI) by inhibiting the ubiquitin ligase activity of the rice RING E3 ubiquitin ligase APIP6 (Park et al., 2012). *Pwl* effectors, another *Avr* gene family (*Pwl1*, *Pwl2*, *Pwl3* and *Pwl4*) are small, glycine-rich proteins that are present in this fungal pathogen (Kang et al., 1995). Though most of the *AVR* genes have been cloned from *M. oryzae* isolates from rice, *Avr1- CO39* was however cloned from a weeping love grass isolate (Peyyala and Farman, 2006). *Avr-Pia*, *Avr-Pik/km/kp*, and *Avr-Pii* were identified in a strain, Ina168 by using whole genome sequencing approach (Yoshida et al., 2009). The *M. oryzae* genome also encodes large suites of enzymes involved in secondary metabolism, including 23 polyketide synthases, several non-ribosomal peptide synthases and cytochrome P450 mono-oxygenases, consistent with the fungus having a significant capacity for secondary metabolite production. The precise function of such metabolites in pathogenesis is not well established, but interestingly one of the PKS-encoding genes *ACE1*, has been identified as an avirulence gene (de Wit et al., 2009). Therefore, it seems that secondary metabolites produced by the fungus play significant roles within the plant during establishment of disease. These effector molecules, including secondary metabolites, may be delivered from appressoria, mediated by transporters like ABC3, an ATP-binding cassette (ABC) type transporter (Sun et al., 2006). In addition, a few secreted proteins that are required for pathogenicity, *viz.* MPG1 (Talbot et al., 1993), EMP1 (Ahn et al., 2004), MHP1 (Kim et al., 2005), MSP1 (Jeong et al., 2007), MC69 (Saitoh et al., 2012), and Slp1 (Mentlak et al., 2012), and four biotrophy associated secreted proteins, BAS1 to BAS4 (Mosquera et al., 2009) have also been characterized. However, the majority of *M. oryzae* secreted proteins have not been experimentally validated for their functions in pathogenicity. Whether these predicted genes are expressed in infected rice plants remains largely unknown (Oh et al.,2008). Therefore, it is necessary to characterize all the predicted genes, especially those are expressed *in planta* to provide full insights into fungal pathogenesis.

Initiation of gene expression studies has been started with the advent of Sanger sequencing by decoding of cDNA and its libraries prepared from the rice blast fungus at different growth stages and during infection processes to rice plant (Takano et al., 2003). Initially, such studies were conducted by the help of differential display method. Using this method, Kim et al. (2000) characterized many defense-related genes expressed in rice against a treatment of *M. oryzae* elicitor. Suppression subtractive hybridization (SSH) is a rapid and effective method to isolate differentially expressed genes (Kim et al., 2001 and 2005; Xiong et al., 2001; Lu et al., 2004; Han et al., 2004). However, the high level of sequence redundancy in SSH libraries limits its ability to identify a large number of differentially expressed genes from rice infected by *M. oryzae*. The expressed sequence tag (EST) is another approach used after SSH to study a set of differentially expressed genes. It was used by Kim et al. (2001) and Jantasuriyarat et al. (2005) to identify defense transcripts on a large scale from rice. *In planta* expressed genes of *M. oryzae* were also identified in the same study using an infected rice cDNA library (Kim et al., 2001). By employing this approach, a Class II hydrophobin MHP1 gene was found preferentially expressed inside host plant during late stages of infection and showed full involvement in virulence of rice blast fungus (Kim et al., 2005). Numa et al. (2009) sequenced 35,189 ESTs obtained from a highly destructive strain of rice blast fungus. These ESTs were mapped on *M. oryzae* genome and they validated only 4480 (~37 %) predicted genes of this fungus as protein coding genes. This study indicates that the evidence based gene prediction is essential for accurate gene-finding from any organism. A total 100 novel genes identified and reported in *M. oryzae* during a compatible rice and blast fungus interaction at later stages of infection using combined strategies of EST and SSH library analysis (Kim et al., 2010). Microarray and RNA-Seq based technologies overcome many of the limitations of using EST analysis and further take the gene expression studies at throughput level. Among microarray and RNA-Seq, the later technique scores many advantages over the earlier one (Table 2), but RNA-Seq is technically more demanding and more tedious than microarray technology (Figure 2).

Molecular analysis of Rice-*Magnaporthe* interaction

Transcriptome profile of Magnaporthe oryzae

Transcriptomics using microarray identified genes that were differentially expressed during infection-related morphogenesis of conidia and appressoria formation in *M. oryzae* (Tanako et al., 2003). Subsequently many studies have been conducted by various research groups to identify differentially expressed genes during compatible and incompatible rice- *M. oryzae* interactions (Table 3). The whole genome oligo microarray chip containing over 13,000 *M. oryzae* elements representing 10,176 predicted genes, during spore germination and appressorium formation on both an inductive hydrophobic surface and in response to cAMP (Oh et al., 2008). They suggested that *M. oryzae* employs a number of backup systems, such as functional redundancy and compensatory processes in order to protect appressorium formation from being de-regulated. Using microarray analysis, 262 fungal genes were found induced more than two fold during biotrophic invasion of *M. oryzae*. The numerous effector candidates identified and described distinctive *in planta* secretion patterns of *M. oryzae*, (*BAS*1-4 genes) which provides a valuable tool for assessing EIHM compartment integrity in individual invaded rice cells (Mosquera et al., 2009). To understand the conidiogenesis in *M. oryzae*, Kim and Lee (2012) measured global gene expression patterns in conidiation process using a whole genome oligonucleotide microarray. Approximately 4.42% and 4.08% genes were significantly up-regulated and down-regulated, respectively, during conidiation process, which shows that the process of conidiation is regulated both positively and negatively. However, the array based methods have several limitations. Dependency upon existing knowledge about genome sequence and high background levels owing to cross-hybridization are some main problems persistence with these technologies. These limitations have been overcome by tag-based methods that include serial analysis of gene expression (SAGE; Gowda et al., 2004) and massively parallel signature sequencing (MPSS; Nobuta et al., 2007). Irie et al. (2003) reported that about hundred statistically significant different tags were obtained from cAMP treated and non-treated SAGE libraries and equal numbers of the tags were induced and -repressed by cAMP. Using RL-SAGE libraries, Gowda et al. (2007) identified a large set of distinct tags (83,832) from the mixed tissues of rice and *M.oryzae* infected plants. They found that nucleotide conversions rates were high in the identified transcript tags and led to proportionate increase in the rate of mismatches in the tags prepared from the infected libraries. Many antisense transcripts were identified from rice and observed their enhanced expression in the infected rice leaves, considered as an evidence for the involvement of RNA variation and antisense transcript expression during plant-fungal interactions. An integrative approach by combining MPSS, robust long-SAGE (RL-SAGE) and oligoarray methods, Gowda et al. (2006) analyzed the mycelium and appressorium transcriptomes from *M .oryzae*. In this study, they identified a total of 2,430 mycelial genes and 1,886 appressorial genes. These differentially expressed

Table 2. Comparison of Microarray and RNA-Seq techniques in relation to transcriptome analysis.

Technology Specifications	Microarray	RNA Sequencing
Principle	Hybridization	High throughput sequencing
Resolution	Several to 100 bp	Single Base
Background Noise	High	Low
Trancriptome Data capture of a gene.	~20% of gene	Entire transcript
Dynamic range to quantify gene expression level.	Up to few hundred fold	>8,000 fold
Ability to distinguish between different isoforms.	Limited	Yes
Ability to distinguish allelic expression.	Limited	Yes
Required amount of RNA.	High	Low
For less abundant transcripts.	Less sensitive	More sensitive
Allele specific (SNP) expression.	No	Yes
Quantitation of splicing.	No	Yes
Novel alternative splicing and Novel genes	Limited	Yes

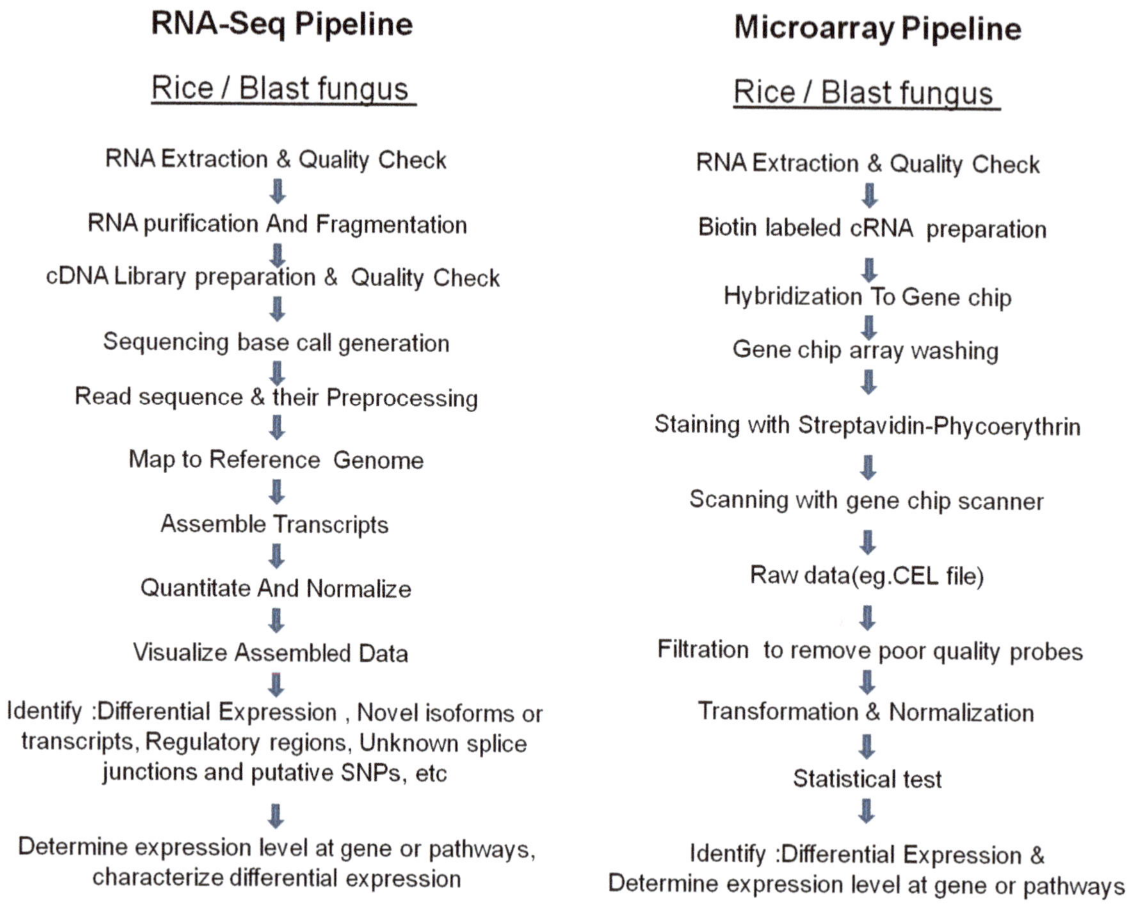

Figure 2. Comparative pipeline of microarray and RNA-Seq techniques.

Table 3. Different transcriptome studies in *M. oryzae* during its interaction with rice.

Time interval	Tissues	Technology	Reference
8 hi 24 hi	Appressoria Conidia	Microarray	Takano et al., 2003
24 hpi 96 hpi	Appressoria Mycelia	Microarray	Gowda et al., 2006
24 hpi 96 hpi	Appressoria Mycelia	MPSS	Gowda et al., 2006
96 hpi	Mycelia	RL-SAGE	Gowda et al., 2006
7 hi 12 hi	Conidia Appressoria	Microarray	Oh et al., 2008
36 hpi	Compatible interaction	Microarray	Mosquera et al., 2009
48 hpi	Conidia	Microarray	Kim and Lee, 2012
24 hpi	Compatible and Incompatible interaction	RNA-Seq	Kawahara et al., 2012
96 hpi	Compatible interaction	RL-SAGE	Chen et al., 2013
3, 6, 12, 24, 48 hpi	Incompatible and Compatible interaction	MPSS	Chen et al., 2013
6, 12, 24 hpi	Incompatible and Compatible interaction	SBS	Chen et al., 2013

transcripts, especially those specific to appressoria, represent a genomic resource useful for gaining a better understanding of the molecular basis of *M. oryzae* pathogenicity. A similar comparative transcriptome study was performed between rice and *M. oryzae* interaction in early and later interaction stages using microarray method (Wang et al., 2014). A total 608 genes were differentially expressed in response to compatible and incompatible interactions with *M. oryzae* fungus. Out of them 231 genes were highly induced in incompatible than compatible interaction. Most of the genes related to the metabolic process, response to stimuli, and cellular processes were highly expressed indicating their direct link with *M. oryzae* infection. These results suggest that a large number of both host and pathogen genes are required in the attack and counter-attack between host-pathogen in their incompatible interaction (Wang et al., 2014).

Advancement in sequencing technologies and its affordability has sorted out the problems related to the exiting methods used for mapping and quantifying transcriptomes. The whole genome transcriptome sequencing (RNA-Seq) is known to have a wider dynamic range, higher technical reproducibility, and provide a better estimate of absolute expression levels and is most commonly used technique for transcriptomic studies (Fu et al., 2009; Marioni et al., 2008). In next generation sequencing (NGS) technologies, RNA-Seq and high throughput-SuperSAGE (HT-SuperSAGE), provide almost similar sensitivity and accuracies of the measured expression levels, however HT-SuperSAGE is much more cost effective than RNA-Seq (Soanes et al., 2012). The later NGS technology has a major limitation as it require the prior information of gene structure to convert tag counts to corresponding transcript expression levels (Chen et al., 2013a). Using RNA-Seq, Kawahara et al. (2012) revealed the expression profiles of both rice and *M. oryzae* simultaneously in rice-infected leaf blades at the initial infection stage (24 hpi). They found that 16,048 rice and 889 fungal transcripts, including 432 and 17 unannotated transcripts, respectively, showed infection-responsive expression. Using two different types of fungal strains, they revealed the differential expression patterns between the compatible and incompatible interactions and also that drastic responsive reaction was common among the incompatible interaction of both rice and *M. oryzae* at the initial stages of infection. Recently, a total of 851 *in planta* expressed genes encoding predicted effector proteins have been identified by a genome-wide transcriptome analysis of *M. oryzae* infected rice leaves through RL-SAGE, MPSS and sequencing by synthesis (SBS) technologies (Chen et al., 2013a). In this study, they identified five novel effectors *MoCDIP1* to *MoCDIP5* from 42 *in planta*-expressed putative secreted proteins that induce plant cell death in rice. These genes are expressed during infection stages, especially 96 hpi in rice plant and share similar physiological phenotypes, such as the response to light, to inhibitors of calcium channel, and to Bcl-x1-mediated cell death suppression. Moreover, the CBD domain of MoCDIP4 protein served as PAMP elicitor that is a functionally conserved domain of the elicitor found among different plant-pathosystems. These results hypothesised that some of these cell death-inducing effectors may facilitate the colonization of *M. oryzae* during the late necrotrophic phase of the *M. oryzae* infection. Initially, several expression studies have been carried out using cDNAs libraries constructed from *M. oryzae* infected rice leaves to identify the interacting molecules between host and pathogen (Kim et al., 2001; Rauyaree et al., 2001). However, such studies are far from satisfactory to obtain a significant level of gene expression changes in *M. oryzae* during rice blast disease development (Chen et al., 2013a). This is possibly due to very less amount of expressed genes of invading fungus as compared to the level of indigenous expressing genes of host that present in the infected plant. Recent advances in genome sequencing technologies have led to a rapid discovery of numerous effectors in *M. oryzae* and have provided a wealth of information on their structure and function. In this pathogen, about 12% of the annotated genes (1546) coding for putative secreted proteins has been predicted from the reference genome of *M. oryzae* (Dean et al., 2005; Yoshida et al., 2009). Variability in effector molecules, especially in *M. oryzae* is mainly by their localization on unstable chromosomal regions and their linkage with transposable element. In several reports, this trend was experimentally observed and hypothesised that the presence of transposons nearby effectors, including *Avr* genes control the effectors to be gained or lost during course of evolution of the pathogen (Kang et al., 1995; Orbach et al., 2000; Yoshida et al., 2009; Singh et al., 2014). By analysis of transcriptome of a FJ81278 strain of *M. oryzae*, a total 256 candidate effectors were found through RNA-seq technique (Chen et al., 2013b). Moreover, 134 candidate effectors were identified from a genome of *M. oryzae* isolate 98-06 employing the same technique (Dong et al., 2015). By functional characterization of isolate specific genes, they also revealed that three genes, IUG6, IUG9 and IUG18 played critical role in pathogenicity of rice blast fungus. Therefore, it is essential to validate the candidate effectors and elucidate their role in rice- *M. oryzae* interaction.

Transcriptome profile of rice

Study of transcriptome revealing the levels of different transcripts and their probable functions in a given temporal and spatial conditions have been extensively studied in case of rice-*Magnaporthe* system. During the hemi-biotrophic interaction between rice and *M. oryzae*, the transcriptome of rice is subjected to stress, leading to constant fluctuations of transcripts levels during various phases of *M. oryzae* infection. In recent years many attempts have been made to explore the expression profiles of different rice lines during both compatible and incompatible rice-*M. oryzae* interactions (Table 4). These studies, though not yet comprehensive, have given a broad understanding about the probable molecular mechanisms during rice-*M. oryzae* interaction.

Transgenic approach helps understand the basic mechanism of disease resistance by over expressing the gene responsible for conditioning disease resistance. Analysis of rice transgenic lines over expressing TF, OsWRKY13 provided important cues of disease resistance

Table 4. Different transcriptome studies in rice during its interaction with *M. oryzae*.

Rice variety	Hours Post inoculation (hpi)	Technology	Reference
LTH, two NILs (IRBL18 and IRBL22)	24	Microarray	Wei et al., 2013
YY, NILs (R), NILs.(S) , LTH	0,12, 24, 36, 48, 60, 72, 96	Microarray	Huang et al.,2014
HR-12	6	Microarray	Vijayan et al., 2013
Nipponbare	84, 96, 120	Sequencing using ABI	Kim et al., 2011
NIL, H7R (resistant with Pi-k/Pir1) and H7S (susceptible)	0, 24	Microarray	Li et al., 2006
Nipponbare	72, 96	Microarray	Ribot et al., 2007
Transgenic rice TP309 (TP-Pi54) and non transgenic TP309	72	Microarray	Gupta et al. 2012
Jinheung	12, 24, 48, 72	Microarray	Wang et al., 2014
Nipponbare (carrying *Pia*)	24	RNAseq	Kawahara et al., 2012
Gigante Vercelli and Vialone Nano	24	RNAseq	Bagnaresi et al., 2012
LTH and its NIL (IRBL 12-M)	24, 48	Microarray	Kato et al., 2009

conditioning. The microarray analysis of two independent transgenic and uninfected rice lines (D11UM1-1 and D11UM7-2) overexpressing OsWRKY13 and also wild-type rice plants Mudanjiang 8, susceptible to *M. grisea*, identified a total of 460 and 445 upregulated and 478 and 605 down regulated genes specific to D11UM1-1 and D11UM7-2, respectively (Qiu et al., 2008). Microarray analysis in compatible and incompatible rice cultivars at 24 hpi suggests that the transcriptional profiles among these two interactions are mostly similar (Wei et al., 2013). Therefore, like in other systems, the incompatible and compatible rice-*M. oryzae* interactions differ mainly quantitatively (Wen et al, 2003; Lu et al., 2004). The WRKY47 transcription factor gene which showed significant upregulation during incompatible interaction was overexpressed in transgenic rice and the resulted plants could show increased resistance to rice blast fungus (Wei et al., 2013). In another RNA-seq experiment, gene ontology (GO) enrichment in compatible and incompatible interaction is similar, whereas the genes sets contributing to each GOs were dissimilar (Bagnaresi et al., 2012). On the contrary, there was a more drastic change in the transcriptome of rice in incompatible interaction compared with that of compatible (Kawahara et al., 2012). Together these studies indicate that the magnitude of change in the transcriptome of rice, during its interaction with *M. oryzae* is largely determined by the genotype of the rice plants and also probably the virulence nature of the blast fungus and the responsive reactions that are involved in the plant defense and pathogen attack are more active in the incompatible interaction (Kawahara, 2012). These correlations can be further supported with other studies also. In one such effort, 15,616 (30.0%) and 872 (5.2%) transcripts were found differentially expressed in the annotated rice and fungal transcripts, respectively (Kawahara et al., 2012). In our studies transcriptome analysis of the transgenic rice plants harbouring major rice blast resistance gene, *Pi54* identified a total of 1154 differentially expressed DEGs in TP-Pi54 plants (incompatible interaction) of which 587 were upregulated, whereas 567 genes were down regulated in case of incompatible interaction (TP-Pi54) in comparison to those in compatible interaction (TP309). Meta-analysis of the genes (FC≥2 and p value<.05) being upregulated in rice transcriptome 24 hours after infection with *M. oryzae* across different studies is demonstrated. Studies using RNA sequencing detected higher number of upregulated genes involved in Rice-*M. oryzae* interaction in comparison to those using microarray analysis. This shows the technological advantages of RNA- seq over microarray for transcriptome analysis. Overall we observed higher number of genes being significantly upregulated during incompatible interaction than compatible interaction in both microarray and RNA seq studies (Figure 3).

Reprogramming of cell wall associated genes

Rice cell wall is the first level of physical barrier for penetrating *M. oryzae* appressorium during early stage of infection. Callose (1, 3-β-glucan) is an important component of plant defense response. It is involved in the blockage of plasmodesmata and thereby acts as a physical barrier for the penetration of fungal mycelia into neighbouring plant cells (Beffa et al., 1996; Hammond-Kosack and Jones, 2000). Role of callose in rice blast resistance has already been reported in the case of blast resistance gene *Pi54* (Rai et al., 2011). Our own work reveals that the upregulation of two genes coding for callose biosynthesis was comparatively higher in the transgenic rice line TP-Pi54 in comparison to non transgenic control TP lines following *M. oryzae* infection (Gupta et al., 2012).

Phenylalanine ammonia lyase (PAL) is a first enzyme in the phenyl propanoid pathway. Previous study has reported the transcriptional activation of the phenylpropanoid pathway genes for various biotic stresses in different systems (Caldo et al., 2004). PAL is involved in phytoalexins and lignin biosynthesis (Dixon et al., 2001). The early accumulation of phenylpropanoids in case of rice- *M. oryzae* interaction has been an important determinant of resistance of rice to *M. oryzae* infection (Parker et al., 2009 and Wei et al., 2013). In resistance responses, the early activation of this pathway leads to production of both anti-microbial secondary metabolites as well as the precursors of lignin or suberin for cell-wall strengthening (Kawasaki et al., 2006). During compatible interaction between rice- *M.*

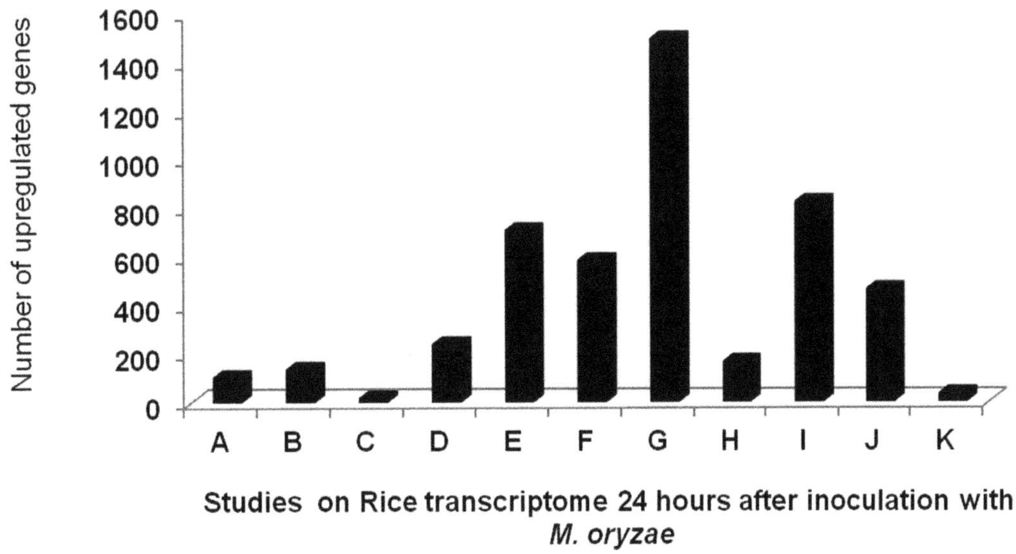

Figure 3. Comparative analysis of the genes upregulated in rice transcriptome after 24 h of infection with *M. oryzae* across. Incompatible interactions in rice line IRBL18 (A) and IRBL22 (B) (Wei et al., 2013); C: Compatible Interaction (Wei et al., 2013); D: After BTH treatment in rice plant (Shimono et al., 2007); E and F: Incompatible and compatible interaction (Bagnaresi et al., 2012); G and H: Incompatible and compatible interaction (Kawahara et al., 2012); I and J: Incompatible and compatible interaction (Priyanka et al., Unpublished); K: Compatible interaction (Vijayan et al., Unpublished).

grisea, the obvious accumulation these metabolites has been correlated with the absence of primary cell-wall thickening at the sites of fungal penetration (Huckelhoven, 2007). In this regard, previous studies, including our own study (unpublished data) reports that a class III peroxidase gene (LOC_Os07g48010), which was found to be 16.94-fold up-regulated in the transgenic line TP-Pi54 (Gupta et al., 2012) was earlier found to be involved in physiological functions like lignification and pathogen defence, and also in lignin biosynthesis (Quiroga et al., 2000; Almagro et al., 2009; Marjamaa et al., 2009). A similar trend in the expression profile of cell wall, beta-glucanase, and proteolysis related gene has been reported to be highly induced in response to *M. oryzae* infection (Bennett and Wallsgrove 1994; Wang et al., 2014).

Peroxidases and laccases are known to catalyze the polymerization of ROS-mediated lignin monomers and variation in their expression level may determine the structure of lignin formed. Beside the expression profile, the scale of expression might also determine the efficiency of lignifications. In one such analysis involving compatible interaction, our experimental results indicate the differential expression of peroxidase genes at different time intervals in response to *M. oryzae* infection. *M. oryzae* trigger early transcription of peroxidases and the number of peroxidase members expressed increases considerably with progress in infection.

Using the publicly available data we performed analysis for different functional categorisation of the upregulated genes (Figure 4). We observed significant differential regulation of cell wall related genes among different studies. The comparison of cell wall related genes being upregulated across different studies also shows that higher numbers of genes are upregulated in incompatible interaction compared to compatible interaction. This analysis indicates that during the initial phases of interaction, when *M. oryzae* penetrates in to the cell cytoplasm, rice plants enforce the defence mechanism by cell wall fortification.

Compromised photosynthesis and fluctuating primary metabolites

Photosynthesis is the source of energy for living organisms. The fungus *M. oryzae*, which infects rice, is the major competitor for photosynthates and it siphons out larger chunk of them leading to confrontation between the rice and *M. oryzae*. The fact that during the infection of rice with *M. oryzae* the rate of photosynthesis is compromised was known to scientific community way back 1993 (Bastiaans and Kropff, 1993). But, the molecular aspects of the process is being explored recently (Vergne et al., 2007; Ribot et al., 2008). During compatible rice- *M. oryzae*, most of the genes involved in photosynthesis were found be repressed, but this phenomenon is a common response of both susceptible and resistant rice plants (Vergne et al., 2007, Ribot et al., 2008). Similar findings by various authors and also our own studies have also revealed the down regulation of photosynthesis related genes post blast infection in rice (Vergne et al., 2007; Li et al., 2006; Yun et al., 2000; Bagnaresi et al., 2012; Vijayan et al, unpublished data). Li et al. (2006) further reported the upregulation of

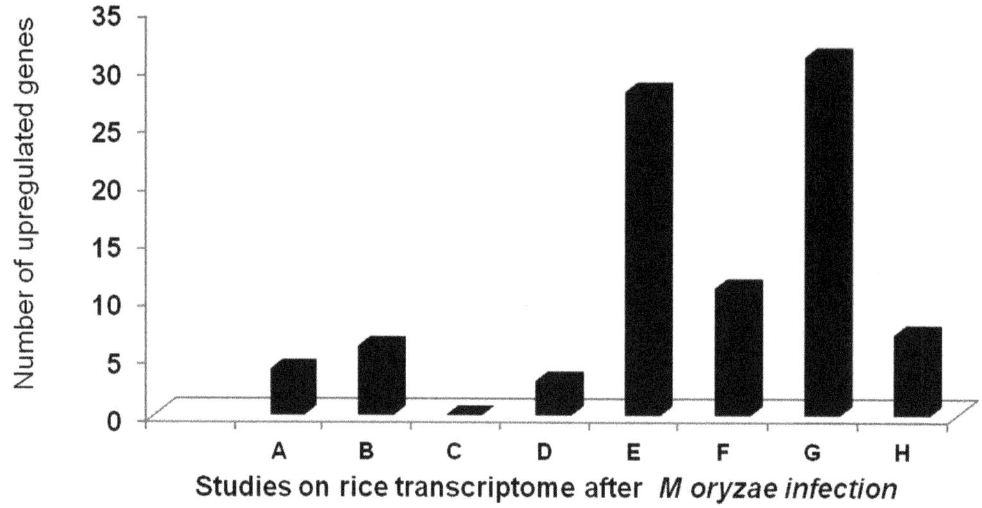

Figure 4. Comparative analysis of cell wall related genes upregulated in rice transcriptome after 24 h of infection with *M. oryzae*. Incompatible interactions in rice line IRBL18 (A) and IRBL22 (B) (Wei et al., 2013); C: Compatible Interaction (Wei et al., 2013); D: After BTH treatment in rice plant (Shimono et al., 2007); E and F: Incompatible and compatible interaction (Bagnaresi et al., 2012); G and H: Incompatible and compatible interaction (Kawahara et al., 2012).

peroxidase, NADH dehydrogenase subunit and protein synthesis-related genes in resistant interaction. Therefore, they concluded that those differentially expressed genes might represent a set of genes with different expression levels between incompatible and compatible disease reactions. Besides, many genes related to photosynthesis and metabolism were down regulated during rice-*M. grisea* interaction, signifying that energy metabolism in rice play an important role in its defense against blast fungus (Li et al., 2006).

Our study reveals the fluctuating expression of photosynthesis related genes between 12 hpi to 72 hpi (unpublished data). The initial down regulation of photosynthesis post pathogen invasion gives plants the much needed flexibility to divert the energy and resources to counter the invading pathogen (Bolton, 2009). Besides the findings reveal the increased expression of cell wall invertases in *M. oryzae* infected rice plants (Berger et al., 2004; Swarbrick et al., 2006). Cho et al. (2005) reported the up-regulations of rice *OsCIN1*, *OsCIN4*, and *OsCIN5*, post infecting with *M. grisea* and reported that these genes may play a role during a switch in metabolism to resist the invading *M. oryzae*.

Proteins play vital role in keeping plant cells viable and functional. Researchers have found that many genes related to protein synthesis were activated and protein degradation related genes were repressed in incompatible response with rice blast, suggesting that protein process might be associated with rice resistance (Li et al., 2006). In a compatible interaction between rice- *M. oryzae*, we also found the differential regulation of genes related to protein modification and degradation and they constituted around 10% of the total differentially expressed genes during early stage of infection (unpublished data). It is very much expected that infection of rice with *M. oryzae* triggers host of differentially expressed genes (DEGs) involved in host metabolism. Wang et al. (2014) analysed and found that, during incompatible interaction, GO terms associated host metabolic process showed a significant differential regulation of 35.37%, 30.19%, and 35.83% in clusters 1, 2, and 4, respectively. They suggested that infection with blast pathogen greatly alter the metabolic processes inside rice plants. Another study dealing with different time intervals (24-120 hpi) of an incompatible interaction also revealed the greater enrichment of transcripts related to carbon fixation,Glycolysis/gluconeogenesis and photosynthesis (Huang et al., 2014). Taken together, these data suggest that rice reprograms metabolic and biological processes related to energy metabolism in response to infection with *M. oryzae*.

Impaired hormonal balance and altered signalling

Plant hormonal pathways that are important regulators of defence-gene expression are Salicylic acid (SA), jasmonic acid (JA) and ethylene (ET) pathways. JA and ET pathways are involved in resistance response against necrotrophs while SA pathway is mainly involved in resistance to biotrophic and hemibiotrophic pathogens (Robert et al., 2011). There is considerable overlap between these pathways during different stress conditions. In Pathogen Triggered Immunity (PTI) response, JA-ET and SA pathways act synergistically. So blocking just one component can perturb overall plant stress response. Therefore, many pathogen effectors suppress the PTI response by interacting with different target proteins. In Effector Triggered Immunity (ETI) response, redundant activities of JA-ET and SA pathways are involved. So in the absence of SA signaling, JA-ET pathway contribute to maintain pathogen resistance in plants (Dodds et al., 2010).

Previous reports reveal that altered JA expression in rice plants acts as a powerful mediator of resistance against hemibiotroph *M. oryzae* and bacterial pathogen *Xanthomonas oryzae pv. oryzae* (Deng et al., 2012; Riemann et al., 2013; Tao et al., 2009; Xiao et al., 2009; Yamada et al., 2012 and Mei et al., 2006). The transcriptome analysis of rice- *M. oryzae* in resistant near isogenic line and susceptible control showed upregulation of enzymes involved in JA biosynthesis (Wei et al., 2013). Similarly, Wang et al. (2014) reported the role of ET and JA in resistance response of rice leaves infected with incompatible *M. oryzae.* During incompatible rice- *M. oryzae* interaction, upregulation of five genes related with ethylene, as well as the genes related with salicylic acid (SA) and jasmonic acid (JA) signaling were reported at 48 hpi (Wang et al., 2014). They also reported the induction of those genes related to JA and SA signaling, oxidation-reduction as well as calcium regulation, MAP kinase, and phosphoinositides. Beside this, our study also confirms the early induction of ET biosynthesis genes and in general gradual increase in the number DEGs related to ABA, auxin, cytokinin, SA, and GA from 24 hpi (Vijayan et al., unpublished data). Together these data indicate a major reprogramming of hormonal pathways and cell signaling after pathogen infection at 24 hpi. Incompatible interaction with *M. oryzae* could also activate ET emissions earlier than compatible interaction (Iwai et al., 2006). Therefore, these results confirm that during early phase of fungal infection there is drastic reprogramming of genes related to hormonal pathways involved in defence mechanism against rice blast fungus.

Plant signal perception and activation of downstream responses is crucial during innate immunity. Cellular events which are common in ETI and PTI mediated response in rice after of *M. oryzae* infection is well deciphered in rice. These include depolarization of plasma membrane, activity of ion channel is modified and is burst of metabolism occurs which leads to generation of reactive oxygen species (ROX) and antimicrobial compounds, mitogen activated protein kinase (MAPK) cascades or calcium dependent protein kinase (CDPKs) are activated, pathogen responsive genes are transcribed and lignin and callose are deposited at plant cell wall (Dodds et al., 2010; Meng., 2013). Other important molecules involved in defence signalling include chitin elicitor binding protein (CEBiP), chitin elicitor receptor kinase (OsCERK1), receptor-like kinases (RLKs), and wall-associated kinases (WAK). Genes involved in signaling pathways within rice after *M. oryzae* infection are highly upregulated in near isogenic line carrying blast resistance gene *Pi9* (Wei et al., 2013) and transgenic line carrying blast resistance gene *Pi54* compared to susceptible control (Gupta et al., 2012) . This shows ETI mediated response to be quantitatively stronger during resistance response than during susceptible interaction (Jones et al., 2006, Dodds et al., 2010, Tao et al., 2003, Wei et al., 2013). Rice CEBiP is characterized to be involved in sensing and binding of chitin oligosaccharides. Post elicitor binding, CEBiP forms a hetero-oligomers with OsCERK1, and this complex through transphosphorylation might trigger downstream signalling cascade (Kaku et al., 2006; Shimizu et al., 2010). The transcriptome analysis of during rice infection with *M. oryzae* has revealed that *CEBiP* and *OsCERK1* genes were highly upregulated during incompatible interaction in comparison to compatible reaction and also additional OsLysM-RLK genes were found induced only in incompatible interaction (Bagnaresi et al., 2012).

In signaling pathways receptor kinase play important role and are upregulated in near isogenic line carrying blast resistance gene *Pi9* (Wei et al., 2013). Receptor kinase category includes receptor like kinase, receptor like cytoplasmic kinase and wall associated kinase which are responsible for internal and external signal perception. (Wei et al., 2013). Wall associated kinases (WAKs) are known to perceive the fungal cell wall associated oligogalacturonides during pathogen infection, and trigger innate immune response in plants (Brutus et al., 2010). The RNA- seq analysis of resistant cv. GV and the susceptible cv. VN showed significant upregulation of 15 WAK transcripts in GV and only 4 WAKs in VN (Bagnaresi et al., 2012). Through microarray analysis, very high upregulation of *OsWAKY71* and *OsWAK25* has been reported during incompatible rice interaction with *M. oryzae* (Wei et al., 2013). OsWAK 25 which is reported to enhance resistance to bacterial pathogen *Xanthomonas oryzae pv. oryzae* (Seo et al., 2011) was also found upregulated in near isogenic line carrying blast resistance gene *Pi9* (Wei et al., 2013). A putative WAK family gene was also found down regulated in compatible disease response during our analysis (Vijayan et al., 2013). Together these data suggest that signal perception and the activation of downstream signalling cascade involves a complex set of signalling molecules and is a vital in providing resistance during rice incompatible reaction with *M. oryzae.*

MAP kinases are the signalling molecules involve in signaling downstream of receptor kinase involved in biosynthesis of photoalexin camalexin in *Arabidopsis* (Ahuja et al., 2012). Four MAPK transcripts including isoforms of MAP3K.3 and MAP3K.1 were found upregulated in resistant line GV but not in susceptible. In an interconnecting signalling cascade, MAPKs OsMPK3, MAPKK OsMKK4 and OsMPK6 were found to play a key role in amplifying chitin elicitor signal for defense responses (Hamel et al., 2006).

Transcription factors in Rice M. oryzae interaction

Transcription factors play a pivotal role in arming the plants with the required phenotypic plasticity which is very critical under various biotic and abiotic stresses. During rice-*M.oryzae* interaction there has been an extensive reprogramming of their respective transcriptomes leading to the adaptive plasticity of plants or establishing compatible interaction by fungus in highly variable environments. This plasticity is mainly achieved by network of transcription factors like *WRKY, NAC, Dof* Zinc finger, *MAD* box and *bZIP* (Gupta et al., 2012). Transcriptome analysis during interaction of rice with *M. oryzae* has revealed that transcripts coding for WRKY family TFs are most commonly reprogrammed and they are crucial regulators of disease resistance (Eulgem, 2005; Ryu et al., 2006; Shimono et al., 2007; Delteil et al., 2012; Bagnaresi et al., 2012; Kawahara et al., 2012; Wei et al., 2013). Among the WRKY family, transcripts coding for *WRKY22*, *WRKY30*, WRKY45, *WRKY45-2*, *WRKY47*, *WRKY53*, *WRKY55/WRKY31* and *WRKY104/WRKY89* are the positive regulators of rice resistance to *M. oryzae*, whereas those coding for *WRKY13*, *WRKY28, WRKY42* and *WRKY76* are negative regulators (Shimono et al., 2007; Cheng and Wang, 2014; Chujo et al., 2013; Qiu et al., 2008; Cheng et al., 2015). Using RNA-seq and microarray approaches, significant upregulation of WRKY family TFs has been reported during rice- *M. oryzae* interaction at 24 hpi (Shimono et al., 2007; Gupta et al., 2011; Kawahara et al., 2012; Bagnaresi et al., 2012; Yokatani et al., 2013; Vijayan et al., 2013; Chujo et al., 2007; Zhang et al., 2008; Wei et al., 2013) .WRKY gene activate several pathogenesis related genes (Chujo et al., 2007 and Peng et al., 2010). Higher number of WRKY were upregulated in blast resistant genotype GV and near isogenic line carrying blast resistance gene *Pi9* compared to susceptible control (Bagnaresi et al., 2012,Wei et al., 2013). These studies show that WRKY as an important regulator in rice blast resistance.

The other important TF gene family genes coding for ethylene response factors (ERF) are differentially regulated by various signalling molecules like ethylene, abscisic, salicylic, jasmonic, gibberellic acids and their interactions during *M. oryzae* interaction with rice. (Sakuma et al., 2002, Grennan, 2008). Besides these, *OsDREB* and *NAC4* TF were also induced during the early phase of infection at 12 and 24 hpi (Kaneda et al., 2007; Kawahara et al., 2012; Vijayan et al., 2013). The expression of *DREB* TF can be attributed to the finding that DREB helps in overcoming the rice susceptibility to blast caused by drought and cold stress (Koga et al., 2004a). The MYB and MAD box TF genes too have been reported to be induced during infection of blast fungus *Magnaporthe* during both compatible and incompatible interactions (Gupta et al., 2011; Vijayan et al., 2013; Wang et al., 2014). These TFs play an important role in activation of plant defense signaling (Ramalingam et al., 2003; Rasmussen et al., 2012). The microarray and Differential Display-PCR analysis have revealed the induction of AP2/EREBP family TF genes, which is required for ethylene responsiveness during rice- *M. oryzae* (Kim et al., 2000; Wei et al., 2013). NAC family transcription factors play a positive role in rice plant defense response. *OsNAC4,* OsNAC6 have been found upregulated in many of the incompatible interaction with *OsNAC* being induced at the early stage of infection (Mosquerel et al., 2009; Gupta et al., 2012; Kawahara et al., 2012). This report signifies the role of *Os NAC4*, which plays an important role in the initial hypersensitive cell death in rice plants after the recognition of flagellin in rice cells (Kaneda et al., 2007).

Our Meta analysis of transcription factors coding genes indicated upregulation of this category of genes in different studies (A to H) and also shows that higher number of genes coding for TF was found upregulated in the incompatible interaction when compared to compatible interaction (Figure 5).

Secondary metabolites and pathogenesis response genes

Secondary metabolites play an important role in PAMPs and Effector Triggered Immunity (PTI/ ETI). Rice have developed various defense techniques like variety of pathogens receptors, generation of reactive oxygen species (ROS) and use of anti-microbial secondary metabolites such as phytoalexins (Jwa et al., 2006). The rice reactive oxygen species (ROS) are involved in cross-linking and host cell wall enforcement during *M. oryzae* infection. ROS levels are well related with *M. oryzae* infection in rice (Chi et al., 2009; Mittler et al., 2004). Further, levels of peroxidase and glutathione-S-transferase

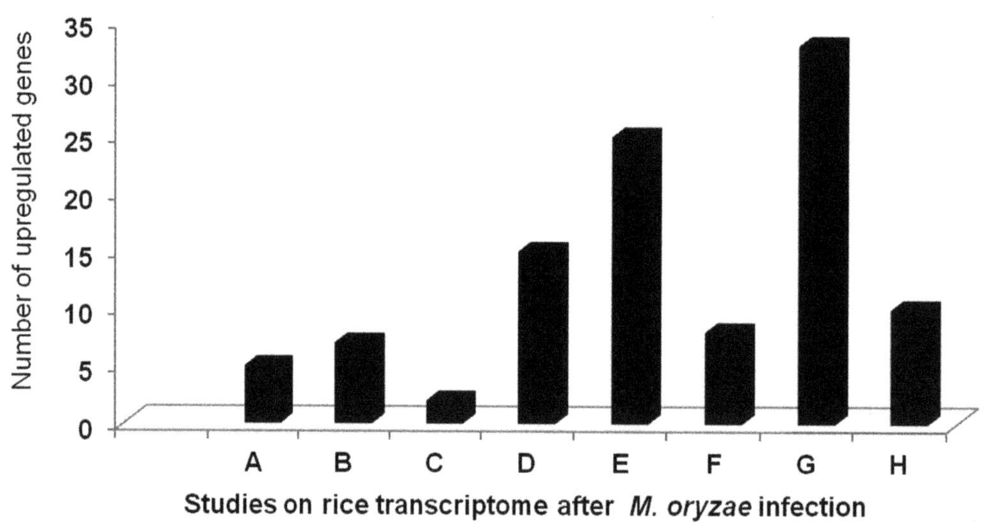

Figure 5. Comparative analysis of transcription factor related genes upregulated in rice transcriptome after 24 h infection with *M. oryzae*. Incompatible interactions in rice line IRBL18 (A) and IRBL22 (B) (Wei et al., 2013); C: Compatible Interaction (Wei et al., 2013); D: After BTH treatment in rice plant (Shimono et al., 2007); E and F: Incompatible and compatible interaction (Bagnaresi et al., 2012); G and H: Incompatible and compatible interaction (Kawahara et al., 2012).

were comparatively more up-regulated at 48 hpi than at 12 hpi suggesting increased level of ROS scavengers in infected tissues at later stage of infection (Wang, 2014). These findings suggest elevation of ROS in *M.oryzae* infected tissues. The two peroxidases detected in an early stage of infection suggest that ROS signalling is activated in the early stage of *M. oryzae* infections (Wang et al., 2014). On the other hand, transcriptome analysis has also revealed the upregulation of transcripts coding for ROS detoxifying enzymes like peroxidases and mono-oxygenases during only incompatible interactions (Qiu et al., 2008; Bagnaresi et al., 2012). Out of the 20 induced ROS homeostasis genes, two transcripts were coding for putative glutaredoxin family proteins (Qiu et al., 2008).

Malate is known to serve as a substrate for NADPH production, which in turn is utilized by NADPH oxidase for the generation of ROS intermediates. NADP-ME (Malic enzyme) activity, which coverts malate into NADPH has been found reduced in susceptible host-pathogen interactions (Parker et al., 2009). The recent study performed in our lab also indicates a similar trend with the down regulation of NADP-ME at the early stages after *M. oryzae* infection, but subsequent up regulation at later stages of infection in rice (Vijayan, unpublished data). The late up regulation of a NADP-ME gene suggests the delayed induction of these ROS producing genes in susceptible host genotypes.

We also performed a meta analysis of genes upregulated at 24 hpi and the findings show significant differential regulation of respiratory burst related genes (peroxidises and glutathione S-transferase) in the incompatible interactions when compared to the compatible interaction among different studies (Figure 6).

Up regulation of secondary metabolite expression is a common trend in response to stress (Bennett and Wallsgrove, 1994). Down regulation of carotenoids has been previously reported as a result of *M. oryzae* infection (Bagnaresi et al., 2013). The repression of genes of the carotenoid biosynthesis pathway might suggest the redirection of substrates to produce defensive constituents of isoprenoid pathway probably in response to pathogen infection (Boba et al., 2011).

Secondary metabolism genes were found upregulated in resistant and susceptible rice line after *M. oryzae* infection. In phenylalanine and shikimate biosynthesis as well as downstream of phenylpropanoid biosynthesis, large number of enzymes show upregulation in resistant near isogenic line (NIL) carrying blast resistance gene *Pi9* (Wei et al., 2013) and transgenic line carrying blast resistance gene *Pi54* (Gupta et al., 2012). Phenyl propanoid pathway plays an important role as phenylpropanoids are important antimicrobial compound. This pathway is involved in synthesizing lignin and phytoalexin that prevent invasion of pathogen in host. To counteract *M. oryzae* infection, rice leaves accumulate diterpene phytoalexins phytocassanes, oryzalexins and momilactones (Okada,2011; Dillon et al. 1997; Umemura et al., 2003). The important genes in diterpene phytoalexin, momilactone and phytocassanases biosynthesis pathways were upregulated in blast resistant genotype GV after *M. oryzae* infection (Bagnaresi et al., 2012). Similarly diterpene phytoalexin biosynthetic (DPB) geneOsKSL8 (LOC_Os11g28530; oryzalexin S synthesis) shows higher expression in incompatible interaction than in compatible reaction after *M. oryzae* infection (Bagnaresi et al., 2012). These results therefore highlight the importance of diterpene phytoalexin biosynthesis genes during rice-*M. oryzae* interaction.

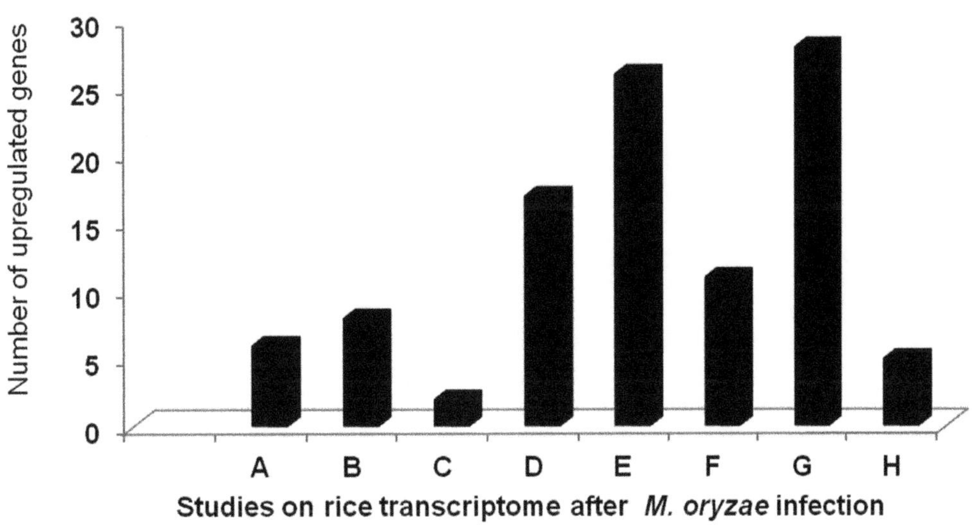

Figure 6. Comparative analysis respiratory burst related genes found in different studies related to the transcriptomes of *M. oryzae* - rice interaction after 24 h of infection. Incompatible interactions in rice line IRBL18 (A) and IRBL22 (B) (Wei et al., 2013); C: Compatible Interaction (Wei et al., 2013); D: After BTH treatment in rice plant (Shimono et al., 2007); E and F: Incompatible and compatible interaction (Bagnaresi et al., 2012); G and H: Incompatible and compatible interaction (Kawahara et al., 2012).

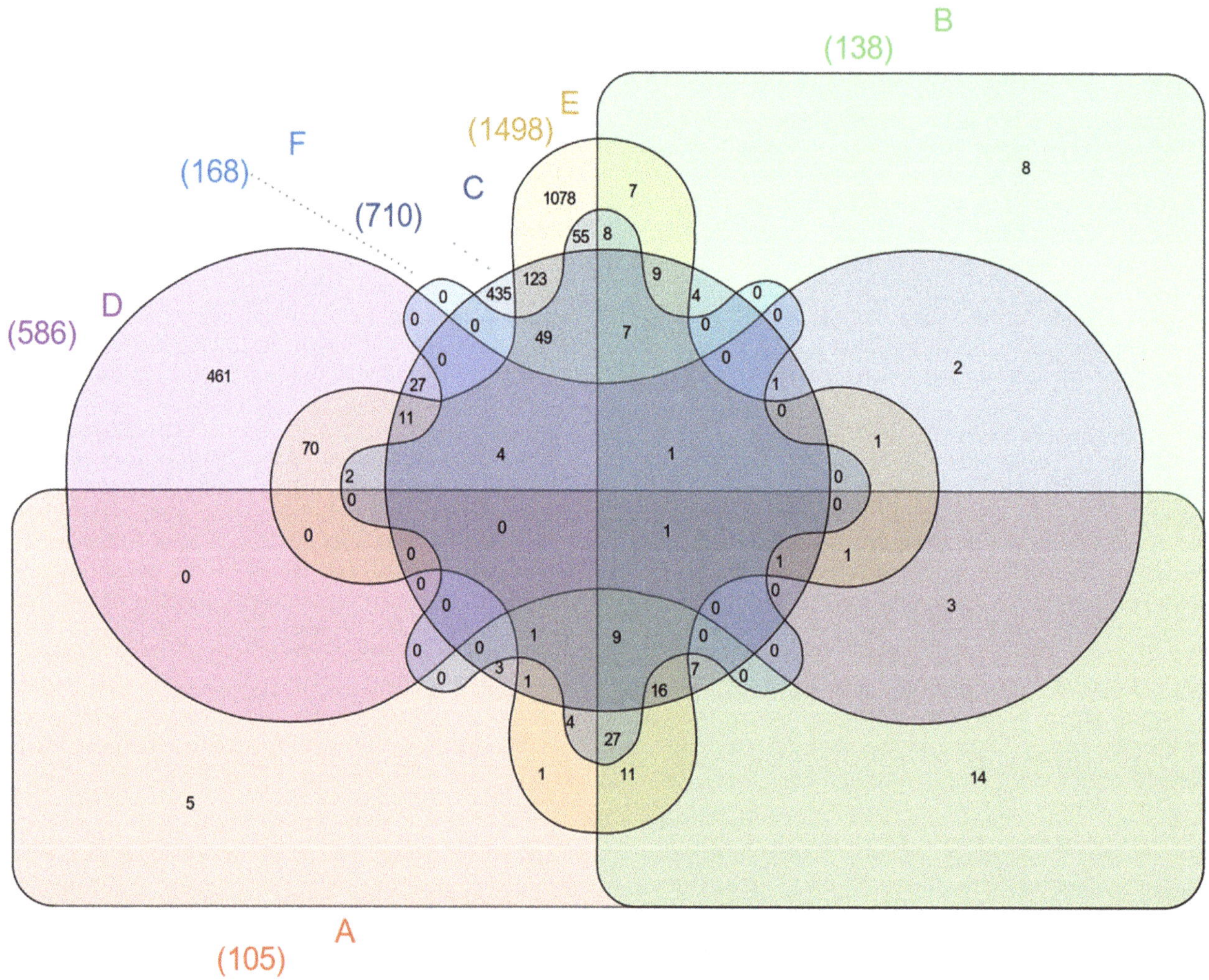

Figure 7. Venn diagram to compare upregulated genes across different studies on rice transcriptome 24 hours after *M. oryzae* inoculation. Incompatible interactions in rice line IRBL18 (A) and IRBL22 (B) (Wei et al., 2013); C and D: Incompatible and compatible interaction (Bagnaresi et al., 2012); E and F: Incompatible and compatible interaction (Kawahara et al., 2012).

Pathogenesis related (PR) genes are those, which are involved in the process of plant defense response and in suppressing pathogen infection. The commonly known PR genes are thaumatin-like proteins and chitinases (Sels et al., 2008). In a resistance interaction, plant immune signaling was triggered rapidly through the recognition of rice receptors by their elicitors/effectors, and PR genes were more rapidly accumulated in incompatible than in compatible interaction (Wang et al., 2014; Kawahara et al., 2012. Transcriptional expression analysis of PR genes also confirmed that an early defense response rapidly occurred in the incompatible interaction at 12 hpi, with high accumulation at 48 hpi (Wang et al., 2014). Similarly, Kawahara et al., 2012 found the higher level of expression of PR genes like PR1, PR10, POX22 and PR2 family transcripts like GH17 during incompatible rice- *M. oryzae* interaction. Moreover, their study reveals the essentiality of fungal penetration for the activation of defense PR genes.

The upregulated genes found in above eight interactions was compared using InteractiVenn (www.interactivenn.net). The findings of this analysis show that there is no common gene observed across these different studies. But when we compared the data from upregulated genes of six interaction studies, it was observed that a single gene coding for cytochrome P450 (LOC_Os03g12500.1) is found to be common among these interactions (Figure 7). Therefore, the cytochrome P450 genes, found to be common among the six studies has a critical role to play during rice- *M. oryzae* interaction. Hence, the application of transcriptome data has huge potential in the identification of novel disease response genes of rice and also novel effector and *Avr* genes of *Magnaporthe oryzae* (Vergne et al., 2007; Chen et al., 2013a). The whole genome transcriptome analysis also helps in understanding the molecular mechanism of rice resistance and *Magnaporthe oryzae* pathogenicity.

Conclusions

Rice-*M.oryzae* interaction has emerged as a model plant-pathosystem over the years. Sequence information of both rice and *M. oryzae* genomes have largely contributed to this development. Transcriptome analysis during rice- *M. oryzae* interaction reveals differentially expressed transcripts at a given phase of infection process. These transcripts provide important information for identification, cloning and characterization of underlying genes for the functional correlation. Interestingly, transcriptome profile of rice during compatible and incompatible interactions differs only quantitatively , otherwise the set of genes differentially regulated remain more or less similar in both the cases, cytochrome P450 being one such molecule. Moreover, different resistant rice lines activate significantly different set of genes in response to *M. oryzae* infection. At the initial stages of infection there is a perturbance in the transcripts related to the respiratory burst, cell wall associated genes, early signalling genes in case of rice and those related to cellular metabolism, conidiation and appressorium formation in case of *Magnaporthe oryzae*. At later stages of infection transcripts coding for effectors are found with elevated expression. Besides effector molecules, transcripts related TF, kinases, hormones, photosynthesis, secondary metabolite and defence response gene were found up-regulated. The transcriptome analysis has a great potential to the identification of novel defence responsive genes and the genes mediating the resistance response in case of rice and also effector and *Avr-* genes contributing to the pathogenesis in case of *M. oryzae*. The transcriptome being the first step in gene regulation provide important cues for understanding the arms race between the host and the pathogen. A sustainable strategy for disease control can only be devised once we understand the metabolic and regulatory pathways involved in host-pathogen interaction.

Acknowledgements

TRS is thankful to Indian Council of Agricultural Research (ICAR) and Department of Biotechnology (DBT), Government of India, for financing Projects related to rice blast disease. He is also grateful to the Department of Science and Technology, Govt. of India for the award of JS Bose Fellowship.

References

Ahn, N., Kim, S., Choi, W., Im, K.H., and Lee, Y.H. (2004). Extracellular matrix protein gene, EMP1, is required for appressorium formation and pathogenicity of the rice blast fungus, *Magnaporthe grisea*. Mol. Cells, 17, 166-173.

Ashikawa, I., Hayashi, N., Yamane, H., Kanamori, H. and Wu, J. (2008). Two adjacent nucleotide-binding site-leucine-rich repeat class genes are required to confer Pikm-specific rice blast resistance. Genetics 180, 2267-2276

Balhadere, P.V., and Talbot, N.J. (2001). PDE1 encodes a P-type ATPase involved in appressorium-mediated plant infection by the rice blast fungus *Magnaporthe grisea*. Plant Cell, 13, 1987-2004.

Beckerman, J.L., and Ebbole, D.J. (1996). MPG1, a gene encoding a fungal hydrophobin of *Magnaporthe grisea*, is involved in surface recognition. Mol. Plant Microbe Interact. 6, 450-456.

Bhadauria, V., Popescu, L., Zhao, W. S. and Peng, Y. L. (2007). Fungal transcriptomics. Microbiol. Res. 162, 285-298.

Bhambra, G.K., Wang, Z.Y., Soanes, D.M., Wakley, G.E., and Talbot, N.J. (2006). Peroxisomal carnitine acetyl transferase is required for elaboration of penetration hyphae during plant infection by *Magnaporthe grisea*. Mol. Microbiol. 61, 46-60.

Bohnert, H.U., Fudal, I., Dioh, W., Tharreau, D., Notteghem, J.L., and Lebrun, M.H. (2004). A putative polyketide synthase/peptide synthetase from *Magnaporthe grisea* signals pathogen attack to resistant rice. Plant Cell, 16, 2499-2513.

Bryan, G.T., Wu, K.S., Farrall, L., Jia, Y., Hershey, H.P., McAdams, S.A., Faulk, K.N., Donaldson, G.K., Tarchini, R., and Valent, B. (2000). A single amino acid difference distinguishes resistant and susceptible alleles of the rice blast resistance gene Pi-ta. Plant Cell 12, 2033-2045.

Chen, S., Songkumarn, P., Venu, R.C., Gowda, M., Bellizzi, M., Hu, J., Liu, W., Ebbole, D., Meyers, B., Mitchell, T., et al. (2013a). Identification and characterization of in-planta expressed secreted effectors from *Magnaporthe oryzae* that induce cell death in rice. Mol. Plant-Microbe. Interact. 26, 191-202.

Chen, C., Lian, B., Hu, J., Zhai, H., Wang, X., et al. (2013b). Genome comparison of two *Magnaporthe oryzae* field isolates reveals genome variations and potential virulence effectors. BMC Genomics, 14, 887.

Chen, J., Shi, Y., Liu, W., Chai, R., Fu, Y., Zhuang, J. and Wu, J. (2011). A Pid3 allele from rice cultivar Gumei2 confers resistance to *Magnaporthe oryzae*. J. Gen. Genomics 38(5), 209-216.

Chen, X. W., Shang, J., Chen, D., Lei, C., Zou, Y., Zhai, W., Liu, G., Xu, J., Ling, Z., Cao, G., Ma, B., Wang, Y., Zhao, X., Li, S. and Zhu, L. (2006). A B-lectin receptor kinase gene conferring rice blast resistance. Plant J. 46, 794-804.

Chen, X., and Ronald, P.C. (2011). Innate immunity in rice. Trends Plant Sci. 16, 451-459.

Couch, B. C., Fudal, I., Lebrun, M. H., Tharreau, D., Valent, B., van Kim, P., Notteghem, J. L. and Kohn, L. M. (2005). Origins of host-specific populations of the blast pathogen Magna-porthe oryzae in crop domestication with subsequent expan-sion of pandemic clones on rice and weeds of rice. Genetics 170, 613-630.

Couch, B.C., and Kohn, L.M. (2002). A multilocus gene genealogy concordant with host preference indicates segregation of a new species, *Magnaporthe oryzae*, from M. grisea. Mycologia. 94, 683-693.

Das, A., Soubam, D., Singh, P. K., Thakur, S., Singh, N. K., and Sharma, T. R. (2012). A novel blast resistance gene, Pi54rh cloned from wild species of rice, Oryza rhizomatis confers broad spectrum resistance to *Magnaporthe oryzae*. Funct. Integr. Genomics 12, 215-228. doi, 10.1007/s10142-012-0284-1

de Wit, P., Mehrabi, R., van den Burg, H.A., and Stergiopoulos, I. (2009). Fungal effector proteins: past, present and future. Mol. Plant Pathol. 10, 735-747.

Dean, R.A., Talbot, N.J., Ebbole, D.J., Farman, M.L., Mitchell, T.K., et al. (2005). The genome sequence of the rice blast fungus *Magnaporthe grisea*. Nature 434, 980-986.

Devanna, N. B., Vijayan, J. and Sharma, T. R. (2014). The Blast Resistance Gene Pi54of Cloned from Oryza officinalis Interacts with Avr-Pi54 through Its Novel Non-LRR Domains. PLoS ONE 9(8): e104840. doi:10.1371/journal.pone.0104840

Dixon, K.P., Xu, J.R., Smirnoff, N., and Talbot, N.J. (1999). Independent signaling pathways regulate cellular turgor during hyperosmotic stress and appressorium-mediated plant infection by *Magnaporthe grisea*. Plant Cell 11, 2045-2058.

Dong, Y., Li, Y., Zhao, M., Jing, M., Liu, X., Liu, M., et al. (2015). Global genome and transcriptome analyses of *Magnaporthe oryzae* epidemic isolate 98-06 uncover novel effectors and pathogenicity- related genes, revealing gene gain and lose dynamics in genome evolution. PLoS Pathog. 11, e1004801.

Fang, E.G., and Dean, R.A. (2000). Site-directed mutagenesis of the magB gene affects growth and development in *Magnaporthe grisea*. Mol. Plant Microbe Interact. 13, 1214-1227.

Flor, H.H. (1971). Current status of the gene-for-gene concept. Annu. Rev. Phytopathol. 9, 275-296.

Fu, X., Fu, N., Guo, S., Yan, Z., Xu, Y., et al. (2009). Estimating accuracy of RNA-Seq and microarrays with proteomics. BMC Genomics, 10, 161.

Fukuoka, S., Saka, N., Koga, H., Ono, K., Shimizu, T. et al. (2009). Loss of function of a proline-containing protein confers durable disease resistance in rice. Science 325(5943), 998-1001.

Fukuoka, S., Yamamoto, S. I., Mizobuchi, R., Yamanouchi, U., Ono, K., Kitazawa, N., Yasuda, N. and Fujita, Y. (2014). Multiple functional polymorphisms in a single disease resistance gene in rice enhance durable resistance to blast. Scientific Reports 4, 4550. DOI, 10.1038/srep04550

Gowda, M., Jantasuriyarat, C., Dean, R. A., and Wang, G-L. 2004. Robust-LongSAGE (RL-SAGE): A substantially improved LongSAGE method for gene discovery and transcriptome analysis. Plant Physiol. 134, 890-897.

Gowda, M., Venu, R.C., Li, H., Jantasuriyarat, C., Chen, S., Bellizzi, M., Pampanwar, V., Kim, H.R., Dean, R.A., Stahlberg, E., Wing, R., Soderlund, C., and Wang, G.L. (2007). *Magnaporthe grisea* infection triggers RNA variation and antisense transcript expression in rice. Plant Physiol. 144, 524-33.

Gowda, M., Venu, R.C., Raghupathy, M.B., Nobuta, K., Li, H., Wing, R., Stahlberg, E., Couglan, S., Haudenschild, C.D., Dean, R., et al. (2006). Deep and comparative analysis of the mycelium and appressorium transcriptomes of *Magnaporthe grisea* using MPSS, RL-SAGE, and oligoarray methods. BMC Genomics 7, 310.

Greer, C. A. and Webster, R. K. (2001). Occurrence, distribution, epidemiology, cultivar reaction, and management of rice blast disease in California. Plant Dis. 85, 1096-1102.

Gupta, S.K., Rai, A.K., Kanwar, S.S., Chand, N.K. Singh, D., Sharma, T.R. (2012) A single blast resistance gene Pi54 activates complex defense mechanism in rice. J Exp. Bot. 63, 757-772.

Hamer, J.E., and Talbot, N.J. (1998). Infection-related development in the rice blast fungus *Magnaporthe grisea*. Current Opinion Microbiol. 1, 693-697.

Han, C.U., Lee, C.H., Jang, K.S., Choi, G.J., Lim, H.K., Kim, J.C., Ahn, S.N., Choi, J.E., Cha, J.S., Kim, H.T., et al. (2004). Identification of rice genes induced in a rice blast-resistant mutant. Mol. Cells 17, 462-468.

Hayashi, K., Yasuda, N., Fujita, Y., Koizumi, S. and Yoshida, H. (2009). Identification of the blast resistance gene Pit in rice cultivars using functional markers. Theor. Appl. Genet. 121(7), 1357-1367.

Hayashi, N., Inoue, H., Kato, T., Funao, T., Shirota, M., Shimizu, T., Kanamori, H., Yamane, H., Hayano-Saito, Y., Matsumoto, T., Yano, M. and Takatsuji, H. (2010). Durable panicle blast-resistance gene Pb1 encodes an atypical CC-NBS-LRR protein and was generated by acquiring a promoter through local genome duplication. Plant J. 64, 498-510.

Howard, R. J. and Valent, B. (1996). Breaking and entering, Host penetration by the fungal rice blast pathogen *Magnaporthe grisea*. Annu. Rev. Microbiol. 50, 491-512.

Howard, R.J., Ferrari, M.A., Roach, D.H., and Money N.P. (1991). Penetration of hard substrates by a fungus employing enormous turgor pressures. Proc. Natl. Acad. Sci. USA 88, 11281-11284.

Hua., L., Wu, J., Chen, C., He, X., Lin, F., Wang, F. et al. (2012). The isolation of Pi1, an allele at the Pik locus which confers broad spectrum resistance to rice blast. Theor. Appl. Genet. 125, 1047-1055.

Idnurm, A., and Howlett, B.J. (2001). Pathogen city Genes of Phytopathogenic Fungi. Mol. Plant Pathol. 2, 241-255.

International Rice Genome Sequencing Project (2005). The Map-Based Sequence of the Rice Genome. Nature 436, 793-800.

Irie, T., Matsumura, H., Terauchi, R., and Saitoh, H. (2003). Serial analysis of gene expression (SAGE) of *Magnaporthe grisea*: Genes involved in appressorium formation. Mol. Genet. Genomics 270, 181-189.

Jacob, F., Vernaldi, S. and Maekawa, T. (2013). Evolution and conservation of plant NLR functions. Front. Immunol. 4, 297. 10.3389/fimmu.2013.00297

Jantasuriyarat, C., Gowda, M., Haller, K., Hatfield, J., Lu, G., Stahlberg, E., Zhou, B., Li, H., Kim, H., Yu, Y., et al. (2005). Large-scale identification of expressed sequence tags involved in rice and rice blast fungus interaction. Plant Physiol. 138, 105-115.

Jeong, J.S., Mitchell, T.K. and Dean, R.A. (2007). The *Magnaporthe grisea* snodprot1 homolog, MSP1, is required for virulence. FEMS Microbiol. Lett. 273, 157-165.

Jia Y., McAdams, S.A., Bryan. G.T., Hershey, H.P. and Valent, B. (2000). Direct interaction of resistance gene and avirulence gene products confers rice blast resistance. EMBO J. 19, 4004-4014.

Jones, J.D.G. and Dangl, J.L. (2006). The plant immune system. Nature 444, 323-329.

Jwa, N. S., Agrawal, G. K., Tamogami, S., Yonekura, M., Han, O., Iwahashi, H. and Rakwal, R. (2006). Role of defense/stress-related marker genes, proteins and secondary metabolites in defining rice self-defense mechanisms. Plant Physiol. Biochem. 44, 261-273.

Kamakura, T., Yamaguchi, S., Saitoh, K., Teraoka, T. and Yamaguchi, I. (2002). A novel gene, CBP1, encoding a putative extracellular chitin-binding protein, may play an important role in the hydrophobic surface sensing of *Magnaporthe grisea* during appressorium differentiation. Mol. Plant Microbe Interact. 15, 437-444.

Kang, S.C., Sweigard, J.A., and Valent, B. (1995). The PWL host specificity gene family in the blast fungus *Magnaporthe grisea*. Mol. Plant Microbe Interact. 8, 939-948.

Kawahara, Y., Oono, Y., Kanamori, H., Matsumoto, T., Itoh, T., et al. (2012). Simultaneous RNA-Seq analysis of a mixed transcriptome of rice and blast fungus interaction. PLoS ONE 11, e49423. doi:10.1371/journal.pone.0049423.

Kawasaki, S. (2004). Proceedings of the 3rd International Rice Blast Conference, Rice Blast, Interaction with rice and control. Dordrecht, The Netherlands, Kluwer Academic Publishers.

Khang, C.H., Park, S.Y., Lee, Y.H., Valent, B. and Kang, S. (2008). Genome organization and evolution of the AVR-Pita avirulence gene family in the *Magnaporthe grisea* species complex. Mol. Plant Microbe Interact. 21, 658-670.

Khush, G.S. and Jena K.K. (2009). Current status and future prospects for research on blast resistance in rice (Oryza sativa L.), in Advances in Genetics, Genomics and Control of Rice Blast Disease, eds Wang, G. L., Valent B., editors. (Dordrecht, Springer;), 1-10 10.1007/978-1-4020-9500-9_1

Kim, S., Ahn, I.P., and Lee, Y.H. (2001). Analysis of genes expressed during rice-*Magnaporthe grisea* interactions. Mol. Plant-Microbe Interact. 14, 1340-1346.

Kim, S., Ahn, I.P., Rho, H.S., and Lee, Y.H. (2005). MHP1, a *Magnaporthe grisea* hydrophobin gene, is required for fungal development and plant colonization. Mol. Microbiol. 57, 1224-1237.

Kim, S., Il-Pyung, A. and Lee, Y. H. (2001). Analysis of genes expressed during Rice-*Magnaporthe grisea* interactions. Mol. Plant-Microbe Interact. 14, 1340-1346.

Kim, S., Park, J., Park, S.Y., Mitchell, T.K., and Lee, Y.H. (2010). Identification and analysis of in-planta expressed genes of *Magnaporthe oryzae*. BMC Genomics 11, 104.

Kim, S.K., and Lee, Y.H. (2012). Gene expression profiling during conidiation in the rice blast pathogen *Magnaporthe oryzae*. PLoS ONE, 7, e43202.

Kumari, A., Das, A., Devanna, B. N., Thakur, S., Singh, P. K., et al. (2013). Mining of rice blast resistance gene *Pi54* shows effect of single nucleotide polymorphisms on phenotypic expression of the alleles. Eur. J. Plant Pathol. 137, 55-65.

Langmead, B., Trapnell, C., Pop, M. and Salzberg, S. (2009). Ultrafast and memory efficient alignment of short DNA sequences to the human genome. Genome Biol. 10, R25.

Lee, S. K., Song, M. Y., Seo, Y. S., Kim, H. K. and Ko, S. (2009). Rice Pi5 mediated resistance to *Magnaporthe oryzae* requires the presence of two CC-NB-LRR Genes. Genetics 181, 1627-1638.

Lee, S., Costanzo, S., Jia, Y., Olsen, K.M. and Caicedo, A.L. (2009). Evolutionary dynamics of the genomic region around the blast resistance gene Pi-ta in AA genome Oryza species. Genetics 183, 1315-1325.

Lee, S., Jia, Y., Jia, M., Gealy, D.R. and Olsen, K.M. (2011). Molecular evolution of the rice blast resistance gene Pi-ta in invasive weedy rice in the USA. PLoS ONE 6(10), e26260.

Lee, Y.-H., and Dean, R.A. (1993). cAMP regulates infection structure formation in the plant pathogenic fungus *Magnaporthe grisea*. Plant Cell 5, 693-700.

Leong, S.A. (2008). The ins and outs of host recognition of *Magnaporthe oryzae*. Genomics of Disease, 24, 199-216.

Li, W., Wang, B., Wu, J., Lu, G., Hu, Y., Zhang X., Zhang, Z., Zhao, Q., Feng, Q., and Zhang, H. (2009) The *Magnaporthe oryzae* avirulence gene AvrPiz-t encodes a predicted secreted protein that triggers the immunity in rice mediated by the blast resistance gene Piz-t. Mol Plant-Microbe Interact. 22, 411-420.

Lin, F., Chen, S., Que, Z., Wang, L. and Liu, X. (2007). The blast resistance gene Pi37 encodes an NBS-LRR protein and is a member of a resistance gene cluster on rice chromosome 1. Genetics 177, 1871-1880.

Liu, X., Yang, Q., Lin, F., Hua, L., Wang, C., Wang, L. and Pan, Q. (2007). Identification and fine mapping of Pi39(t), a major gene conferring the broad-spectrum resistance to *Magnaporthe oryzae*. Mol. Gen. Genomics 278, 403-410.

Lu, G., Jantasuriyarat, C., Zhou, B., and Wang, G.L. (2004). Isolation and characterization of novel defense response genes involved in compatible and incompatible interactions between rice and *Magnaporthe grisea*. Theor. Appl. Genet. 108, 525-534.

Marioni, J.C., Mason, C.E., Mane, S.M., Stephens, and M., Gilad, Y. (2008). RNA-seq: an assessment of technical reproducibility and comparison with gene expression arrays. Genome Res. 18, 1509-1517.

Matsumura, H., Reich, S., Ito, A., Saitoh, H., Kamoun, S., Winter, P., Kahl, G., Reuter, M., Kruger, D.H. and Terauchi, R.O. (2003). Gene expression analysis of plant host-pathogen interactions by SuperSAGE. Proc. Natl. Acad. Sci. USA. 100(26), 15718-15723.

Matsushima, N. and Miyashita H. (2012). Leucine-Rich Repeat (LRR) domains containing intervening motifs in plants. Biomolecules 2, 288-311. 10.3390/biom2020288

Mentlak, T.A., Kombrink, A., Shinya, T., Ryder, L.S., Otomo, I., Saitoh, H., Terauchi, R., Nishizawa, Y., Shibuya, N., Thomma, B.P., and Talbot, N.J. (2012). Effector-mediated suppression of chitin-triggered immunity by *Magnaporthe oryzae* is necessary for rice blast disease. Plant Cell 24, 322-335.

Miki, S., Matsui, K., Kito, H., Otsuka, K., Ashizawa, T., Yasuda, N., Fukiya, S., Sato, J., Hirayae, K., Fujita, Y., et al. (2009). Molecular cloning and characterization of the AVR-Pia locus from a Japanese field isolate of *Magnaporthe oryzae*. Mol. Plant Pathol. 10, 361-374.

Monosi, B., Wisser, R.J., Pennill, L. and Hulbert, S.H. (2004). Full-genome analysis of resistance gene homologues in rice. Theor. Appl. Genet. 109, 1434-1447. 10.1007/s00122-004-1758-x

Mosquera, G., Giraldo, M. C., Khang, C. H., Coughlan, S., and Valent, B. (2009). Interaction transcriptome analysis identifies *Magnaporthe oryzae* BAS1-4 as biotrophy-associated secreted proteins in rice blast disease. Plant Cell 21, 1273-1290.

Nobuta, K., Venu, R. C., Lu, C., Beló, A., Vemaraju, K., Kulkarni, K., Wang, W., Pillay, M., Green, P. J., Wang, G-L., and Meyers, B. C. (2007). An expression atlas of rice mRNAs and small RNAs. Nat. Biotechnol. 25, 473-477.

Numa, H., Nishimura, M., Tanaka, T., Kanamori, H., Yang, C.C., Matsumoto, T., Nagamura, Y., and Itoh, T. (2009). Genome-wide validation of *Magnaporthe grisea* gene structures based on transcription evidence. FEBS Letters, 583, 797-800.

Oh, Y., Donofrio, N., Pan, H., Coughlan, S., Brown, D.E., Meng, S., Mitchell, T., and Dean, R.A. (2008). Transcriptome analysis reveals new insight into appressorium and function in the rice blast fungus *Magnaporthe oryzae*. Genome Biol. 9, R85.

Okuyama, Y., Kanzaki, H., Abe, A., Yoshida, K., Tamiru, M., Saitoh, H., Fujibe, T., Matsumura, H., Shenton, M., Galam, D. C., Undan, J., Ito, A., Sone, T. and Terauchi, R. (2011) A multifaceted genomics approach allows the isolation of the rice Pia blast resistance gene consisting of two adjacent NBS-LRR protein genes. Plant J. 66, 467-479.

Orbach, M.J., Farrall, L., Sweigard, J.A., Chumley, F.G. and Valent, B. (2000). A telomeric avirulence gene determines efficacy for the rice blast resistance gene Pi-ta. Plant Cell 12, 2019-2032.

Osmani, S.A., Engle, D.B., Doonan, J.H., and Morris, N.R. (1988). Spindle formation and chromatin condensation in cells blocked at interphase by mutation of a negative cell cycle control gene. Cell 52, 241-251.

Ou, S.H. (1985). Rice diseases. Kew, Surrey, Commonwealth Mycological Institute.

Park, C., Chen, S., Shirsekar, G., Zhou, B., Khang, C., et al. (2012). The *Magnaporthe oryzae* effector AvrPiz-t targets the ring E3 ubiquitin ligase apip6 to suppress pathogen-associated molecular pattern-triggered immunity in rice. Plant Cell 24, 4748-4762.

Park, G., Xue, C., Zhao, X., Kim, Y., Orbach, M., and Xu, J. R. (2006). Multiple upstream signals converge on the adaptor protein Mst50 in *Magnaporthe grisea*. Plant Cell 18, 2822-2835.

Peyyala, R., and Farman, M.L. (2006). *Magnaporthe oryzae* isolates causing gray leaf spot of perennial ryegrass possess a functional copy of the AVR1-CO39 avirulence gene. Mol. Plant Pathol.. 7, 157-165.

Qu, S., Liu, G., Zhou, B., Bellizzi, M., Zeng, L., Dai, L., Han, B. and Wang, G. (2006). The broad-spectrum blast resistance gene Pi9 encodes a nucleotide-binding site-leucine-rich repeat protein and is a member of a multigene family in rice. Genetics 172(3), 1901-1914.

Ramos-Pamplona, M. and Naqvi, N.I. (2006). Host invasion during rice-blast disease requires carnitine-dependent transport of peroxisomal acetyl-CoA. Mol. Microbiol. 61, 61-75.

Rao, Z.M., Dong, H.T., Zhuang, J.Y., Chai, R.Y., Fan, Y.Y., Li, D.B. and Zheng, K.L. (2002). Analysis of gene expression profiles during host-*Magnaporthe grisea* interactions in a pair of near isogenic lines of rice. Yi Chuan Xue Bao. 29, 887-893.

Rauyaree, P., Choi, W., Fang, E., Blackmon, B., and Dean, R.A. (2001). Genes expressed during early stages of rice infection with the rice blast fungus *Magnaporthe grisea*. Mol. Plant Pathol. 2, 347-354.

Rossman, A.Y., Howard, R.J. and Valent, B. (1990) Pyricularia oryzae the correct name for the rice blast fungus. Mycology 82, 545-557.

Saitoh, H., Fujisawa, S., Mitsuoka, C., Ito, A., Hirabuchi, A., et al. (2012). Large-scale gene disruption in *Magnaporthe oryzae* identifies MC69, a secreted protein required for infection by monocot and dicot fungal pathogens. PLoS Pathogen. 8, e1002711.

Sesma, A. and Osbourn, A.E. (2004). The rice leaf blast pathogen undergoes developmental processes typical of root infecting fungi. Nature 431, 582-586.

Shang, J., Tao, Y., Chen, X., Zou, Y., Lei, C., Wang, J., Li, X., Zhao, X., Zhang, M., Lu, Z., Xu, J., Cheng, Z., Wan, J. and Zhu, L. (2009). Identi-fication of a new rice blast Resistance gene, Pid3, by genome wide comparison of paired nucleotide-binding site-leucine-rich repeat genes and their pseudogene alleles between the two sequenced rice genomes. Genetics 182, 1303-1311.

Shang, J.,Tao, Y., Chen, X., Zou,Y., Lei, C., Wang, J., Li, X., Zhao, X., Zhang, M., Lu, Z., Xu, J., Cheng, Z., Wan, J. and Zhu, L. (2009). Identification of a new rice blast Resistance gene, Pid3, by genome wide comparison of paired nucleotide-binding site-leucine-rich repeat genes and their pseudogene alleles between the two sequenced rice genomes. Genetics 182, 1303-1311.

Sharma, T.R., Das, A., Thakur, S., and Jalali, S. (2014). Recent understanding on structure, function and evolution of plant disease resistance genes. Proc. Indian Nat. Sci. Acad. 1, 83-93.

Sharma, T.R., Madhav, M.S., Singh, B.K., Shanker, P., Jana, T. K., Dalal, V., Pandit, A., Singh, A., Gaikwad, K., Upreti, H.C. and Singh, N.K. (2005). High-resolution mapping, cloning and molecular characterization of the Pi-kh gene of rice, which confers resistance to *Magnaporthe grisea*. Mol. Gen. Genomics 274, 569-578.

Sharma, T. R., Rai, A.K., Gupta, S. K. and Singh, N. K. (2010). Broad spectrum blast resistance gene *Pi-kh* designated as *Pi-54*. J. Plant Biochem. Biotechnol. 19, 987-998.

Sharma, T.R., Rai, A.K., Gupta, S.K., Vijayan, J., Devanna, B.N., and Ray, S. (2012). Rice blast management through host-plant resistance, retrospect and prospects. Agri. Res. 1, 37-52. doi, 10.1007/s40003-011-0003-5

Sharma, T. R., Shanker, P., Singh, B. K., Jana, T. K., Madhav, M. S., Gaikwad, G., Singh, N. K., Plaha, P. and Rathour, R. (2005). Molecular mapping of rice blast resistance gene Pi-kh in rice variety Tetep. J. Plant Biochem. Biotechnol. 14, 127-133.

Singh, N. K., Dalal, V., Batra, K., Singh, B. K., Chitra, G., Singh, A., Ghazi, I. A., Yadav, M., Pandit, A., Dixit, R., Singh, P. K., Singh, H., Koundal, K. R., Gaikwad, K., Mohapatra, T. and Sharma, T. R. (2007). Single-copy genes define a conserved order between rice and wheat for understanding differences caused by duplication, deletion, and transposition of genes. Funct. Integr. Genomics 7, 17-35.

Singh, P.K., Thakur, S., Rathour, R., Variar, M., Prashanthi, S.K., Singh, A.K., Singh, U.D., Sharma, V., Singh, N.K., and Sharma, T.R. (2014). Transposon-based high sequence diversity in Avr-Pita alleles increases the potential for pathogenicity of *Magnaporthe oryzae* populations. Funct. Integr. Genomics, 2, 419-29.

Singh, S., Chand, S., Singh, N.K., Sharma, T.R. (2015) Genome-wide distribution, organisation and functional characterization of disease resistance and defence response genes across rice species. PLoS One. 22; 10(4):e0125964. doi: 10.1371/journal.pone.0125964.

Skamnioti, P., and Gurr, S.J. (2009). Against the grain: safeguarding rice from rice blast disease. Trends Biotechnol. 27, 141-150.

Slot, K.A.E. and Knogge, W. (2002). A dual role of microbial pathogen-derived proteins in plant disease and resistance. CRC Crit. Rev. Plant Sci. 21, 229-271.

Soanes, D.M., Chakrabarti, A., aszkiewicz, K.H., Dawe, A.L. and Talbot, N.J. (2012). Genome-wide transcriptional profiling of appressorium development by the rice blast fungus *Magnaporthe oryzae*. PLoS Pathog. 8, e1002514.

Soundararajan, S., Jedd, G., Li, X.L., Ramos-Pamplona, M., Chua, N.H., and Naqvi, N.I. (2004). Woronin body function in *Magnaporthe grisea* is essential for efficient pathogenesis and for survival during nitrogen starvation stress. Plant Cell, 16, 1564-1574.

Sun, C.B., Suresh, A., Deng, Y.Z. and Naqvi, N.I. (2006). A multidrug resistance transporter in Magnaporthe is required for host penetration and for survival during oxidative stress. Plant Cell, 18, 3686-3705.

Sweigard, J.A., Carroll, A.M., Kang, S., Farrall, L., Chumley, F.G., and Valent, B. (1995). Identification, cloning, and characterization of PWL2, a gene for host species specificity in the rice blast fungus. Plant Cell, 7, 1221-1233.

Sweigard, J.A., Chumley, F.G., and Valent, B. (1992). Cloning and analysis of Cutl, a cutinase gene from *Magnaporthe grisea*. Mol. Gen. Genet. 232, 174-182.

Takahasi, A., Hayashi, N., Miyao, A. and Hirochika, H. (2010). Unique features of the rice blast resistance Pish locus revealed by large scale retrotransposon tagging. BMC Plant Biol. 10, 175-189.

Takano, Y., Choi, W., Mitchell, T.K., Okuno, T., and Dean, R.A. (2003). Large scale parallel analysis of gene expression during infection-related morphogenesis of *Magnaporthe grisea*. Mol. Plant Pathol. 4, 337-346.

Talbot, N.J. (2003) On the Trail of a Cereal Killer, Exploring the biology of *Magnaporthe grisea*. Ann. Rev. Microbiol. 57(1), 177-202.

Talbot, N.J. (1995). Having a blast: exploring the pathogenicity of *Magnaporthe grisea*. Trends Microbiol. 3, 9-16.

Talbot, N.J., Ebbole, D.J., and Hamer, J.E. (1993). Identification and characterization of MPG1, a gene involved in pathogenicity from the rice blast fungus *Magnaporthe grisea*. Plant Cell 5, 1575-1590.

Tang, J., Zhu, X., Wang, Y., Liu, L., Xu, B., Li. F., Fang, J. and Chu, C. (2011). Semi-dominant mutations in the CC-NB-LRR-type R gene, NLS1, lead to constitutive activation of defense responses in rice. The Plant J. 66, 996-1007.

Teng, P.S. (1994). The epidemiological basis for blast management. Wallingford, Oxon, UK, CAB International and IRRI.

Thakur, S., Gupta, Y. K., Singh, P. K., Rathour, R., Variar, M., Prashanthi, S. K., et al. (2013a). Molecular diversity in rice blast resistance gene Pi-ta makes it highly effective against dynamic population of *Magnaporthe oryzae*. Funct. Integr. Genomics 13, 309-322. doi, 10.1007/s10142-013-0325-4

Thakur, S., Singh, P. K., Das, A., Rathour, R., Variar, M., Prashanthi, S. K. et. al (2015). Extensive sequence variation in rice blast resistance gene Pi54 makes it broad spectrum in nature. Front. Plant Sci. 6, 345. doi: 10.3389/fpls.2015.00345

Thakur, S., Singh, P. K., Rathour, R., Variar, M., Prashanthi, S. K., Singh, A. K., et al. (2013b). Positive selection pressure on rice blast resistance allele Pizt makes it divergent in Indian land races. J. Plant Interact. 8, 34-44. doi, 10.1080/17429145.2012.721523

Thakur, S., Singh, P. K., Rathour, R., Variar, M., Prashanthi, S. K., Gopalakrishnan, S., et al. (2014). Genotyping and development of single-nucleotide polymorphism (SNP) markers associated with blast resistance genes in rice using GoldenGate assay. Mol. Breeding 34, 1449-1463. doi, 10.1007/s11032-014-0129-9

Thines, E., Weber, R.W.S. and Talbot, N.J. (2000). MAP kinase and protein kinase Adependent mobilisation of triacylglycerol and glycogen during appressorium turgor generation by *Magnaporthe grisea*. Plant Cell 12, 1703-1718.

Trapnell, C., Pachter, L. and Salzberg, S. L. (2009). TopHat: discovering splice junctions with RNA-Seq. Bioinformatics 25, 1105-1111.

Trapnell, C., Williams, B. A., Pertea, G., Mortazavi, A., Kwan, G., et al. (2010) Transcript assembly and quantification by RNA-Seq reveals unannotated transcripts and isoform switching during cell differentiation. Nat. Biotechnol. 28, 511-515.

Valent, B., and Khang, C.H. (2010). Recent advances in rice blast effector research. Curr. Opin. Plant. Biol. 13, 434-441.

Veneault-Fourrey, C., Barooah, M., Egan, M., Wakley, G., and Talbot, N.J. (2006). Autophagic fungal cell death is necessary for infection by the rice blast fungus. Science 312, 580-583.

Vijayan, J., Jain, S., Jain, N., Devanna, B.N., Rathour, R., Variar, M., Prashanthi, S.K., Singh, A.K., Singh, U.D., Singh, N.K., and Sharma T.R. (2013) Identification of differentially expressed genes in rice during its early phases of interaction with Magnaporthe oryzae. Indian J. Genet., 73, 233-243.

Wang, Z., Gerstein, M. and Snyder, M (2009). RNA-Seq: a revolutionary tool for transcriptomics. Nat. Rev. Genet. 10, 57-63.

Wang, C., Yang, Y., Yuan, X., Xu, Q., Feng, Y., Yu, H., et al. (2014). Genome-wide association study of blast resistance in indica rice. BMC Plant Biol. 14, 6. doi, 10.1186/s12870-014-0311-6

Wang, X., Jia, Y., Shu, Q.Y. and Wu, D. (2008). Haplotype diversity at the Pi-ta locus in cultivated rice and its wild relatives. Phytopathology 98, 1305-1311.

Wang, Y., Kwon, S.J., Wu, J., Choi, J., Lee,Y.-H., Agrawal, G.K., Tamogami, S., Rakwal, R., Park, S.-R., Kim, B.-G., Jung, K.-H., Kang, K.Y., Kim, S.G., and Kim, S.T. (2014). Transcriptome analysis of early responsive genes in rice during *Magnaporthe oryzae* infection. Plant Pathol. J. 30, 1-12.

Wang, Z. X., Yano, M., Yamanouchi, U., Iwamoto, M., Monna, L., Hayasaka, H., Katayose, Y. and Sasaki Takuji. (1999). The Pib gene for rice blast resistance belongs to the nucleotide binding and leucine-rich repeat class of plant disease resistance genes. Plant J. 19, 55-64.

Wang, Z.Y., Soanes, D.M., Kershaw, M.J., and Talbot, N.J. (2007). Functional analysis of lipid metabolism in the rice blast fungus *Magnaporthe grisea* reveals a role for peroxisomal β-oxidation in appressorium-mediated plant infection. Mol. Plant Microbe Interact. 20, 475-491.

Wilson, R.A. and Talbot, N.J. (2009). Under pressure: investigating the biology of plant infection by *Magnaporthe oryzae*. Nat. Rev. Microbiol. 7, 185-195.

Xiong, L., Lee, M.W., Qi, M., and Yang, Y. (2001). Identification of defense-related rice genes by suppression subtractive hybridization and differential screening. Mol. Plant Microbe Interact. 14, 685-692.

Xu, J.R. (2000). Map kinases in fungal pathogens. Fungal Genet. Biol. 31, 137-152.

Xu, J.R., Staiger, C.J., and Hamer, J.E. (1998). Inactivation of the mitogen-activated protein kinase Mps1 from the rice blast fungus prevents penetration of host cells but allows activation of plant defense responses. Proc. Natl. Acad. Sci. USA 95, 12713-12718.

Yokotani, N., Sato, Y., Tanabe, S., Chujo T., Shimizu, T., Okada, K., Yamane, H., Shimono, M., Sugano, S., Takatsuji, H., Kaku, H., Minami, E., and Nishizawa, Y. (2013) WRKY76 is a rice transcriptional repressor playing opposite roles in blast disease resistance and cold stress tolerance. J. Exp. Bot. 64 (16), 5085-97.

Yoshida, K. and Miyashita, N. T. (2009). DNA polymorphism in the blast disease resistance gene Pi-ta of the wild rice Oryza rufipogon and its related species. Genes Genet. Syst. 84, 121-136.

Yoshida, K., Saitoh, H., Fujisawa, S., Kanzaki, H., Matsumura, H., et al. (2009). Association genetics reveals three novel avirulence genes from the rice blast fungal pathogen *Magnaporthe oryzae*. Plant Cell 21, 1573-1591.

Yuan, B., Zhai, C., Wang, W., Zeng, X., Xu, X. et al. (2011). The Pik-p resistance to *Magnaporthe oryzae* in rice is mediated by a pair of closely linked CC-NBS-LRR genes. Theor. Appl. Genet. 1225, 1017-1028.

Zhai, C., Lin, F., Dong, Z., He, X., Yuan, B., Zeng, X., Wang, L. and Pan, Q. (2011). The isolation and characterization of Pik, a rice blast resistance gene which emerged after rice domestication. New Phytol. 189, 321-334.

Zheng, X.W., Chen, X.W., Zhang, X.H., Lin, Z. Z., Shang, J. J., Xu, J. C., Zhai, W. X. and Zhu, L. H. (2004). Isolation and identification of a gene in response to rice blast disease in rice. Plant Mol. Biol. 54(11), 99- 109.

Zhou, B., Qu, S., Liu, G., Dolan, M. and Sakai, H. (2006). The eight amino-acid differences within three leucine-rich repeats between Pi2 and Piz-t resistance proteins determine the resistance specificity to *Magnaporthe grisea*. Mol. Plant Microbe Interact. 19, 1216-1228.

Other Publications of Interest

Xu, J.-P. (2014). Next-generation Sequencing: Current Technologies and Applications. Caister Academic Press, U.K. ISBN: 978-1-908230-33-1

He, Z. (2014). Microarrays: Current Technology, Innovations and Applications. Caister Academic Press, U.K. ISBN: 978-1-908230-49-2

Poptsova, M.S. (2014). Genome Analysis: Current Procedures and Applications. Caister Academic Press, U.K. ISBN: 978-1-908230-29-4

Fuentes, M. and LaBaer, J. (2014). Proteomics: Targeted Technology, Innovations and Applications. Caister Academic Press, U.K. ISBN: 978-1-908230-46-1

Murillo, J., Vinatzer, B.A., Jackson, R.W. and Arnold, D.L. (2015). Bacteria-Plant Interactions: Advanced Research and Future Trends. Caister Academic Press, U.K. ISBN: 978-1-908230-58-4

Caranta, C., Aranda, M.A., Tepfer, M. and Lopez-Moya, J.J. (2011). Recent Advances in Plant Virology. Caister Academic Press, U.K. ISBN: 978-1-904455-75-2

Nannipieri, P., Pietramellara, G. and Renella, G. (2014). Omics in Soil Science. Caister Academic Press, U.K. ISBN: 978-1-908230-32-4

Herold, K.E. and Rasooly, A. (2009). Lab-on-a-Chip Technology: Fabrication and Microfluidics. Caister Academic Press, U.K. ISBN: 978-1-904455-46-2

Transcriptomic Analyses on the Role of Nitric Oxide in Plant Disease Resistance

Capilla Mata-Pérez[1], Juan C. Begara-Morales[1], Francisco Luque[2], María N. Padilla[1], Jaime Jiménez-Ruiz[1], Beatriz Sánchez-Calvo[1], Jesús Fierro-Risco[1] and Juan B. Barroso[1,2]*

[1]Group of Biochemistry and Cell Signaling in Nitric Oxide, Department of Experimental Biology, Faculty of Experimental Sciences, Campus Universitario "Las Lagunillas" s/n, University of Jaén, E-23071, Jaén, Spain
[2]Center for Advanced Studies in Olive Grove and Olive Oils, Department of Experimental Biology, Campus Universitario "Las Lagunillas" s/n, University of Jaén, E-23071, Jaén, Spain
*Corresponding author. Email: jbarroso@ujaen.es
Tel. 34-953-212764; Fax. 34-953-211874.

Abstract
Nitric oxide (NO) is a gaseous molecule having key roles in many physiological processes such as germination, growth, development and senescence. It has been also shown the important role of NO as a signaling molecule in the response to a wide variety of stress situations, including both biotic and abiotic stress conditions. In the last few years, a growing number of studies have focused on NO-cell targets by several approaches such as transcriptomic and proteomic analyses. This review is centered on offering an update about the principal medium- and large-scale transcriptomic analyses performed with several NO donors including microarray, cDNA-amplification fragment length polymorphism (AFLP) and high throughput sequencing (RNA-seq technology) approaches mainly focused on the role of this reactive nitrogen species in relation to plant disease resistance. Different putative NO-responsive genes have been identified in different plant tissues and plant species by application of several NO donors suggesting the implication of NO-responsive genes with plant adaptive responses to biotic stress processes. Finally, it is also provided an overview about common transcription factor-binding sites of NO-responsive genes and the need to further analyze the different NO-targets by other omics studies.

Introduction
Nitric oxide (NO) is a small and gaseous molecule reportedly involved in a wide array of functions in vegetal systems. In this sense, NO has been shown to be related to defense response (Delledonne et al., 1998; Durner et al., 1998), growth and development (Beligni and Lamattina, 2000; Pagnussat et al., 2002), senescence (Begara-Morales et al., 2013; Beligni and Lamattina, 2001) and in the response to abiotic (Begara-Morales et al., 2014a; Corpas and Barroso, 2013; Corpas et al., 2008; Chaki et al., 2011a; Chaki et al., 2011b; Valderrama et al., 2007) and biotic stress conditions (Feechan et al., 2005; Lindermayr et al., 2005). NO and NO-derived molecules are referred as reactive nitrogen species (RNS) and among them are included peroxynitrite ($ONOO^-$) or nitrogen dioxide (NO_2^-). These RNS are involved in different stressful situations since they can mediate several post-translational modifications through mechanisms such as *S*-nitrosylation and protein tyrosine nitration (Begara-Morales et al., 2013; Corpas and Barroso, 2013; Chaki et al., 2009; Lozano-Juste et al., 2011; Valderrama et al., 2007). Furthermore, it has been demonstrated that NO is important in plant-pathogen interaction being involved in disease resistance (Huang et al., 2002) and having shown that this RNS is quickly produced after pathogen attack and being involved in cell signaling during hypersensitive response (HR) in plants (Delledonne, 2005).

Otherwise, pathogen recognition triggers a rapid production of reactive oxygen species (ROS) such as hydrogen peroxide (H_2O_2) or superoxide (O_2^-) (Lamb and Dixon, 1997). This oxidative burst promotes several defense mechanisms like cross-linking of the cell wall (Bradley et al., 1992) and the expression of genes involved in cellular protection and defense (Levine et al., 1994). In this regard, it has been shown that ROS are necessary for the initiation of hypersensitive response (HR) and thereby limiting damage of the infection site (Lamb and Dixon, 1997). Furthermore, it has been demonstrated that a synergistic production of ROS and NO potentiates the induction of hypersensitive cell death in soybean cells therefore showing that NO plays a key role in disease resistance in plants (Delledonne et al., 1998; Delledonne et al., 2001). Accordingly, activation of immune responses in plants constitutes an imbricated response associated with a parallel burst of both ROS and NO.

In recent years, significant advances have been achieved in understanding the molecular basis of the role of nitric oxide in plant-pathogen interactions and in disease resistance. For this proposal, it has been carried out several transcriptomic studies by medium- and large-scale approaches including cDNA-amplification fragment length polymorphism (AFLP) (Ferrarini et al., 2008; Huang et al., 2002; Polverari et al., 2003) and microarray methods (Ahlfors et al., 2009; Badri et al., 2008; Parani et al., 2004; Zeidler et al., 2004) with gaseous NO or different NO-donors. In the last few years, new high-throughput sequencing methods, called massively parallel sequencing or RNA-seq, have emerged as a useful tool that could replace and improve upon existing methods given their advantages as compared to array-based methods (Van Verk et al., 2013; Wilhelm and Landry, 2009). In this regard, it has recently analyzed the gene expression profile by RNA-seq technology of *Arabidopsis thaliana* plants incubated with S-nitrosoglutathione -GSNO- (Begara-Morales et al., 2014b) which is considered a biological reservoir of NO both in animal (Zhang and Hogg, 2004) and vegetal systems (Wang et al., 2006).

In this review, we summarize recent results obtained by different transcriptomic approaches to update the understanding of NO-responsive genes involved in plant disease resistance. Furthermore, we propose other "omic" experiments to complete and well-define the results obtained by these transcriptional studies to improve our knowledge in the imbricated network in which NO is involved.

Nitric oxide in physiological and abiotic stress processes

Several studies about the role of nitric oxide in plant physiology have been carried out. In this regard, it is well known that NO has an important role in plant growth and development (Corpas et al., 2006; Ribeiro Jr et al., 1999) due to it is involved in different physiological processes such as germination and alleviation of seed dormancy (Beligni and Lamattina, 2000; Bethke et al., 2007; Bethke et al., 2006; Libourel et al., 2006; Zheng et al., 2009). Furthermore, this RNS is implicated in the regulation of plant metabolism and senescence (Begara-Morales et al., 2013; Guo and Crawford, 2005; Leshem et al., 1998), promoting cell death (Pedroso and Durzan, 2000), regulating stomatal movement (Garcia-Mata and Lamattina, 2007) and photosynthesis (Takahashi and Yamasaki, 2002), gravitropism (Hu et al., 2005), regulation of flowering time (He et al., 2004), flower development and apical dominance (Kwon et al., 2012; Lee et al., 2008) and orientation of the pollen tube growth (Prado et al., 2004).

To date, few transcriptional studies have been focused on the role of NO in several physiological aspects. In this sense, the application of different NO-donors like sodium nitroprusside (SNP) in *Arabidopsis thaliana* roots increased the exudation of phytochemicals (Badri et al., 2008). With this respect, several transport systems likely to be involved in the root exudation including MATE, ABC, MFS and other transporters changed their expression in response to NO-treatment. Furthermore and very recently, Monzón et al. (2014) (Monzón et al., 2014) have been determined by microarray technology that NO is required for regulating root architecture and lignin composition in sunflower. In this study, the application of cPTIO (2-(4-carboxyphenyl)-4,4,5,5-tretramethylimidazoline-1-oxyl-3-oxide), a NO scavenger, showed that a fine adjustment of NO levels in the roots would be critical to regulate lignin composition and therefore branching, root architecture and plant growth. Besides, the role of NO in plant physiology has been also related to plant hormones such as jasmonates or auxins showing an interaction at multiple and diversified levels (Huang et al., 2004; Sanz et al., 2014). In this regard, scarce transcriptional analyses with NO donors have been performed in relation to plant hormone study. One of them consists on a microarray study in *Arabidopsis thaliana* plants with salicylic acid- (SA), methyl jasmonate- (MeJA) and NO-treatments. This research showed a limited overlap at different levels such as in the phenylpropanoid biosynthesis or glucosinolate pathways, being the specific genes different in each treatment (Badri et al., 2008). In sum, there is limited information focused on the role of NO in plant physiology by transcriptional approaches, which could be really interesting to understand the functioning of NO in vegetal systems.

Otherwise, most of the transcriptional studies with NO donors have focused on the involvement of this RNS in relation to biotic stress (Ahlfors et al., 2009; Badri et al., 2008; Ferrarini et al., 2008; Polverari et al., 2003). However, inside these studies there are different descriptions about the participation of NO in different target genes involved in several abiotic stress conditions. In this regard, Parani et al. (2004) (Parani et al., 2004) observed the induction on the transcript levels of several dehydration responsive element-binding (DREB1 and DREB2) and late embryogenesis abundant (LEA) proteins by a NO donor as is SNP. The increase on the expression of these genes is related to the tolerance to drought, cold and salinity which could be associated to the drought tolerance observed after NO treatment in wheat (García-Mata and Lamattina, 2001).

Analysis of transcriptomic data about NO-responsive genes related to plant disease resistance

At present, there are scarce transcriptomic studies evidencing the role of NO in plant-pathogen interaction. In this respect, it has been recently performed a differential expression analysis by RNA-seq technology during nodulation of *Medicago truncatula* plants in which NO was removed by incubation with cPTIO (Boscari et al., 2013). This NO depletion provoked a down-regulation in some genes related to nodulation process such as HAP3-like gene, suggesting these results that NO plays a key role in repressing the defense reaction in symbiotic conditions, thereby favoring establishment of the beneficial plant-microbe interaction and differing this action from the signaling function of NO in pathogenic interactions. Moreover, it has been suggested in a recent RNA-seq analysis that in susceptible mutant to *Fusarium* head blight (FHB) disease, NAUH117, both ROS and NO were accumulated while not in resistant mutant (Wangshuibai) variety (Xiao et al., 2013). Authors suggest that this behavior may contribute to the necrotrophic phase during *Fusarium* infection through PCD. Thus, the repression of ROS and NO bursts would lead to enhanced FHB resistance in wheat.

Besides these results, most transcriptional studies that have identified NO-modulated genes were carried out by using pharmacological approaches through the employment of several NO donors like SNP, NOR-3 or GSNO among others. These studies have used medium- and large-scale transcriptional analyses using microarray, cDNA-AFLP and RNA-seq methodologies. First studies were conducted by cDNA microarray containing 200 defense-related genes and 50 genes related to primary metabolism (Huang et al., 2002). Treatment with NOR-3, a NO donor, in *Arabidopsis thaliana* cell suspension cultures highly induced the expression of alternative oxidase 1a (AOX1a), several pathogenesis-related (PR) proteins and a set of peroxidases and glutathione *S*-transferases. These authors suggest that the induction of AOX1a is SA-independent and that this enzyme might fulfill an important function for cellular homeostasis under NO stress, thereby counteracting the toxicity of NO. Also using microarray technology with 700 defense-related genes, Zeidler et al. (2004) (Zeidler et al., 2004) observed that *Arabidopsis thaliana* leaves react to different lipopolysaccharides (LPS) with a rapid burst of NO and a concomitant activation of defense genes. Among them, they found that LPS induces a broad of defense or stress-associated genes including glutathione *S*-transferases, cytochrome P450 and different PR proteins, altogether contributing to the activation of plant defense responses. Besides these studies, several transcript profiling genome-scale studies were carried out to achieve a global vision of NO-involvement. In this regard, Parani et al. (2004) (Parani et al., 2004) conducted a genome array covering over 24,000 genes in *Arabidopsis thaliana* plants irrigated with 0.1 mM or 1 mM of SNP. This study showed that NO-responsive genes were related to plant defense response, protection against oxidative stress, iron homeostasis, signal transduction and gene expression control through transcription factors such as WRKY members. Moreover, another whole genome microarray with 26,090 genes was used to study and identify root targets modulated by the NO donor SNP in 18-day-old *Arabidopsis thaliana* plants (Badri et al., 2008). In addition to SNP-treatment, it was also performed the incubation with different phytohormones such as methyl jasmonate (MeJA) and salicylic acid (SA) which may control specific signaling pathways different from NO-treatment. It was also characteristic that the number of repressed genes was higher than those up-regulated genes by NO and, finally, only a member of NAC transcription factor family, NAC3, seems to be regulated by the three signaling compounds analyzed. Following with whole-genome microarray analysis, an SNP-treatment of 21-day-old *Arabidopsis thaliana* plants provoked the modulation of 614 genes mostly related to phytohormone-treatments and to the response to biotic or abiotic stresses (Ahlfors et al., 2009). Remarkably, plants sprayed with SNP and ozone (O_3) at the same time promoted the expression of a similar subset of genes indicating that both molecules act synergistically. Finally, one of the main findings was that NO appears to attenuate the induction of SA biosynthesis and signaling genes, thereby suggesting that NO could play a role in the initiation of cell death. On the other hand and in addition to microarray studies, changes in NO-responsive genes were also analyzed by cDNA-AFLP transcriptomic profiling. In this regard, it was observed in *Arabidopsis thaliana* leaves infiltrated with SNP the modulation of 120 genes mainly related to signal transduction, defense response and cell death, removal or production of toxic oxygen species, photosynthesis and energy transfer, cellular trafficking and in basic metabolic pathways (Polverari et al., 2003). Moreover, roots from 4-week-old *Medicago truncatula* plants were treated with SNP or GSNO to assess the effect of different reactive nitrogen molecules (Ferrarini et al., 2008) and they were subjected to cDNA-AFLP analysis with a coverage of 55 % of the *Medicago truncatula* transcriptome. Despite SNP and GSNO are NO donors, only 11 % of genes in leaves and 1.6 % in roots exerted the same modulatory effect, suggesting a modulation of gene expression strictly dependent on the form of NO and on the cellular response in different tissues. On the basis of these results, 999 modulated-genes were used to construct a microarray to monitor the expression of NO-responsive genes in this species during an incompatible interaction and during a symbiotic interaction, being 275 and 290 NO-modulated genes, respectively. In the incompatible interaction, several resistance-related genes and those involved in reactive oxygen species generation or in plant tolerance to oxidative stresses, biosynthesis of JA and defense signaling, proteasome degradation pathway and signal transduction cascades were NO-modulated. On the other hand and related to symbiotic interactions, several genes involved in flavonoid biosynthesis, proteasome degradation pathway, redox signaling and reprogramming of plant cell primary metabolism were processes regulated by NO during this symbiotic interaction. Therefore, this study highlights that NO-responsive genes behave differently depending on the plant organ and on the type of interaction and suggest the role of NO in the HR during pathogenic interaction and might regulate important processes of symbiotic nodule development and functioning. Finally, due to the large number of advantages of massively parallel sequencing or RNA-seq, it has been recently carried out a growing number of NO-related studies using this approach. In this regard, Begara-Morales et al. (2014) (Begara-Morales et al., 2014b) have demonstrated the participation of *S*-nitrosoglutathione (GSNO), which is consider as a biological reservoir of NO, in the response to different stress conditions by an RNA-seq analysis. This methodology has a large number of advantages due to it is not necessary to know the transcribed region previously and permit gene expression quantification in a single experiment among other important advantages (Wilhelm and Landry, 2009). Therefore, GSNO-treatment provoked the induction of different heat shock proteins and wound-responsive proteins in *Arabidopsis thaliana* roots probably involved in the response to a wide array of abiotic stress conditions. Furthermore, due to the synergistic production of reactive oxygen species (ROS) during different abiotic stress situations (Apel and Hirt, 2004; Miller et al., 2008), several oxidative stress-related genes were induced by GSNO-treatment such as mitochondria alternative oxidase 3 and 1a, several peroxidases and members of glutathione transferase proteins. Remarkably, the high induction on the expression of methionine sulfoxide reductase b7 gene supports the protection against oxidative damage from RNS by GSNO as it has been shown in other systems like *Escherichia coli* and *Mycobacterium tuberculosis*. In this sense, it is important to note that although most of GSNO-responsive genes have been previously described, they have identified target genes of GSNO such as MSRB7 gene, which had not been previously detected by array-based methods, supporting the advantages of using massively parallel sequencing or RNA-seq. Finally, an important result obtained from this study was the differential gene expression between leaves and roots, suggesting an organ-specific modulation of gene expression by NO. Consequently, this is the first transcriptomic study with a biological NO donor as is

GSNO by a high throughput technology such as RNA-seq methodology. Moreover, Zeng et al. (2014) (Zeng et al., 2014) have analyzed the transcriptional involvement of NO in birch cells (*Betula platyphylla*) by RNA-seq technology. In this regard, NO-treatment identified 403 up- and 971 down-regulated genes after 12 h SNP-treatment which was then confirmed by qRT-PCR. The results of this transcriptome analysis showed that NO is involved in several processes such as protection against ROS by the regulation of key enzymes such as ascorbate peroxidase or different glutathione *S*-transferases, involvement in carbohydrate metabolism and cell wall biosynthesis, terpenoid biosynthesis and growth regulation. Based on these results, authors indicated that birch cell exposure to NO has a significant effect on many key plant processes as it has been previously described for other plant systems.

Promoter analysis of NO-responsive genes

Recent advances had been done about how NO might regulate gene expression. In this sense, several studies have investigated promotor regions of NO-responsive genes. One of these analyses was focused on 28 common up-regulated genes in different experiments with several NO donors like NOR-3 and gaseous NO (Palmieri et al., 2010). In this regard, eight families of transcription factor binding sites (TFBS) occurred at least 15 % more often in the promotor regions of NO-target genes. Among them, authors found the basic region/leucine zipper motif (bZIP) involved in the regulation of many different physiological processes such as biotic and abiotic stress signaling or seed maturation, WRKY family which are members also involved in pathogen defense, senescence and trichome development and MYCL (myelocytomatosis viral oncogene homolog L) transcription factors. These findings suggest an involvement of NO-responsive genes in different processes such as pathways related to the response to biotic or abiotic stress situations and with several physiological processes. Interestingly, NO-treatment promotes the up-regulation of different JA-related genes like 12-oxophytodienoate reductases (OPR1, OPR2 and OPR3) and two lipoxygenases (LOX3 and a putative lipoxygenase protein) suggesting a close connection between NO and JA-associated processes. Although these results should be experimentally validated, they may be a first step to understand the regulatory networks in which NO is involved. Furthermore, Begara-Morales et al. (2014) (Begara-Morales et al., 2014b) have recently observed two core oligomers (AATTAT and AAAACA) found on the promoters of GSNO-inducible genes. Both have been described as GT-1 factor-binding sites in *Phaseolus vulgaris* bean embryo which are located in the upstream promoter region of light-responsive genes. Among the GSNO-inducible promoters and related to AAAACA core oligomer, was found JAZ10 which is a negative regulator of JA signaling whilst, in relation to AATTAT core oligomer, it was related to a DNA-binding transcription factor and with LSU3, a receptor-like kinase. In sum, the most remarkable fact is that a greater part of GSNO-induced genes share these elements on the promoters and therefore this could be a good starting point to understand or to extend our knowledge about GSNO as a signaling molecule.

Concluding remarks

The literature contains numerous studies reporting the effect of exogenously applied NO on plant processes. However, given the diverse nature of NO-functions, it is more informative to perform large-scale gene expression studies in order to understand the signaling networks in which NO is involved. The development of medium- and large-scale transcriptional analysis from a part or whole-genome such as microarray, cDNA-AFLP or RNA-seq, it has become a very useful tool for the study of the role of NO in plant physiology. However, within the existing transcriptional analyses is sometimes difficult to make an efficient comparison due to the different NO donors employed (SNP, SNAP, GSNO, gaseous NO, NOR-3...) and the different model systems studied (*Arabidopsis thaliana*, *Medicago truncatula*, *Glycine max*, *Betula platyphyilla*...) at distinct developmental stages (cell suspension cultures, seedlings, plantlets, leaves, roots...). In this sense and based on the comparison made by Besson-Bard et al. (2009) (Besson-Bard et al., 2009), NO-responsive genes are usually involved in the plant responses to biotic, abiotic stimuli and stresses. Among these genes, they found partially in common several enzymes involved in detoxification of ROS such as different peroxidases and glutathione *S*-transferases, various genes involved in mitochondrial energy linked-processes, tryptophan synthesis and other genes involved in defense response. These data strengthen the emerging picture that NO functions as a signal involved in the plant adaptive response to plant disease resistance. In spite of the advantages of transcriptomic studies, it is important to bear in mind the need to complement them with other "omic" studies like proteomics. In this regard, it is sometimes slightly comparable the induction of gene expression with the induction of protein expression. Furthermore, proteins are really susceptible to NO-post translational modifications (NO-PTM) such as tyrosine nitration or *S*-nitrosylation which could produce conformational changes and therefore affecting the protein activity. For this reason, with the aid of novel techniques such as TMT (tandem mass tags) isobaric labeling coupled to mass spectrometry is possible to identify and quantify protein expression in different conditions (Kozuka-Hata et al., 2013). Among the advantages of this approach are included an increased multiplex relative quantitation, increased sample throughput and fewer missing quantitative values among samples with a higher identification rates which could be very helpful to highlight the role of NO in several aspects like the involvement in different post-translational modifications. Moreover, it is also important to note that transcriptomic studies do not consider three dimensional (3D) structures of proteins which could be a key feature in NO-proteomic studies. In this regard, Chaki et al. (2014) (Chaki et al., 2014) have compared the prediction of candidate proteins to *S*-nitrosylation by several approaches including biotin switch, mass spectrometry and different computer-assisted methods. This study showed that computational analyses are a useful tool for this prediction; however, further development, such as including the three dimensional structure of proteins in such analyses, would improve the identification of *S*-nitrosylation sites. For this reason, the combination of computational prediction and experimental verification represents a good approach to better understand the molecular mechanisms and the regulatory functions in which NO is involved.

Acknowledgments
CMP would like to thank the University of Jaén for funding the Ph.D. fellowship. This study was supported by an ERDF grant co-financed by the Ministry of Economy and Competitiveness (projects BIO2012-33904 and P10-AGR-6038) and the Junta de Andalucía (group BIO286) in Spain.

References

Ahlfors, R., Brosché, M., Kollist, H., and Kangasjärvi, J. (2009). Nitric oxide modulates ozone-induced cell death, hormone biosynthesis and gene expression in Arabidopsis thaliana. The Plant Journal *58*, 1-12.

Apel, K., and Hirt, H. (2004). Reactive oxygen species: metabolism, oxidative stress, and signal transduction. Annu Rev Plant Biol *55*, 373-399.

Badri, D.V., Loyola-Vargas, V.M., Du, J., Stermitz, F.R., Broeckling, C.D., Iglesias-Andreu, L., and Vivanco, J.M. (2008). Transcriptome analysis of Arabidopsis roots treated with signaling compounds: a focus on signal transduction, metabolic regulation and secretion. New Phytologist *179*, 209-223.

Begara-Morales, J.C., Chaki, M., Sánchez-Calvo, B., Mata-Pérez, C., Leterrier, M., Palma, J.M., Barroso, J.B., and Corpas, F.J. (2013). Protein tyrosine nitration in pea roots during development and senescence. Journal of experimental botany *64*, 1121-1134.

Begara-Morales, J.C., Sánchez-Calvo, B., Chaki, M., Valderrama, R., Mata-Pérez, C., López-Jaramillo, J., Padilla, M.N., Carreras, A., Corpas, F.J., and Barroso, J.B. (2014a). Dual regulation of cytosolic ascorbate peroxidase (APX) by tyrosine nitration and S-nitrosylation. Journal of experimental botany, ert396.

Begara-Morales, J.C., Sánchez-Calvo, B., Luque, F., Leyva-Pérez, M.O., Leterrier, M., Corpas, F.J., and Barroso, J.B. (2014b). Differential transcriptomic analysis by RNA-seq of GSNO-responsive genes between Arabidopsis roots and leaves. Plant and Cell Physiology, pcu044.

Beligni, M.a.V., and Lamattina, L. (2001). Nitric oxide: a non-traditional regulator of plant growth. Trends in plant science *6*, 508-509.

Beligni, M.V., and Lamattina, L. (2000). Nitric oxide stimulates seed germination and de-etiolation, and inhibits hypocotyl elongation, three light-inducible responses in plants. Planta *210*, 215-221.

Besson-Bard, A., Astier, J., Rasul, S., Wawer, I., Dubreuil-Maurizi, C., Jeandroz, S., and Wendehenne, D. (2009). Current view of nitric oxide-responsive genes in plants. Plant science *177*, 302-309.

Bethke, P.C., Libourel, I.G., Aoyama, N., Chung, Y.-Y., Still, D.W., and Jones, R.L. (2007). The Arabidopsis aleurone layer responds to nitric oxide, gibberellin, and abscisic acid and is sufficient and necessary for seed dormancy. Plant Physiology *143*, 1173-1188.

Bethke, P.C., Libourel, I.G., and Jones, R.L. (2006). Nitric oxide reduces seed dormancy in Arabidopsis. Journal of Experimental Botany *57*, 517-526.

Boscari, A., del Giudice, J., Ferrarini, A., Venturini, L., Zaffini, A.-L., Delledonne, M., and Puppo, A. (2013). Expression dynamics of the Medicago truncatula transcriptome during the symbiotic interaction with Sinorhizobium meliloti: which role for nitric oxide? Plant physiology *161*, 425-439.

Bradley, D.J., Kjellbom, P., and Lamb, C.J. (1992). Elicitor- and wound-induced oxidative cross-linking of a proline-rich plant cell wall protein: a novel, rapid defense response. Cell *70*, 21-30.

Corpas, F.J., and Barroso, J.B. (2013). Peroxynitrite (ONOO-) is endogenously produced in arabidopsis peroxisomes and is overproduced under cadmium stress. Annals of botany, mct260.

Corpas, F.J., Barroso, J.B., Carreras, A., Valderrama, R., Palma, J.M., León, A.M., Sandalio, L.M., and del Río, L.A. (2006). Constitutive arginine-dependent nitric oxide synthase activity in different organs of pea seedlings during plant development. Planta *224*, 246-254.

Corpas, F.J., Chaki, M., Fernández-Ocaña, A., Valderrama, R., Palma, J.M., Carreras, A., Begara-Morales, J.C., Airaki, M., del Río, L.A., and Barroso, J.B. (2008). Metabolism of reactive nitrogen species in pea plants under abiotic stress conditions. Plant and cell physiology *49*, 1711-1722.

Chaki, M., Kovacs, I., Spannagl, M., and Lindermayr, C. (2014). Computational prediction of candidate proteins for S-nitrosylation in Arabidopsis thaliana. PloS one *9*, e110232.

Chaki, M., Valderrama, R., Fernández-Ocaña, A.M., Carreras, A., Gómez-Rodríguez, M.V., Pedrajas, J.R., Begara-Morales, J.C., Sánchez-Calvo, B., Luque, F., and Leterrier, M. (2011a). Mechanical wounding induces a nitrosative stress by down-regulation of GSNO reductase and an increase in S-nitrosothiols in sunflower (Helianthus annuus) seedlings. Journal of experimental botany *62*, 1803-1813.

Chaki, M., Valderrama, R., Fernández-Ocaña, A.M., Carreras, A., López-Jaramillo, J., Luque, F., Palma, J.M., Pedrajas, J.R., Begara-Morales, J.C., and Sánchez-Calvo, B. (2009). Protein targets of tyrosine nitration in sunflower (Helianthus annuus L.) hypocotyls. Journal of experimental botany *60*, 4221-4234.

Chaki, M., Valderrama, R., Fernández-Ocaña, A.M., Carreras, A., Gómez-Rodríguez, M.V., López-Jaramillo, J., Begara-Morales, J.C., Sánchez-Calvo, B., Luque, F., and Leterrier, M. (2011b). High temperature triggers the metabolism of S-nitrosothiols in sunflower mediating a process of nitrosative stress which provokes the inhibition of ferredoxin-NADP reductase by tyrosine nitration. Plant, cell and environment *34*, 1803-1818.

Delledonne, M. (2005). NO news is good news for plants. Current opinion in plant biology *8*, 390-396.

Delledonne, M., Xia, Y., Dixon, R.A., and Lamb, C. (1998). Nitric oxide functions as a signal in plant disease resistance. Nature *394*, 585-588.

Delledonne, M., Zeier, J., Marocco, A., and Lamb, C. (2001). Signal interactions between nitric oxide and reactive oxygen intermediates in the plant hypersensitive disease resistance response. Proceedings of the National Academy of Sciences *98*, 13454-13459.

Durner, J., Wendehenne, D., and Klessig, D.F. (1998). Defense gene induction in tobacco by nitric oxide, cyclic GMP, and cyclic ADP-ribose. Proceedings of the National Academy of Sciences *95*, 10328-10333.

Feechan, A., Kwon, E., Yun, B.-W., Wang, Y., Pallas, J.A., and Loake, G.J. (2005). A central role for S-nitrosothiols in plant disease resistance. Proceedings of the National Academy of Sciences *102*, 8054-8059.

Ferrarini, A., De Stefano, M., Baudouin, E., Pucciariello, C., Polverari, A., Puppo, A., and Delledonne, M. (2008). Expression of Medicago truncatula genes responsive to nitric oxide in pathogenic and symbiotic conditions. Molecular plant-microbe interactions *21*, 781-790.

Garcia-Mata, C., and Lamattina, L. (2007). Abscisic acid (ABA) inhibits light-induced stomatal opening through calcium-and nitric oxide-mediated signaling pathways. Nitric Oxide *17*, 143-151.

García-Mata, C., and Lamattina, L. (2001). Nitric oxide induces stomatal closure and enhances the adaptive plant responses against drought stress. Plant Physiology *126*, 1196-1204.

Guo, F.-Q., and Crawford, N.M. (2005). Arabidopsis nitric oxide synthase1 is targeted to mitochondria and protects against oxidative damage and dark-induced senescence. The Plant Cell Online *17*, 3436-3450.

He, Y., Tang, R.-H., Hao, Y., Stevens, R.D., Cook, C.W., Ahn, S.M., Jing, L., Yang, Z., Chen, L., and Guo, F. (2004). Nitric oxide represses the Arabidopsis floral transition. Science *305*, 1968-1971.

Hu, X., Neill, S.J., Tang, Z., and Cai, W. (2005). Nitric oxide mediates gravitropic bending in soybean roots. Plant Physiology *137*, 663-670.

Huang, X., Stettmaier, K., Michel, C., Hutzler, P., Mueller, M.J., and Durner, J. (2004). Nitric oxide is induced by wounding and influences jasmonic acid signaling in Arabidopsis thaliana. Planta *218*, 938-946.

Huang, X., von Rad, U., and Durner, J. (2002). Nitric oxide induces transcriptional activation of the nitric oxide-tolerant alternative oxidase in Arabidopsis suspension cells. Planta *215*, 914-923.

Kozuka-Hata, H., Oyama, M., and Goto, Y. (2013). Phosphoproteomics-Based Characterization of Cancer Cell Signaling Networks (INTECH Open Access Publisher).

Kwon, E., Feechan, A., Yun, B.-W., Hwang, B.-H., Pallas, J.A., Kang, J.-G., and Loake, G.J. (2012). AtGSNOR1 function is required for multiple developmental programs in Arabidopsis. Planta *236*, 887-900.

Lamb, C., and Dixon, R.A. (1997). The oxidative burst in plant disease resistance. Annual review of plant biology *48*, 251-275.

Lee, U., Wie, C., Fernandez, B.O., Feelisch, M., and Vierling, E. (2008). Modulation of nitrosative stress by S-nitrosoglutathione reductase is critical for thermotolerance and plant growth in Arabidopsis. The Plant Cell Online *20*, 786-802.

Leshem, Y.a.Y., Wills, R.B., and Ku, V.V.-V. (1998). Evidence for the function of the free radical gas—nitric oxide (NO$^{\bullet}$)—as an endogenous maturation and senescence regulating factor in higher plants. Plant Physiology and Biochemistry *36*, 825-833.

Levine, A., Tenhaken, R., Dixon, R., and Lamb, C. (1994). H_2O_2 from the oxidative burst orchestrates the plant hypersensitive disease resistance response. Cell *79*, 583-593.

Libourel, I.G., Bethke, P.C., De Michele, R., and Jones, R.L. (2006). Nitric oxide gas stimulates germination of dormant Arabidopsis seeds: use of a flow-through apparatus for delivery of nitric oxide. Planta *223*, 813-820.

Lindermayr, C., Saalbach, G., and Durner, J. (2005). Proteomic identification of S-nitrosylated proteins in Arabidopsis. Plant Physiology *137*, 921-930.

Lozano-Juste, J., Colom-Moreno, R., and León, J. (2011). In vivo protein tyrosine nitration in Arabidopsis thaliana. Journal of experimental botany, err042.

Miller, G., Shulaev, V., and Mittler, R. (2008). Reactive oxygen signaling and abiotic stress. Physiologia Plantarum *133*, 481-489.

Monzón, G.C., Pinedo, M., Di Rienzo, J., Novo-Uzal, E., Pomar, F., Lamattina, L., and de la Canal, L. (2014). Nitric oxide is required for determining root architecture and lignin composition in sunflower. Supporting evidence from microarray analyses. Nitric Oxide *39*, 20-28.

Pagnussat, G.C., Simontacchi, M., Puntarulo, S., and Lamattina, L. (2002). Nitric oxide is required for root organogenesis. Plant Physiology *129*, 954-956.

Palmieri, M.C., Lindermayr, C., Bauwe, H., Steinhauser, C., and Durner, J. (2010). Regulation of plant glycine decarboxylase by S-nitrosylation and glutathionylation. Plant physiology *152*, 1514-1528.

Parani, M., Rudrabhatla, S., Myers, R., Weirich, H., Smith, B., Leaman, D.W., and Goldman, S.L. (2004). Microarray analysis of nitric oxide responsive transcripts in Arabidopsis. Plant Biotechnology Journal 2, 359-366.

Pedroso, M., and Durzan, D. (2000). Effect of different gravity environments on DNA fragmentation and cell death in Kalanchoe leaves. Annals of Botany *86*, 983-994.

Polverari, A., Molesini, B., Pezzotti, M., Buonaurio, R., Marte, M., and Delledonne, M. (2003). Nitric oxide-mediated transcriptional changes in Arabidopsis thaliana. Molecular Plant-Microbe Interactions *16*, 1094-1105.

Prado, A.M., Porterfield, D.M., and Feijó, J.A. (2004). Nitric oxide is involved in growth regulation and re-orientation of pollen tubes. Development *131*, 2707-2714.

Ribeiro Jr, E.A., Cunha, F.Q., Tamashiro, W., and Martins, I.S. (1999). Growth phase-dependent subcellular localization of nitric oxide synthase in maize cells. FEBS letters *445*, 283-286.

Sanz, L., Fernández-Marcos, M., Modrego, A., Lewis, D.R., Muday, G.K., Pollmann, S., Dueñas, M., Santos-Buelga, C., and Lorenzo, O. (2014). Nitric Oxide Plays a Role in Stem Cell Niche Homeostasis through Its Interaction with Auxin. Plant physiology *166*, 1972-1984.

Takahashi, S., and Yamasaki, H. (2002). Reversible inhibition of photophosphorylation in chloroplasts by nitric oxide. FEBS letters *512*, 145-148.

Valderrama, R., Corpas, F.J., Carreras, A., Fernández-Ocaña, A., Chaki, M., Luque, F., Gómez-Rodríguez, M.V., Colmenero-Varea, P., del Río, L.A., and Barroso, J.B.

(2007). Nitrosative stress in plants. Febs Letters *581*, 453-461.

Van Verk, M.C., Hickman, R., Pieterse, C.M., and Van Wees, S.C. (2013). RNA-Seq: revelation of the messengers. Trends in plant science *18*, 175-179.

Wang, Y., Yun, B.-W., Kwon, E., Hong, J.K., Yoon, J., and Loake, G.J. (2006). S-nitrosylation: an emerging redox-based post-translational modification in plants. Journal of experimental botany *57*, 1777-1784.

Wilhelm, B.T., and Landry, J.-R. (2009). RNA-Seq—quantitative measurement of expression through massively parallel RNA-sequencing. Methods *48*, 249-257.

Xiao, J., Jin, X., Jia, X., Wang, H., Cao, A., Zhao, W., Pei, H., Xue, Z., He, L., and Chen, Q. (2013). Transcriptome-based discovery of pathways and genes related to resistance against Fusarium head blight in wheat landrace Wangshuibai. BMC genomics *14*, 197.

Zeidler, D., Zähringer, U., Gerber, I., Dubery, I., Hartung, T., Bors, W., Hutzler, P., and Durner, J. (2004). Innate immunity in Arabidopsis thaliana: lipopolysaccharides activate nitric oxide synthase (NOS) and induce defense genes. Proceedings of the National Academy of Sciences of the United States of America *101*, 15811-15816.

Zeng, F., Sun, F., Li, L., Liu, K., and Zhan, Y. (2014). Genome-Scale Transcriptome Analysis in Response to Nitric Oxide in Birch Cells: Implications of the Triterpene Biosynthetic Pathway. PloS one *9*, e116157.

Zhang, Y., and Hogg, N. (2004). The mechanism of transmembrane S-nitrosothiol transport. Proceedings of the National Academy of Sciences of the United States of America *101*, 7891-7896.

Zheng, C., Jiang, D., Liu, F., Dai, T., Liu, W., Jing, Q., and Cao, W. (2009). Exogenous nitric oxide improves seed germination in wheat against mitochondrial oxidative damage induced by high salinity. Environmental and Experimental Botany *67*, 222-227.

Other Publications of Interest

Xu, J.-P. (2014). Next-generation Sequencing: Current Technologies and Applications. Caister Academic Press, U.K. ISBN: 978-1-908230-33-1

He, Z. (2014). Microarrays: Current Technology, Innovations and Applications. Caister Academic Press, U.K. ISBN: 978-1-908230-49-2

Poptsova, M.S. (2014). Genome Analysis: Current Procedures and Applications. Caister Academic Press, U.K. ISBN: 978-1-908230-29-4

Fuentes, M. and LaBaer, J. (2014). Proteomics: Targeted Technology, Innovations and Applications. Caister Academic Press, U.K. ISBN: 978-1-908230-46-1

Murillo, J., Vinatzer, B.A., Jackson, R.W. and Arnold, D.L. (2015). Bacteria-Plant Interactions: Advanced Research and Future Trends. Caister Academic Press, U.K. ISBN: 978-1-908230-58-4

Caranta, C., Aranda, M.A., Tepfer, M. and Lopez-Moya, J.J. (2011). Recent Advances in Plant Virology. Caister Academic Press, U.K. ISBN: 978-1-904455-75-2

Nannipieri, P., Pietramellara, G. and Renella, G. (2014). Omics in Soil Science. Caister Academic Press, U.K. ISBN: 978-1-908230-32-4

Herold, K.E. and Rasooly, A. (2009). Lab-on-a-Chip Technology: Fabrication and Microfluidics. Caister Academic Press, U.K. ISBN: 978-1-904455-46-2

An Overview of Proteomics Tools for Understanding Plant Defense Against Pathogens

Carolina Grandellis§, Cecilia V. Vranych§, Ainelén Piazza, Betiana S. Garavaglia, Natalia Gottig and Jorgelina Ottado*

Instituto de Biología Molecular y Celular de Rosario, Consejo Nacional de Investigaciones Científicas y Técnicas (IBR-CONICET). Ocampo y Esmeralda, Rosario, 2000, Argentina
§These authors contributed equally to this work
*Corresponding author. Email: ottado@ibr-conicet.gov.ar
Tel/Fax: +54 341 4237500.

Abstract
Plant diseases are responsible for important losses in crops and cause serious impacts in agricultural production. In the last years, proteomics has been used to examine plant defense responses against pathogens. Such studies may be pioneer in the generation of crops with enhanced resistance. In this review, we focus on proteomics advances in the understanding of host and non-host resistance against pathogens.

Introduction
The term "Proteomics" was first introduced by Marc Wilkins in 1994, as a data-rich discipline that uses Mass Spectrometry (MS) and bioinformatics to characterize the proteome of a cell (Rossignol, 2001; Wilkins et al., 2008; Wilkins, 2009). Unlike genomic tools, proteomics are useful to know the identity, abundance, turnover rate, Post-Translational Modifications (PTMs) and interactions of different proteins in a mixture (Hirano et al., 2004). Protein identification has become possible by coupling gel-based techniques to MS in which samples are converted into gaseous ions and detected by a mass spectrometer according to their mass to charge ratio (m/z) (Quirino et al., 2010; Lodha et al., 2013). With a second mass spectrometer (MS/MS) the amino acid sequence can be inferred (Quirino et al., 2010). Also, gel-free approaches have been recently developed such as Liquid Chromatography (LC) (Quirino et al., 2010).

Nowadays, proteomics assays can be solved by two different approaches referred to as 'bottom-up' and 'top-down'. In bottom-up proteins are digested with a protease prior to MS identification (Tipton et al., 2011). On the other hand, top-down proteomics is applied to intact proteins (Catherman et al., 2014; Kiss et al., 2014; Tran et al., 2011).

Worldwide, plants are affected by pathogens and insects attacks, which are serious threats to agricultural industries and biodiversity conservation. To protect themselves, plants have developed innate immune and resistance systems. Once the first barrier (cell wall) is overcome by the pathogen, the plant detects conserved pathogen structures named Pathogen or Microbial Associated Molecular Patterns (PAMPs or MAMPs, respectively) through their recognition with specific plant receptors named Pattern Recognition Receptors (PRRs). These interactions trigger PAMP-Triggered Immunity (PTI) that constitutes the first step of active plant defense. PTI response involves signaling through the second messenger cGMP, Mitogen-Activated Protein Kinases (MAPKs), Ca^{2+} and H^+ influx, early accumulation of Reactive Oxygen Species (ROS), cell wall thickening, and enhanced expression of Pathogenesis-Related (PR) genes (Lannoo and Van Damme, 2014). Successful pathogens suppress and overcome PTI responses by the secretion of effectors resulting in Effector-Triggered Susceptibility (ETS). This interaction is named compatible and causes plant disease. However, some plants have evolved Resistance (R) proteins which are typically intracellular NB-LRR (Nucleotide-Binding Sites-Leucine-Rich Repeat) type receptors that recognize the pathogen effectors activating the so-called Effector-Triggered Immunity (ETI). The ability to trigger ETI is strain or race specific and is associated with programmed cell death, which is referred to as the Hypersensitive Response (HR) and Systemic Acquired Resistance (SAR). This interaction is named incompatible and includes non-host interactions (Jones and Dangl, 2006; Chisholm et al., 2006; Afroz et al., 2013; Zimaro et al., 2011).

In this review, we will specifically discuss current proteomics techniques that constitute valuable approaches to characterize the defense that plants mount during their battle against pathogens.

Protein identification by Mass Spectrometry
Mass spectrometry is crucial for protein identification and specifically Peptide Mass Fingerprinting (PMF) employs different algorithms to compare analyzed proteins spectra with known standards (Li and Assmann, 2000; Quirino et al., 2010; Lodha et al., 2013). Moreover, tandem mass spectrometry (MS/MS) allows 'de novo' sequencing (Hirano, et al., 2004; Quirino et al., 2010). MS includes: an ion source, a mass analyzer and a detector (Hirano et al., 2004). Although there are several ionization techniques, Electrospray Ionization (ESI) and Matrix-Assisted Laser Desorption/Ionization (MALDI) are the most commonly employed (Lodha, et al., 2013; Hirano et al., 2004; Quirino, et al., 2010).

Besides, major technical breakthroughs such as the generation of more complex analyzers (Time-Of-Flight or TOF, and Quadrupole or Q mass analyzer) and the advent of high-resolution mass spectrometers, like Orbitrap, have turned MS in a key player to study the proteome (Yang, et al., 2015). However, it is necessary to select the accurate MS instrument for a particular analysis. In general, Q-MS and Ion Trap-MS (IT-MS) are combined with ESI while TOF-MS is associated with MALDI. TOF has been coupled to quadrupole (Q-TOF), and combinations such as MALDI-TOF/TOF and MALDI-Q-TOF were also developed (Hirano et al., 2004). The evolution of MS has turned it into an essential technology for many proteomics applications in the study of plant resistance. We next describe some of the most used techniques based on gel electrophoresis and also gel-free technologies in plant resistance characterization. In Figure 1, a schematic representation of them is depicted (Figure 1).

Two Dimensional Polyacrylamide Gel Electrophoresis

Two Dimensional PolyAcrylamide Gel Electrophoresis (2D-PAGE) is used to separate proteins from a mixture providing information such as: molecular weight, isoelectric point, presence or absence of proteins in a sample and PTMs (Görg et al., 2004; Quirino et al., 2010; Lodha, et al., 2013).

2D-PAGE methods combined with PMF or MS/MS have been used to study the interaction of rice with *Magnaporthe oryzae* that causes the most devastating disease in cultivated area. Differentially expressed protein spots were observed in rice apoplast after *M. oryzae* infection and a

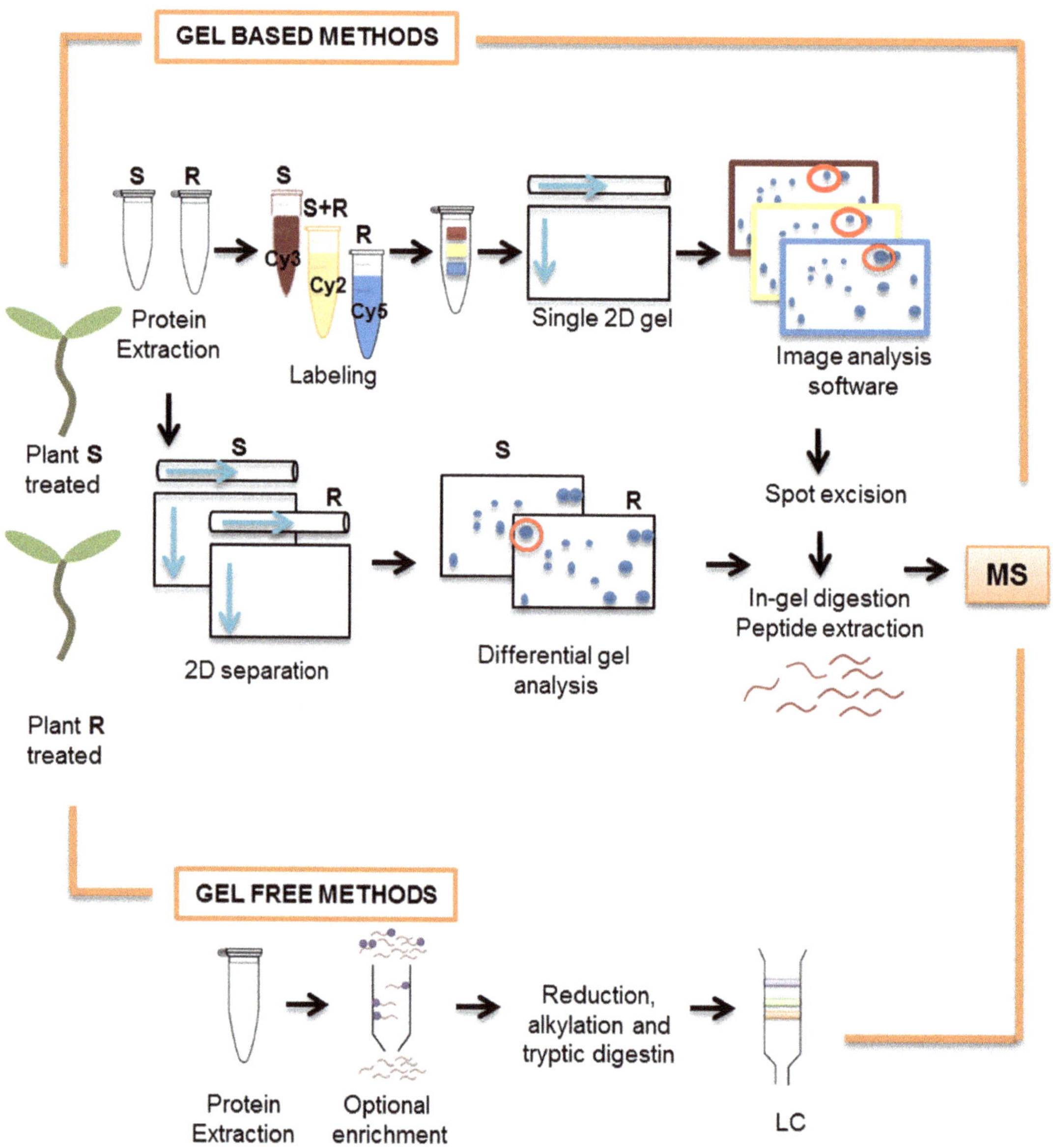

Figure 1. Schematic overview of protocols employed for some of the gel-based and gel-free proteomics techniques described in this review. In this example, protein extracts are obtained from two plants samples: susceptible and resistant denoted S and R, respectively. Differential spots are red highlighted. MS: Mass Spectrometry, LC: Liquid Chromatography.

number of proteins were identified by MALDI-TOF-MS. A number of proteins including: Glycosyl Hydrolase (GH) proteins, PR and antioxidant proteins, subtilisin-like proteases and OsDUF26 were strongly expressed, suggesting that secreted proteolytic enzymes modulate plant resistance in this plant-pathogen interaction (Wu et al., 2014).

MS/MS was used to study corn resistance to the fungus *Curvularia lunata*. Protein markers related to host resistance were identified using four maize inbred lines with different resistance levels. The number of spots presenting quantitative differences was higher in resistant inbred lines than in susceptible ones. Also, a larger amount of up-regulated protein was detected in resistant maize. These proteins were identified as GLPs (Germin-Like Proteins), eIF-5A (Translation Initiation Factor 5A), OEE 1 (Oxygen-Evolving Enhancer Protein 1), OEE 2 (Oxygen-Evolving Enhancer Protein 2), and OEC (Oxygen-Evolving Complex), and they have been associated with host defense and resistance to pathogen infection (Huang et al., 2009).

Non-host resistance has also been studied by 2D-PAGE followed by MS/MS analyzing the non-host response of rice to the rust fungi *Puccina striiformis* f. sp. *tritici*. The proteins related to this non-host resistance were mainly involved in H_2O_2 production and glycerol-3-phosphate signaling (Zhao et al., 2014).

Two-Dimensional Differential in-Gel Electrophoresis

Two Dimensional Differential in-Gel Electrophoresis (2D-DIGE), a 2D-PAGE modification, is a powerful technique that detects reproducibly differences between two protein samples simultaneously on the same gel (Unlü et al., 1997). When performing 2D-DIGE, different samples are covalently labeled with distinct spectrally resolvable fluorophores, combined and then fractionated together on a single 2D gel. Frequently, a third sample known as internal standard, composed of a mixture of equal amounts of experimental samples, is applied in the analysis. This enables to relativize each protein to its corresponding spot in the internal standard in the same gel. Then, changes between proteins abundances are visualized using fluorescence imaging (Marouga et al., 2005). 2D-DIGE and MALDI-TOF-MS/MS were used to detect differential proteins spots in resistant and susceptible maize ecotypes during sugarcane mosaic virus infection. Identified proteins were subdivided into functional categories: stress and defense response, energy and metabolism, photosynthesis, protein synthesis and folding, carbon fixation, structure protein and signal transduction/ transcription. A high proportion of proteins identified as stress and defense related, were up-accumulated in the resistant maize genotype. Furthermore, 17 novel proteins that have not been previously described as virus-responsive were identified, providing novel data into the molecular mechanism of plant-virus interactions (Wu et al., 2013).

Another example in which 2D-DIGE was used led to the identification of rice defense related proteins in the incompatible interaction with the fungus *M. oryzae* when compared to proteins detected in the compatible interaction. Noticeably, similar proteins were identified in both incompatible and compatible interactions. However, the incompatible interaction caused an earlier and higher expression of these proteins in resistant plants in comparison with susceptible ones, which might determine resistance to the pathogen (Kim et al., 2013).

Gel-Free Technologies

Although gel based techniques allow rapid, simple and reproducible separation of proteins, they present several limitations, such as separation of high molecular weight and basic proteins (Rossignol, 2001; Agrawal, 2013; Quirino et al., 2010). These restrictions have been overcome with the application of the "shotgun" method and gel-free approaches like LC-MS/MS (Hirano et al., 2004). Hence, prior to MS analysis, LC separation is performed which is based on Strong Cation eXchange (SCX) and Reversed Phase (RP) chromatography (Hirano et al., 2004; Agrawal et al., 2013). LC-MS/MS was used to study the compatible *Pseudomonas syringae-Arabidopsis thaliana* interaction (Hurley et al., 2014). Virulence of *P. syringae* relies on the Type III Secretion System (T3SS) to inject effector proteins into host cells (Afroz et al., 2013; Hurley et al., 2014). These effectors are able to suppress the initial PTI during compatible plant-pathogen interaction by targeting PRR complexes and downstream signaling components. Recognition of the AvrRpt2 effector by the *A. thaliana* the RPS2 (Resistance to *P. syringae* 2) immune receptor activates ETI. Nevertheless, HopF2 effector suppresses this ETI-response. Using LC-MS/MS, the proteins that co-purified with HopF2 in *A. thaliana* were identified, and this analysis was relevant to determine protein functionality. A high molecular weight complex co-purified with HopF2, and members corresponded to membrane associated proteins, that included AHA2 and ERD4 (involved in the closing of stomata in response to *P. syringae)*, and others proteins (AtHIR2 and AtREM1.3) that play a role in pathogen perception (Hurley et al., 2014).

Proteomics methods to analyze post-translational modifications

PTMs of amino acids extend the range of functions of the protein. Enzymes can cleave amino acids or append different functional groups such as acetate, phosphate, various lipids and carbohydrates, changing the activity state, localization, turnover, and interactions with other proteins (Mann and Jensen, 2003). Plants proteins undergo PTMs in response to pathogens and other stimuli. The most intensely studied PTM is phosphorylation (Jones et al., 2013) that is crucial in plant resistance events by fine-tuning regulation of protein function (Yang et al., 2013). Phosphoproteins are usually present in a low copy number, so their detection is challenging. Generally an enrichment step is performed to obtain protein mixtures in which phosphoproteins are abundant.

We next summarize proteomics methods involving the study of PTMs in plant defense. For instance in tomato, *Cladosporium fulvum* resistance genes (Cf) recognize fungal effectors resulting in HR and resistance, and phosphorylation mediates Cf cascades. Leaves extracts from infected and no infected tissue were subjected to phosphopeptide identification and label-free quantification by LC-MS/MS (Matthiesen et al., 2005). Titanium dioxide (TiO_2) enrichment followed by LC-ESI-QTOF allowed the identification of phosphoproteins. TiO_2 possess high affinity for phosphopeptides, which are then eluted using ammonia solution (Yu and Veenstra, 2013). Results showed that photosynthesis is quickly suppressed upon HR and differentially phosphorylated Hsp90 (Heat Shock Protein 90) chaperon isoforms are expressed, suggesting a role in plant defense. Likewise, infected tissue underwent a switch to anaerobic respiration (Stulemeijer et al., 2009). Plant defense related proteins were indentified in wheat resistance to the fungus *Septoria tritici* by comparing resistant and susceptible infected plants (Yang et al., 2013). Protein extracts were subjected to TiO_2 enrichment. Hydrophilic Interaction LIquid Chromatography (HILIC) fractionation and LC-MS/MS analysis were conducted. Differential phosphorylation was observed for GTP-binding proteins, 14-3-3 proteins, and calcium-binding proteins. HILIC allows separation of small polar compounds because they present good solubility in the mobile phase of HILIC, overcoming the poor solubility encountered in LC (Kahsay et al., 2014).

In addition, rice phosphoproteome was studied during the interaction with *M. oryzae*. Changes of phosphoprotein profiles in rice leaves were analyzed during early compatible and incompatible interactions using PEG (Polyethylene Glycol) prefractionation to deplete RuBisCO, and by enrichment of low abundant phosphoproteins through $Al(OH)_3$ Metal-Oxide Affinity Chromatography ($Al(OH)_3$ MOAC), 2D-PAGE, Pro-Q diamond phosphoprotein stain (Pro-Q DPS) and MS. Spots were subjected to MALDI-TOF/TOF-MS and many *M. oryzae*-regulated phosphoproteins involved in plant resistance were found. Among them, signaling-related, microtubule-associated, amino acid synthesis-related proteins and energy-related enzymes were successfully identified (Li et al., 2015).

N-Glycosylation and Lys-Acetylation were analyzed in grapevine proteins in response to *Lobesia botrana* (Melo-Braga et al., 2012). Enrichment in glycopeptides was achieved by Zwitterionic-Hydrophilic Interaction Liquid Chromatography (ZIC-HILIC). This method involves Trifluoroacetic Acid (TFA) and Ion Pairing (IP) to increase the hydrophilicity difference between glycopeptides and non-glycosylated peptides. Also, Lys-Acetylation was studied using an anti-acetyllysine antibody to immunoprecipitate acetylated peptides. At last, about 30% grapevine proteins differed in abundance in infected tissue related to an uninfected control (Melo-Braga et al., 2012). Another type of PTM observed in resistance proteins is the S-acylation (palmitoylation) that consists in the addition of fatty acids to cysteine residues (Kim et al., 2005). To study this PTM, sample preparation must include blocking the free cysteines with N-ethylmaleimide followed by removal of S-acyl groups by hydroxylamine treatment. Exposed cysteines are tagged with biotin and then purified and analyzed by MS. For example, FLS2, the plant receptor for bacterial flagellin, has been found to be S-acylated. Apart from FLS2 and RPS2 previously observed to be palmitoylated (Kim et al., 2005), several other S-acylated proteins have been identified, such as RPS2-containing complexes and the calcium channels ACA8 and ACA10. This information confers an important role for S-acylation in pathogen recognition and plant resistance (Hemsley et al., 2013).

Surveying plant cellular compartments and tissues through proteomics

Subcellular proteomics consist in the study of specific plant fractions such as the apoplast, organelles or plasma membrane. By this approach, the complexity of protein extracts can be reduced to have a major representation in a defined compartment of the plant cell (Delaunois et al., 2013; Quirino et al., 2010).

Membrane fractions of *A. thaliana* following activation of RPS2 have been analyzed by a large scale proteomics approach. Plasma Membrane (PM) proteins enrichment was achieved using the aqueous two-phase partitioning method and they were analyzed by LC-MS/MS. In total 2303 *Arabidopsis* proteins were detected and 22% of the up-regulated proteins were previously characterized in plant immunity or cell death regulation (Elmore et al., 2012). These results suggest that the PM proteome is dynamic during plant immune signaling and the activation of RPS2 initiates ETI signaling and elicits HR.

Many important events take place in the apoplast where pathogen and host interact, thus this compartment has a crucial role in plant defense mechanisms. Apoplastic fluid is usually isolated using Vacuum Infiltration Centrifugation (VIC) method. The tissue is infiltrated with extraction buffer, then the surface is dried and leaves are rolled and inserted into tubes. Finally, centrifugation allows the recovery of apoplastic proteins (Delaunois et al., 2014). In a recent report, grapevine leaves apoplastic fluid showed defense proteins such as osmotin and thaumatin-like proteins, suggesting the presence of preformed defenses. Chitinases and glucanases, that degrade fungal cell walls, were also represented. Furthermore, peroxidases and superoxide dismutase were found actively involved in the plant defense response (Delaunois et al., 2014).

MALDI Imaging Mass Spectrometry

MALDI Imaging Mass Spectrometry (MALDI-IMS) is the combination of molecular mass analysis and spatial information. This most recent technology can be used to detect, identify and localize specific molecules such as drugs, lipids and proteins directly from intact tissue sections. Essentially, MALDI-IMS allows direct measurement of mass spectra from tissue, and the distribution of the detected compounds can be visualized as an image, representing 2D distribution maps of selected components (Amstalden van Hove et al., 2010; Burnum et al., 2008; Goodwin et al., 2008; Rohner et al., 2005). Until today, MALDI-IMS has not been used to identify proteins involved in plant-pathogen interactions, even though recent reports showed the identification and localization of other molecules involved in plant resistance. For instance, the differential accumulation of the flavonoid hesperidin in

Citrus sinensis grafted on *Citrus limonia* cv Pêra using MALDI-IMS was studied with the aim to analyze if the accumulation of this compound and tolerance to citrus variegated chlorosis bacterium were correlated (Soares et al., 2015). Transversal sections of leaves and petioles were analyzed to obtain different ion intensities of hesperidin from infected and healthy tissue, as well as to determine its anatomical distribution in both tissues. By MALDI-IMS, the highest ion intensities of hesperidin were observed from infected plants when compared with healthy tissues in all analyses. In petioles of infected plants, hesperidin was detected in whole tissues, while in leaf transversal sections it was observed mainly in xylem, collenchyma and mesophyll cell types. In healthy plants hesperidin ions were low-abundant and after infection, relocalization occurred (Soares et al., 2015).

MALDI-IMS has also been used to investigate the *in-situ* response to the pathogen *Plasmopara viticola.* This plant produces high amounts of toxic compounds against fungi such as resveratrol, pterostilbene and viniferins. Grapevine infected leaves showed differential presence of compounds in susceptible and resistant varieties. In a susceptible variety, less toxic compounds for fungi were detected on the infection sites, and the viniferins localized too far away from infection sites to play a real antifungal role. Using this technology, locations from a range of compounds in both susceptible to resistant grapevine species is now possible to understand their role in defense (Becker et al., 2014). Even if this technique has not been used to analyze proteins yet, it is challenging to apply this technology to the *in situ* identification of plant defense proteins during both PTI and ETI.

Concluding remarks

A major challenge for scientists is to direct efforts to control plant diseases, and improve resistance to pathogens. Phytopathogens threaten agriculture, and despite efforts to manage crop diseases, pathogens account for losses of 15% of worldwide food production. The understanding of the host molecular events leading to the setting up of resistance mechanisms and its functional characterization is essential to adopt new technologies for plant breeding and crop improvement. In this means, the current omics technologies provide a range of methods which offer new opportunities to develop strategies that are likely to provide durable disease resistance.

Data from genomes states that human beings are 98% chimpanzees, and chimpanzees are 98% human beings as well. According to this, only 2% of the genomic information differentiates both species, but this is not enough to explain the apparent differences. A genome containing 30,000 genes can usually generate 100,000 mRNAs. Moreover, those mRNAs can generate 400,000 proteins that together constitute the proteome in a certain condition or stage. Thus, the genome provides insights about events that could potentially occur in a cell or in an organism and transcriptomic data report transcripts levels that are not always translated into a functional protein. The proteome tells us exactly what is occurring in a specific state of a cell, a certain tissue or an organism and, the interactome, defined by physical interactions between proteins and molecules, allow us to predict what might occur in the living cell or tissue analyzed.

To elucidate plant-pathogen interactions, numerous studies were carried out from plant and pathogen perspectives, revealing that these interactions involve a complex and specific communication. Here, we present evidence of the efficacy and convenience to apply these techniques to the study of plant-pathogen interactions and the consequent leading molecular events at the protein level that ultimately confer resistance to the host plant.

Acknowledgments

This work was supported by grants from the Argentine Federal Government: Agencia Nacional de Promoción Científica y Tecnológica (ANPCyT) [PICT2013-0556] and [PICT2013-0625] and Consejo Nacional de Investigaciones Científicas y Técnicas (CONICET) [PIP2014-2016]. BSG, NG and JO are staff members and CG, CVV and AP are fellows of CONICET.

References

Afroz, A., Zahur, M., Zeeshan, N., and Komatsu, S. (2013). Plant-bacterium interactions analyzed by proteomics. Front. Plant Sci. *4*, 21.

Agrawal, G.K., Sarkar, A., Righetti, P.G., Pedreschi, R., Carpentier, S., Wang, T., Barkla, B.J., Kohli, A., Ndimba, B.K., Bykova, N. V., et al. (2013). A decade of plant proteomics and mass spectrometry: Translation of technical advancements to food security and safety issues. Mass Spectrom. Rev. *32*, 335-365.

Amstalden van Hove, E.R., Smith, D.F., and Heeren, R.M. a (2010). A concise review of mass spectrometry imaging. J. Chromatogr. A *1217*, 3946-3954.

Becker, L., Carré, V., Poutaraud, A., Merdinoglu, D., and Chaimbault, P. (2014). MALDI mass spectrometry imaging for the simultaneous location of resveratrol, pterostilbene and viniferins on grapevine leaves. Molecules *19*, 10587-10600.

Burnum, K.E., Frappier, S.L., and Caprioli, R.M. (2008). Matrix-assisted laser desorption/ionization imaging mass spectrometry for the investigation of proteins and peptides. Annu. Rev. Anal. Chem. (Palo Alto. Calif). *1*, 689-705.

Catherman, A.D., Skinner, O.S., and Kelleher, N.L. (2014). Top Down proteomics: facts and perspectives. Biochem. Biophys. Res. Commun. *445*, 683-693.

Chisholm, S.T., Coaker, G., Day, B., and Staskawicz, B.J. (2006). Host-microbe interactions: shaping the evolution of the plant immune response. Cell *124*, 803-814.

Delaunois, B., Colby, T., Belloy, N., Conreux, A., Harzen, A., Baillieul, F., Clément, C., Schmidt, J., Jeandet, P., and Cordelier, S. (2013). Large-scale proteomic analysis of the grapevine leaf apoplastic fluid reveals mainly stress-related proteins and cell wall modifying enzymes. BMC Plant Biol. *13*, 24.

Delaunois, B., Jeandet, P., Clément, C., Baillieul, F., Dorey, S., and Cordelier, S. (2014). Uncovering plant-pathogen crosstalk through apoplastic proteomic studies. Front. Plant Sci. *5*, 249.

Elmore, J.M., Liu, J., Smith, B., Phinney, B., and Coaker, G. (2012). Quantitative Proteomics Reveals Dynamic Changes in the Plasma Membrane During Arabidopsis Immune Signaling. Mol. Cell. Proteomics *11*, M111.014555-M111.014555.

Goodwin, R.J. a, Pennington, S.R., and Pitt, A.R. (2008). Protein and peptides in pictures: Imaging with MALDI mass spectrometry. Proteomics *8*, 3785-3800.

Görg, A., Weiss, W., and Dunn, M.J. (2004). Current two-dimensional electrophoresis technology for proteomics. Proteomics *4*, 3665-3685.

Hemsley, P. a., Weimar, T., Lilley, K.S., Dupree, P., and Grierson, C.S. (2013). A proteomic approach identifies many novel palmitoylated proteins in Arabidopsis. New Phytol. *197*, 805-814.

Hirano, H., Islam, N., and Kawasaki, H. (2004). Technical aspects of functional proteomics in plants. Phytochemistry *65*, 1487-1498.

Huang, X., Liu, L., Zhai, Y., Liu, T., and Chen, J. (2009). Proteomic comparison of four maize inbred lines with different levels of resistance to Curvularia lunata (Wakker) Boed infection. Prog. Nat. Sci. *19*, 353-358.

Hurley, B., Lee, D., Mott, A., Wilton, M., Liu, J., Liu, Y.C., Angers, S., Coaker, G., Guttman, D.S., and Desveaux, D. (2014). The Pseudomonas syringae Type III Effector HopF2 Suppresses Arabidopsis Stomatal Immunity. PLoS One *9*, e114921.

Jones, J.D.G., and Dangl, J.L. (2006). The plant immune system. Nature *444*, 323-329.

Jones, A.M.E., Monaghan, J., and Ntoukakis, V. (2013). Editorial: Front. Plant Sci. *4*, 64.

Kahsay, G., Song, H., Van Schepdael, A., Cabooter, D., and Adams, E. (2014). Hydrophilic interaction chromatography (HILIC) in the analysis of antibiotics. J. Pharm. Biomed. Anal. *87*, 142-154.

Kim, H.-S., Desveaux, D., Singer, A.U., Patel, P., Sondek, J., and Dangl, J.L. (2005). The Pseudomonas syringae effector AvrRpt2 cleaves its C-terminally acylated target, RIN4, from Arabidopsis membranes to block RPM1 activation. Proc. Natl. Acad. Sci. U. S. A. *102*, 6496-6501.

Kim, S.G., Wang, Y., Lee, K.H., Park, Z.Y., Park, J., Wu, J., Kwon, S.J., Lee, Y.H., Agrawal, G.K., Rakwal, R., et al. (2013). In-depth insight into *in vivo* apoplastic secretome of rice-Magnaporthe oryzae interaction. J. Proteomics *78*, 58-71.

Kiss, A., Smith, D.F., Reschke, B.R., Powell, M.J., and Heeren, R.M.A. (2014). Top-down mass spectrometry imaging of intact proteins by laser ablation ESI FT-ICR MS. Proteomics *14*, 1283-1289.

Lannoo, N., and Van Damme, E.J.M. (2014). Lectin domains at the frontiers of plant defense. Front. Plant Sci. *5*, 397.

Li, J., and Assmann, S.M. (2000). Mass spectrometry. An essential tool in proteome analysis. Plant Physiol. *123*, 807-809.

Li, Y., Ye, Z., Nie, Y., Zhang, J., Wang, G.-L., and Wang, Z. (2015). Comparative phosphoproteome analysis of Magnaporthe oryzae-responsive proteins in susceptible and resistant rice cultivars. J. Proteomics *115*, 66-80.

Lodha, T.D., Hembram, P., and Basak, N.T.J. (2013). Proteomics: A Successful Approach to Understand the Molecular Mechanism of Plant-Pathogen Interaction. Am. J. Plant Sci. *04*, 1212-1226.

Mann, M., and Jensen, O.N. (2003). Proteomic analysis of post-translational modifications. Nat. Biotech *21*, 255-261.

Marouga, R., David, S., and Hawkins, E. (2005). The development of the DIGE system: 2D fluorescence difference gel analysis technology. Anal. Bioanal. Chem. *382*, 669-678.

Matthiesen, R., Trelle, M.B., Højrup, P., Bunkenborg, J., and Jensen, O.N. VEMS 3.0: algorithms and computational tools for tandem mass spectrometry based identification of post-translational modifications in proteins. J. Proteome Res. *4*, 2338-2347.

Melo-Braga, M.N., Verano-Braga, T., Leon, I.R., Antonacci, D., Nogueira, F.C.S., Thelen, J.J., Larsen, M.R., and Palmisano, G. (2012). Modulation of Protein Phosphorylation, N-Glycosylation and Lys-Acetylation in Grape (Vitis vinifera) Mesocarp and Exocarp Owing to Lobesia botrana Infection. Mol. Cell. Proteomics *11*, 945-956.

Quirino, B.F., Candido, E.S., Campos, P.F., Franco, O.L., and Krüger, R.H. (2010). Proteomic approaches to study plant-pathogen interactions. Phytochemistry *71*, 351-362.

Rohner, T.C., Staab, D., and Stoeckli, M. (2005). MALDI mass spectrometric imaging of biological tissue sections. Mech. Ageing Dev. *126*, 177-185.

Rossignol, M. (2001). Analysis of the plant proteome. [Review] [29 refs]. Curr. Opin. Biotechnol. *12*, 131-134.

Soares, M.S., da Silva, D.F., Forim, M.R., da Silva, M.F.D.G.F., Fernandes, J.B., Vieira, P.C., Silva, D.B., Lopes, N.P., de Carvalho, S.A., de Souza, A.A., et al. (2015). Quantification and localization of hesperidin and rutin in Citrus sinensis grafted on C. limonia after Xylella fastidiosa infection by HPLC-UV and MALDI imaging mass spectrometry. Phytochemistry *115*, 161-170.

Stulemeijer, I.J.E., Joosten, M.H. a J., and Jensen, O.N. (2009). Quantitative phosphoproteomics of tomato mounting a hypersensitive response reveals a swift suppression of photosynthetic activity and a differential role for Hsp90 Isoforms. J. Proteome Res. *8*, 1168-1182.

Tipton, J.D., Tran, J.C., Catherman, A.D., Ahlf, D.R., Durbin, K.R., and Kelleher, N.L. (2011). Analysis of intact protein isoforms by mass spectrometry. J. Biol. Chem. *286*, 25451-25458.

Tran, J.C., Zamdborg, L., Ahlf, D.R., Lee, J.E., Catherman, A.D., Durbin, K.R., Tipton, J.D., Vellaichamy, A., Kellie, J.F., Li, M., et al. (2011). Mapping intact protein isoforms in discovery mode using top-down proteomics. Nature *480*, 254-258.

Unlü, M., Morgan, M.E., and Minden, J.S. (1997). Difference gel electrophoresis: a single gel method for detecting changes in protein extracts. Electrophoresis *18*, 2071-2077.

Wilkins, M. (2009). Proteomics data mining. Expert Rev. Proteomics *6*, 599-603.

Wilkins, M.R., Arthur, J.W., Junius, F.K., Harrison, M.J., Traini, M.D., Bizannes, P., Doggett, P.E., and McDonald, W. (2008). Information management for proteomics: a perspective. Expert Rev. Proteomics *5*, 663-678.

Wu, J., Lee, D.Y., Wang, Y., Kim, S.T., Baek, S.-B., Kim, S.G., and Kang, K.Y. (2014). Protein profiles secreted from phylloplane of rice leaves free from cytosolic proteins: Application to study rice-Magnaporthe Oryzae interactions. Physiol. Mol. Plant Pathol. *88*, 28-35.

Wu, L., Han, Z., Wang, S., Wang, X., Sun, A., Zu, X., and Chen, Y. (2013). Comparative proteomic analysis of the plant-virus interaction in resistant and susceptible ecotypes of maize infected with sugarcane mosaic virus. J. Proteomics *89*, 124-140.

Yang, F., Melo-Braga, M.N., Larsen, M.R., Jørgensen, H.J.L., and Palmisano, G. (2013). Battle through signaling between wheat and the fungal pathogen Septoria tritici revealed by proteomics and phosphoproteomics. Mol. Cell. Proteomics *12*, 2497-2508.

Yang, Y., Hu, M., Yu, K., Zeng, X., and Liu, X. (2015). Mass spectrometry-based proteomic approaches to study pathogenic bacteria-host interactions. Protein Cell *6*, 265-274.

Yu, L.-R., and Veenstra, T. (2013). Phosphopeptide enrichment using offline titanium dioxide columns for phosphoproteomics. Methods Mol. Biol. *1002*, 93-103.

Zhao, J., Yang, Y., and Kang, Z. (2014). Proteomic Analysis of Rice Nonhost Resistance to Puccinia striiformis f. sp. tritici Using Two-Dimensional Electrophoresis. Int. J. Mol. Sci. *15*, 21644-21659.

Zimaro, T., Gottig, N., Garavaglia, B.S., Gehring, C., and Ottado, J. (2011). Unraveling plant responses to bacterial pathogens through proteomics. J. Biomed. Biotechnol. *2011*.

Other Publications of Interest

Xu, J.-P. (2014). Next-generation Sequencing: Current Technologies and Applications. Caister Academic Press, U.K. ISBN: 978-1-908230-33-1

He, Z. (2014). Microarrays: Current Technology, Innovations and Applications. Caister Academic Press, U.K. ISBN: 978-1-908230-49-2

Poptsova, M.S. (2014). Genome Analysis: Current Procedures and Applications. Caister Academic Press, U.K. ISBN: 978-1-908230-29-4

Fuentes, M. and LaBaer, J. (2014). Proteomics: Targeted Technology, Innovations and Applications. Caister Academic Press, U.K. ISBN: 978-1-908230-46-1

Murillo, J., Vinatzer, B.A., Jackson, R.W. and Arnold, D.L. (2015). Bacteria-Plant Interactions: Advanced Research and Future Trends. Caister Academic Press, U.K. ISBN: 978-1-908230-58-4

Caranta, C., Aranda, M.A., Tepfer, M. and Lopez-Moya, J.J. (2011). Recent Advances in Plant Virology. Caister Academic Press, U.K. ISBN: 978-1-904455-75-2

Nannipieri, P., Pietramellara, G. and Renella, G. (2014). Omics in Soil Science. Caister Academic Press, U.K. ISBN: 978-1-908230-32-4

Herold, K.E. and Rasooly, A. (2009). Lab-on-a-Chip Technology: Fabrication and Microfluidics. Caister Academic Press, U.K. ISBN: 978-1-904455-46-2

Linking Biomarker and Comparative Omics to Pathogens in Legumes

Marwan Diapari

Southern Crop Protection and Food Research Centre, Agriculture and Agri-Food Canada / Agriculture et Agroalimentaire Canada, 1391 Sandford St., London, ON, N5V 4T3 Canada. Email Marwan.Diapari@agr.gc.ca

Abstract
It is envisioned that a more precise study of the association between the traits and biomarkers will dramatically decrease the time and costs required to bring new improved disease resistance lines to market. The field of omics has an enormous potential to assess diseases more precise, including the identification and understanding of pathogenic mechanisms in legume crops, and have been exemplified by a relatively large number of studies. Recently, molecular genetic studies have accumulated a huge amount of genotypic data, through a more affordable next generation sequencing (NGS) technology, causing the omics approaches to fall behind. In this paper I provide an overview of genomics and proteomics and their use in legume crops, including the use of comparative genomics to identify homologous markers within legume crops.

Introduction
Legumes are economically important, rich source of quality protein for humans and animals, and enrich the soil by producing their own nitrogen in symbiosis with nitrogen-fixing bacteria (Singh et al., 2007). A better knowledge of its physiological characteristics and behaviour within its own environment is imperative in order to increase legume competitiveness and to adapt its production to the new constraints arising from the climatic changes and the occurrence of new diseases (Adam-Blondon, 2007).

Molecular approaches (OMICS) that involve mRNA (transcriptomics), proteins (proteomics), metabolites (metabolomics), are indispensable in the era of functional genomics, and have been exemplified by a relatively large number of studies, such as gene expression, genetic diversity, phylogeny, comparative genomics and epigenetics (Ma et al., 2013; Sabir et al., 2014). OMICS has brought the possibilities for genetic improvement, conservation of biodiversity, and imparting disease resistance in crops for further targeted breeding (Rubiales et al., 2015). Maintaining genetic diversity and capturing genetic improvement are often through structuring populations into a genetic hierarchy, setting population sizes, and managing pedigrees and inbreeding.

Plant diseases can drastically abate the crop yields as the degree of disease outbreak is getting severe around the world. Major diseases are rusts, powdery and downy mildews, ascochyta blights, botrytis gray molds, anthracnose, damping-off, root rots, collar rots, vascular wilts and white molds (Gururani et al., 2012; Rubiales et al., 2015). New disease occurrence can involve gene duplication, transfer or recombination (Hollomon and Brent, 2009). The negative impact of diseases and the genetic mechanisms and gene interactions involved in the resistance response are still poorly understood (Cantu et al., 2013). Comprehensive body of knowledge and understanding of the cellular functions will aid the development of novel practical approaches to enhance disease resistance (Choi and Hwang, 2014).

Legume crops have been exploited by breeding program (Robertson et al., 1996), including for disease purpose breeding program. It is believed that wild relatives are considered as important source of genetic diversity for agriculture, although their gene repertoire remains largely unexplored (Li et al., 2014). Proposed draft legume genomes allow the prediction of the full complement of genes, phylogenetic relationships and chromosomal locations will provide breeders and scientists with a useful tool to identify novel disease resistance traits (Andolfo et al., 2014). It is expected the array will accelerate hypothesis generation and gene discovery in disease defense pathways, responses to abiotic stresses, development, and evolutionary diversity in plants (Close et al., 2004). Directed molecular evolution can be used to develop enabling technologies for plant biologists; how genes can be optimized to generate improved input traits such as those conferring insect tolerance, disease control and herbicide tolerance; and how plant quality can be altered to improve yield (Lassner and McElroy, 2002). Molecular genetics provide an unparalleled opportunity to understand how genes and genetic changes interact with environmental stimuli to either preserve health or cause disease (Schwartz et al., 2004).

Plant breeders strive to stay ahead of the evolving pathogens by releasing cultivars with new resistance genes or gene combinations (Webb and Fellers, 2006). Breeding for disease resistance in crops has mainly been accomplished by incorporating single resistance gene. There are advantages to quantitative resistance in terms of durability. New technologies in genomics and proteomics are providing insights into disease-resistance pathways (Jordan et al., 2006). Genetic improvement of the crop is being actively pursued and numerous functional genomics studies aimed at characterizing gene controlling key agronomic characteristics and providing some insight into the basis of disease resistances (Hulbert and Pumphrey, 2014; Mellor et al., 2012).

The fields of environmental genetics and environmental genomics has enormous potential to affect our ability to accurately assess the risk of developing disease, identify and understand basic pathogenic mechanisms that are critical to disease progression (Schwartz et al., 2004). Genetic studies pathogens during the past decade have made it an excellent system for investigating pathogen-host interactions (Xu and Xue, 2002).

Genetic variation has an important role in determining biologic phenotypes and disease resistance (Sladek and Hudson, 2006). Inter-genomic comparisons identified lineage-specific genes, some of which show evidence of positive selection and may contribute to variation of agronomic traits such as disease resistance (Li et al., 2014). Scarcity of genetic polymorphism makes crops especially vulnerable to a wide variety of pathogens (Ratnaparkhe et al., 2011). Sequence analysis of plant disease resistance genes shows similarity among themselves (Ramachandra et al., 2011) and are present in abundance and majority of these encode proteins with a nucleotide-binding site (NBS) (Basak et al., 2007; Ribas et al., 2011), comprising up to 2% of all genes (Mun et al., 2009).

NBS encoding resistance genes are key plant disease-resistance genes (R-genes) represent the major class of disease resistance genes (Mun et al., 2009; Richly et al., 2002) also play an important role in offering resistance to pathogens (Yu et al., 2014). Numerous plant R-genes have been used with varying degree of success in crop improvement (Gururani et al., 2012). A number of studies have focused on NBS-encoding genes in disease resistant breeding programs for diverse plants (Wei et al., 2013). Plant possesses a large number of R-genes with diverse recognition specificities which are activated in response to a variety of microbial pathogens (Sharma et al., 2014) and have the ability to detect a pathogen attack and facilitate a counter attack against the pathogen (Gururani et al., 2012; Martin et al., 2003) also often show significant similarity amongst themselves in terms of both their DNA sequences and structural motifs present in their protein products (López et al., 2003).

The majority of verified plant disease resistance genes isolated to date are of the NBS-LRR class (Madsen et al., 2003). The NBS-LRR genes are the largest class of disease resistance genes in plants (Wei et al., 2013; Xu et al., 2011). NBS profiling, in particular, targets conserved nucleotide binding site-encoding sequences of resistance gene analogs (RGAs), and is widely used to identify molecular markers for *R* genes (Vossen et al., 2013).

Reactive oxygen species (ROS) signaling network mediating disease resistance in plants information became available, thanks to the complete Arabidopsis (*Arabidopsis thaliana*) and rice (*Oryza sativa*) genome sequences (Kotchoni and Gachomo, 2006). ROS play multiple roles in interactions between plants and microbes (Daub et al., 2013) and in the plant hypersensitive disease resistance response (Cecconi et al., 2009), including induced systemic resistances (Deepak et al., 2008).

Genomic and Proteomics

Recent advances in metabolomics, genomics, transcriptomic and metabolic modelling offer new opportunities to address this question and generate a system-level understanding of metabolic interactions at the host-pathogen interface (Rico et al., 2011). Functional genomics represents a systematic approach to elucidating the function of the novel genes revealed by complete genome sequences. The different approaches are grouped into four domains: genome, transcriptome, proteome, and metabolome (Oliver, 2002). Genomics is a new and promising area, which can broadly be defined as the application of high throughput genomics and functional genomic technologies to the study of pathology in plants (Brown and van der Ouderaa, 2007). Access to complete genomic sequences, coupled with rapidly accumulating data related to RNA and protein expression patterns, has made it possible to determine comprehensively how genes contribute to complex phenotypes (Close et al., 2004).

The huge amount of genomic data now becoming available offers both opportunities and challenges for breeding program (Garrett et al., 2006). Innovative approaches may be required to sequence, assemble, annotate and analyse host and pathogen genomes. Overcoming these challenges will enable scientists to investigate the genes and genome organisation may ultimately lead to new solutions for control diseases legume (Van Zee et al., 2007). The methods utilised provide new opportunities for studying the nature and role of defence mechanisms in plants (Collinge et al., 2008).

Knowledge of structural and functional genomics and genes and comparative genomics with model species has increased our understanding genomic diversity through integration of genetical, evolutionary and structural data (Heslop-Harrison and Schwarzacher, 2007). We will also gain a better understanding of the genes responsible for disease resistance. With a collection of desirable genes in mind, we can again use genomics as a diagnostic tool to search for these genes in the wide variety and follow their transfer by classical breeding (Regan et al., 2006). Following largescale source of sequencing of libraries and bioinformatics searches in public databases pave the way for the design and conduct of large-scale experiments that will help to a better understanding the complex biology of disease resistance (Bernier, 2004) and facilitating the isolation of novel genes, as well as by helping to identify targets for chemical control (Kamoun et al., 2002).

DNA sequencing is now complete for some hosts and several pathogens to form predictions about epidemic features and outcomes and for understanding host resistance and pathogen evolution (Garrett et al., 2006). The use of large-scale complementary DNA library constructions and genome-wide transcript profiles of plants exposed to biotic stress provide the data required to drive hypotheses concerning the function of newly identified genes (Collinge et al., 2008). For example, an extensive programme of functional genomics is being undertaken through the systematic analysis of insertional mutants in Arabidopsis (Delseny and Pelletier, 2001).

Proteins are involved in disease resistance and defense (Mooney et al., 2006). In recent years, proteomics has been used to gain in-depth understanding of many aspects of the host defense against pathogens to identify differentially accumulated proteins a mutant and a wild type (Chen et al., 2013; Zimaro et al., 2011). Fueled by ever growing DNA sequence information, proteomics is providing insight into the mechanisms of biological processes in a high-throughput mode (Sharma et al., 2008). Proteomic approach was employed to investigate the cold stress-responsive proteins in citrus (Long et al., 2012). Various proteomic techniques for isolation and purification of proteins have been developed; each tailored to preserve protein properties relevant to study of a particular disease type (Vo and Palsson, 2007). The application of combined proteomic and RNA interfering analyses is an efficient strategy to identify genes required disease resistance and probably other biological processes in plants (Xu et al., 2012).

The recent development of large scale phenotyping, genome sequencing and analysis of gene, protein and metabolite expressions can be of great help to further decipher plant-pathogen interactions and identify key resistance components that may be introgressed into crop plants through breeding (Rubiales et al., 2015). With the sequencing of entire genomes it has become technically feasible to study transcription on a global scale (Barnett and Fisher, 2006). In 2001, The Medicago Genome Initiative (MGI) is a database of EST sequences of the model legume *Medicago truncatula* (http://www.noble.org), which is taking a global approach in studying the genetic and biochemical events associated with the growth, development and environmental interactions of this model legume (Bell et al., 2001). Genome sequencing of the model legumes, Medicago truncatula and Lotus japonicus provides an opportunity for large-scale sequence-based comparison of two genomes in the same plant family (Cannon et al., 2006). Recent genome sequencing enables mega-base scale comparisons between related genomes. In 2013, chickpea draft sequence genome using next-generation sequencing platforms, bacterial artificial chromosome end sequences and a genetic map has been proposed. The draft genome sequence is expected to facilitate genetic enhancement and breeding to develop improved chickpea varieties (Jain et al., 2013). In the same year whole genome shotgun sequence of CDC Frontier, a kabuli chickpea variety, this contains an estimated 28,269 genes published. Candidate genes for disease resistance and agronomic traits are highlighted, including traits that distinguish the two main market classes of cultivated chickpea--desi and kabuli. These data comprise a resource for chickpea improvement through molecular breeding and provide insights into both genome diversity and domestication (Varshney et al., 2013). The genome of V. faba is exceedingly large (ca. 13 Gb) provides a valuable tool for future transcript profiling and also enriches the genetic resources available for this important legume crop species (Arun-Chinnappa and McCurdy, 2015). Development of bacterial artificial chromosome (BAC) library in White clover, sought a large-scale BAC-end sequencing strategy has the potential to anchor a significant proportion of the genome of white clover onto the gene-space sequence of *M. truncatula (Febrer et al., 2007)*. The first draft of the genome sequence of pigeonpea, and development of markers will be useful for fingerprinting and diversity analysis of pigeonpea germplasm and molecular breeding applications (Dubey et al., 2011; Singh et al., 2012). In 2012, Illumina next-generation sequencing platform to generate 237.2 Gb of sequence applied in pigeonpea (Varshney et al., 2012).

In the whole genome sequencing, genetic map provides an essential framework for accurate and efficient genome assembly and validation. The analysis revealed an overall conserved marker order, although some localized inversions between draft genome assembly and the genetic map were detected (Deokar et al., 2014). Many vectors, especially artificial chromosome vectors, have been developed for genome-scale mapping and sequencing. These data showed that this genomic library was reliable for further molecular research in *Lotus japonicus* (Guo Xi-ZhiGuo et al., 2004).

Comparative Genomics

Within the legume plant family, glimpses of synteny have also been observed. Characterizing syntenic relationships in legumes is important in transferring knowledge from model legumes to crops that are important sources of protein, fixed nitrogen, and health-promoting compounds (Mudge et al., 2005). The Lotus and Medicago genomes share a minimum of 10 large-scale synteny blocks, each with substantial collinearity and frequently extending the length of whole chromosome arms (Cannon et al., 2006). 56 conserved synteny blocks were identified between eggplant and tomato (Hirakawa et al., 2014). The December 22 2011 issue of *Nature* published the result of syntenic relationships between Medicago, Glycine, Lotus and Vitis illustrated in Circos diagram (Young et al., 2011) (Figure 1). Phylogenetic interpretation of DNA sequence data from multiple genomic regions has become the gold standard for species delimitation and population genetics (Kohn, 2004).

The interaction between genomics and quantitative genetics has been a two-way street. Genomics contributed genetic markers and genetic maps making it possible to study quantitative trait loci (QTLs), and quantitative genetics contributed new theories and computational techniques to deal with the data generated by QTL studies (van Buijtenen, 2001). It will be possible to integrate-based genomic tools into approaches as genetic approaches aiming at developing efficient marker-assisted selection strategies for the development of resistant varieties for example, or for designing models for predicting the behaviour of the plant in the field for greater resistance (Adam-Blondon, 2007).

Gene discovery holds enormous promise, but depends heavily on comparative genomics, capitalizing on genomic information (Burdon and Wilcox, 2007). Comparative genomics approach are often used to investigate crops with narrow genetic base of cultivars, and to evaluate the

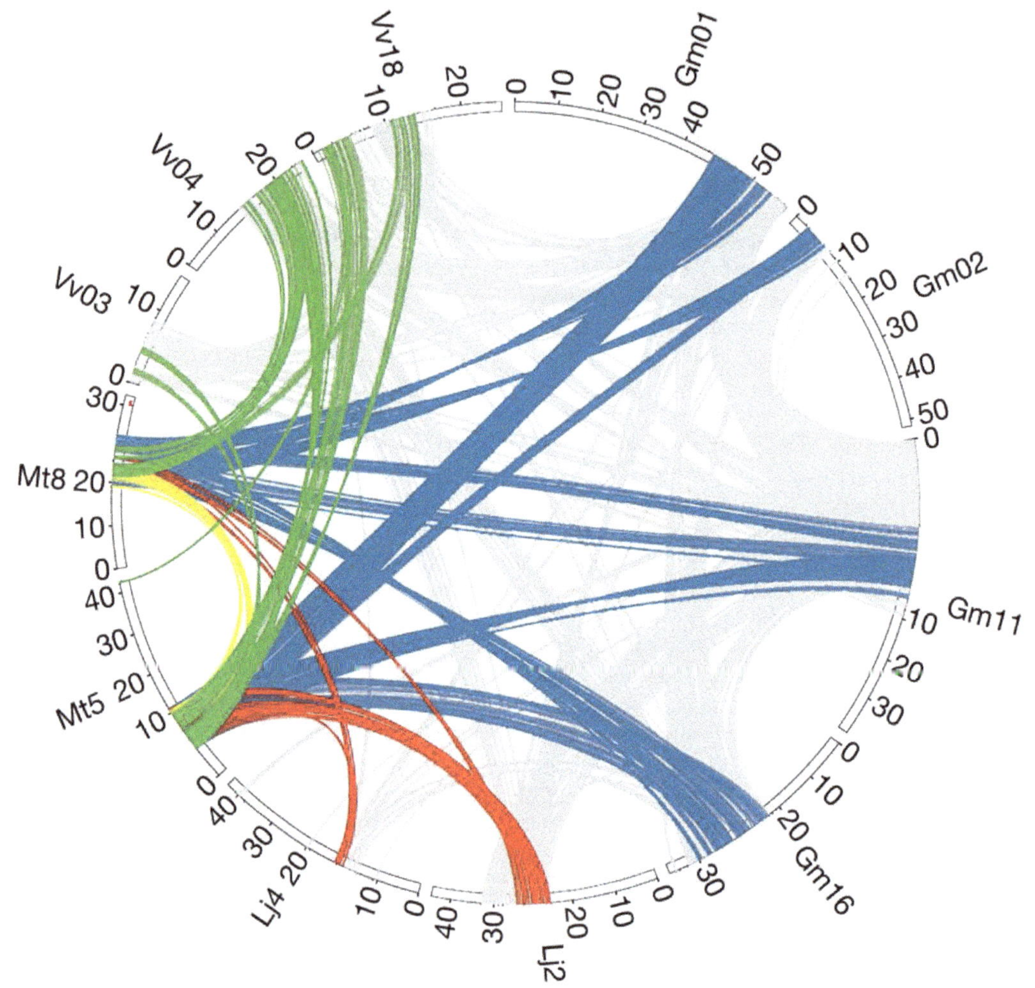

Figure 1. Homologous gene pairs were identified for all pairwise comparisons between *M. truncatula* (Mt), *G. max* (Gm), *L. japonicus* (Lj) and *V. vinifera* (Vv) genomes, syntenic regions are associated with the ancestral whole-genome duplication (WGD) occurred approximately 58 Myr ago (Young et al., 2011).

evolution of a complex gene cluster that is associated with several disease resistance (R) genes of known function (Ashfield et al., 2012; Bertioli et al., 2003; Sharma et al., 2013) Comparative genomics provides a tool to utilize the exponentially increasing sequence information from model plants to clone agronomically important genes from less studied crop species (Huang et al., 2005). Comparative genomics also plays a very important role in identification of genes/QTLs linked to the resistance genes (Babu et al., 2014). With access to sequences various model organisms and many important microbial pathogens, structural biology is on the verge of a dramatic transformation. Structural genomics will yield a large number of experimental protein structures (tens of thousands) and an even larger number of calculated comparative protein structure models (millions) (Burley, 2000).

The emergence of "omic" technologies and the establishment of model legume plants, such as *Medicago truncatula* and *Lotus japonicus* are promising strategies for understanding the molecular genetic basis of stress resistance (Dita et al., 2006). The legume species *M. truncatula* is a model plant to help in the identification of agronomically important genes in crop legumes, e.g. disease resistance (Journet et al., 2001).

QTL mapping and synteny analysis identified genomic loci harboring resistance factors has offered considerable promise and experimental power with varying degrees of success (Colwell, 2002; Muchero et al., 2011). The potential of GWAS approaches likely bearing resistance alleles of significant effect useful in breeding programs (Debibakas et al., 2014). Genome-wide analyses of gene function and gene expression are beginning to yield valuable information in many areas of biological research, and these genomic tools are now being applied to crop pest and disease research (Keon et al., 2003), thus strategy of applying a sequence-to-phenotype (or functional genomics) paradigm applied to the discovery of R-genes (Kamoun et al., 2002).

Conclusions

Current endeavours to find disease-causing genes have received a growing appreciation of the extent of conservation of genes, the extent to which synteny will be a useful tool throughout all legume species. The impact of the discovery of conserved synteny is also most important in the consideration of genomics programs for the major legume genomes. Genomics, through deciphering allelic effects on phenotypes and identifying patterns of adaptive variation at the landscape level, will in the future constitute a useful tool, if cost-effective, to design conservation strategies for plants (Gonzalez-Martinez et al., 2006), including disease resistance breeding program. However, the discovery still needs to be validated in a much larger sample size, where a high hurdle that many putative biomarkers fail to make. Poste (Poste, 2011) explained that although over 150,000 papers claimed to have found disease biomarkers in human, only 100 biomarkers were routinely used in clinic. Perhaps the same case may occurred in legumes, and future research on linking biomarkers comparative omics will help to fulfill this challenge.

References

Adam-Blondon, A.-F. (2007). What benefits can viticulture possibly expect from genomics? Quels benefices la genomique peut-elle apporter a la viticulture? [French]. Cahiers Agricultures *16*, 87-92.

Andolfo, G., Jupe, F., Witek, K., Etherington, G.J., Ercolano, M.R., and Jones, J.D.G. (2014). Defining the full tomato NB-LRR resistance gene repertoire using genomic and cDNA RenSeq. BMC Plant Biol *14*.

Arun-Chinnappa, K.S., and McCurdy, D.W. (2015). De novo assembly of a genome-wide transcriptome map of Vicia faba (L.) for transfer cell research. Front Plant Sci *6*, 217.

Ashfield, T., Egan, A.N., Pfeil, B.E., Chen, N.W.G., Podicheti, R., Ratnaparkhe, M.B., Ameline-Torregrosa, C., Denny, R., Cannon, S., Doyle, J.J., *et al.* (2012). Evolution of a complex disease resistance gene cluster in diploid phaseolus and tetraploid glycine. Plant Physiol *159*, 336-354.

Babu, B.K., Dinesh, P., Agrawal, P.K., Sood, S., Chandrashekara, C., Bhatt, J.C., and Kumar, A. (2014). Comparative genomics and association mapping approaches for blast resistant genes in finger millet using SSRs. PLoS One *9*.

Barnett, M.J., and Fisher, R.F. (2006). Global gene expression in the rhizobial-legume symbiosis. Symbiosis *42*, 1-24.

Basak, J., Kundu, S., and Pal, A. (2007). Phylogenetic analysis of divergent structural organization of nucleotide binding domain encoded by resistance genes and gene homologs in the family Fabaceae. Ind J Biotechnol *6*, 9-17.

Bell, C.J., Dixon, R.A., Farmer, A.D., Flores, R., Inman, J., Gonzales, R.A., Harrison, M.J., Paiva, N.L., Scott, A.D., Weller, J.W., *et al.* (2001). The Medicago Genome Initiative: A model legume database. Nucleic Acids Res *29*, 114-117.

Bernier, L. (2004). Genomics of ophiostomatoid fungi: the first two thousand genes. Phytoprotection *85*, 39-43.

Bertioli, D.J., Leal-Bertioli, S.C.M., Lion, M.B., Santos, V.L., Pappas Jr, G., Cannon, S.B., and Guimarães, P.M. (2003). A large scale analysis of resistance gene homologues in Arachis. Mol Genet Genomics *270*, 34-45.

Brown, L., and van der Ouderaa, F. (2007). Nutritional genomics: food industry applications from farm to fork. Br J Nutr *97*, 1027-1035.

Burdon, R.D., and Wilcox, P.L. (2007). Population management: potential impacts of advances in genomics. New Forests *34*, 187-206.

Burley, S.K. (2000). An overview of structural genomics. Nat Struct Biol *7*, 932-934.

Cannon, S.B., Sterck, L., Rombauts, S., Sato, S., Cheung, F., Gouzy, J., Wang, X., Mudge, J., Vasdewani, J., Scheix, T., *et al.* (2006). Legume genome evolution viewed through the Medicago truncatula and Lotus japonicus genomes. Proc Natl Acad Sci U S A *103*, 14959-14964.

Cantu, D., Yang, B., Ruan, R., Li, K., Menzo, V., Fu, D., Chern, M., Ronald, P.C., and Dubcovsky, J. (2013). Comparative analysis of protein-protein interactions in the defense response of rice and wheat. BMC Genomics *14*.

Cecconi, D., Orzetti, S., Vandelle, E., Rinalducci, S., Zolla, L., and Delledonne, M. (2009). Protein nitration during defense response in Arabidopsis thaliana. Electrophoresis *30*, 2460-2468.

Chen, C., Chen, H., Shan, J.X., Zhu, M.Z., Shi, M., Gao, J.P., and Lin, H.X. (2013). Genetic and physiological analysis of a novel type of interspecific hybrid weakness in rice. Mol Plant *6*, 716-728.

Choi, H.W., and Hwang, B.K. (2014). Molecular and cellular control of cell death and defense signaling in pepper. Planta *241*.

Close, T.J., Wanamaker, S.I., Caldo, R.A., Turner, S.M., Ashlock, D.A., Dickerson, J.A., Wing, R.A., Muehlbauer, G.J., Kleinhofs, A., and Wise, R.P. (2004). A new resource for cereal genomics: 22K Barley GeneChip comes of age. Plant Physiol *134*, 960-968.

Collinge, D.B., Jensen, M.K., Lyngkjaer, M.F., and Rung, J. (2008). How can we exploit functional genomics approaches for understanding the nature of plant defences? Barley as a case study. Eur J Plant Pathol *121*, 257-266.

Colwell, R.R. (2002). Fulfilling the promise of biotechnology. Biotechnol Adv *20*, 215-227.

Daub, M.E., Herrero, S., and Chung, K.R. (2013). Reactive oxygen species in plant pathogenesis: The role of perylenequinone photosensitizers. Antioxid Redox Signal *19*, 970-989.

Debibakas, S., Rocher, S., Garsmeur, O., Toubi, L., Roques, D., D'Hont, A., Hoarau, J.Y., and Daugrois, J.H. (2014). Prospecting sugarcane resistance to Sugarcane yellow leaf virus by genome-wide association. Theor Appl Genet *127*, 1719-1732.

Deepak, S.A., Shibato, J., Ishii, H., Ogawa, Y., Yoshida, Y., Iwahashi, H., Masuo, Y., Agrawal, G.K., and Rakwal, R. (2008). Proteomics approach for investigating the disease resistance using cucumber as model plant. Am J Biochem Biotechnol *4*, 231-238.

Delseny, M., and Pelletier, G. (2001). From Arabidopsis to rice genomics: A survey of French programmes. Comptes Rendus de l'Academie des Sciences Serie III Sciences de la Vie *324*, 1103-1110.

Deokar, A.A., Ramsay, L., Sharpe, A.G., Diapari, M., Sindhu, A., Bett, K., Warkentin, T.D., and Tar'an, B. (2014). Genome wide SNP identification in chickpea for use in development of a high density genetic map and improvement of chickpea reference genome assembly. BMC Genomics *15*, 708.

Dita, M.A., Rispail, N., Prats, E., Rubiales, D., and Singh, K.B. (2006). Biotechnology approaches to overcome biotic and abiotic stress constraints in legumes. Euphytica *147*, 1-24.

Dubey, A., Farmer, A., Schlueter, J., Cannon, S.B., Abernathy, B., Tuteja, R., Woodward, J., Shah, T., Mulasmanovic, B., Kudapa, H., *et al.* (2011). Defining the transcriptome assembly and its use for genome dynamics and transcriptome profiling studies in pigeonpea (Cajanus cajan L.). DNA research : an international journal for rapid publication of reports on genes and genomes *18*, 153-164.

Febrer, M., Cheung, F., Town, C.D., Cannon, S.B., Young, N.D., Abberton, M.T., Jenkins, G., and Milbourne, D. (2007). Construction, characterization, and preliminary BAC-end sequencing analysis of a bacterial artificial chromosome library of white clover (Trifolium repens L). Genome *50*, 412-421.

Garrett, K.A., Hulbert, S.H., Leach, J.E., and Travers, S.E. (2006). Ecological genomics and epidemiology. Eur J Plant Pathol *115*, 35-51.

Gonzalez-Martinez, S.C., Krutovsky, K.V., and Neale, D.B. (2006). Forest-tree population genomics and adaptive evolution. New Phytol *170*, 227-238.

Guo Xi-ZhiGuo, X.-Z., Liu Yao-GuangLiu, Y.-G., and Luo DaLuo, D. (2004). The construction and screening of genomic library based on transformable artificial chromosome (TAC) vector in Lotus japonicus. Zhiwu Shengli yu Fenzi Shengwuxue Xuebao *30*, 234-238.

Gururani, M.A., Venkatesh, J., Upadhyaya, C.P., Nookaraju, A., Pandey, S.K., and Park, S.W. (2012). Plant disease resistance genes: Current status and future directions. Physiol Mol Plant Pathol *78*, 51-65.

Heslop-Harrison, J.S., and Schwarzacher, T. (2007). Domestication, genomics and the future for banana. Ann Bot *100*, 1073-1084.

Hirakawa, H., Shirasawa, K., Miyatake, K., Nunome, T., Negoro, S., Ohyama, A., Yamaguchi, H., Sato, S., Isobe, S., Tabata, S., *et al.* (2014). Draft genome sequence of eggplant (Solanum melongena L.): the representative solanum species indigenous to the old world. DNA research : an international journal for rapid publication of reports on genes and genomes *21*, 649-660.

Hollomon, D.W., and Brent, K.J. (2009). Combating plant diseases - The Darwin connection. Pest Manage Sci *65*, 1156-1163.

Huang, S., Van Der Vossen, E.A.G., Kuang, H., Vleeshouwers, V.G.A.A., Zhang, N., Borm, T.J.A., Van Eck, H.J., Baker, B., Jacobsen, E., and Visser, R.G.F. (2005). Comparative genomics enabled the isolation of the R3a late blight resistance gene in potato. Plant J *42*, 251-261.

Hulbert, S., and Pumphrey, M. (2014). A time for more booms and fewer busts? unraveling cereal-rust interactions. Mol Plant-Microbe Interact *27*, 207-214.

Jain, M., Misra, G., Patel, R.K., Priya, P., Jhanwar, S., Khan, A.W., Shah, N., Singh, V.K., Garg, R., Jeena, G., *et al.* (2013). A draft genome sequence of the pulse crop chickpea (Cicer arietinum L.). The Plant journal : for cell and molecular biology *74*, 715-729.

Jordan, M.C., Cloutier, S., Somers, D., Procunier, D., Rampitsch, C., and Xing, T. (2006). Beyond R genes: dissecting disease-resistance pathways using genomics and proteomics. Can J Plant Pathol *28*, S228-S232.

Journet, E.P., Carreau, V., Gouzy, J., Thoquet, P., Rosenberg, C., Barker, D., Huguet, T., Dénarié, J., and Gamas, P. (2001). Génomique de la légumineuse modèle medicago truncatula : État des lieux et perspectives. OCL Ol Corps Lipides *8*, 478-484.

Kamoun, S., Dong, S., Hamada, W., Huitema, E., Kinney, D., Morgan, W.R., Styer, A., Testa, A., and Torto, T.A. (2002). From sequence to phenotype: Functional genomics of Phytophthora. Can J Plant Pathol *24*, 6-9.

Keon, J., Curtis, R., Cabrera, H., and Hargreaves, J. (2003). A genomics approach to crop pest and disease research. Pest Manage Sci *59*, 143-148.

Kohn, L.M. (2004). Applying comparative genomics to plant disease epidemiology. Phytoprotection *85*, 45-48.

Kotchoni, S.O., and Gachomo, E.W. (2006). The reactive oxygen species network pathways: An essential prerequisite for perception of pathogen attack and the acquired disease resistance in plants. J Biosci *31*, 389-404.

Lassner, M.W., and McElroy, D. (2002). Directed molecular evolution: Bridging the gap between genomics leads and commercial products. OMICS A Journal of Integrative Biology *6*, 153-162.

Li, Y.H., Zhou, G., Ma, J., Jiang, W., Jin, L.G., Zhang, Z., Guo, Y., Zhang, J., Sui, Y., Zheng, L., *et al.* (2014). De novo assembly of soybean wild relatives for pan-genome analysis of diversity and agronomic traits. Nat Biotechnol *32*, 1045-1052.

Long, G., Song, J., Deng, Z., Liu, J., and Rao, L. (2012). Ptcorp gene induced by cold stress was identified by proteomic analysis in leaves of poncirus trifoliata (L.) Raf. Mol Biol Rep *39*, 5859-5866.

López, C.E., Zuluaga, A.P., Cooke, R., Delseny, M., Tohme, J., and Verdier, V. (2003). Isolation of resistance gene candidates (RGCs) and characterization of an RGC cluster in cassava. Mol Genet Genomics *269*, 658-671.

Ma, N.L., Rahmat, Z., and Lam, S.S. (2013). A Review of the Omics Approach to Biomarkers of Oxidative Stress in Oryza sativa. Int J Mol Sci *14*, 7515-7541.

Madsen, L.H., Collins, N.C., Rakwalska, M., Backes, G., Sandal, N., Krusell, L., Jensen, J., Waterman, E.H., Jahoor, A., Ayliffe, M., *et al.* (2003). Barley disease resistance gene analogs of the NBS-LRR class: Identification and mapping. Mol Genet Genomics *269*, 150-161.

Martin, G.B., Bogdanove, A.J., and Sessa, G. (2003). Understanding the Functions of Plant Disease Resistance Proteins. In Annu Rev Plant Biol, pp. 23-61.

Mellor, K.E., Hoffman, A.M., and Timko, M.P. (2012). Use of ex vitro composite plants to study the interaction of cowpea (Vigna unguiculata L.) with the root parasitic angiosperm Striga gesnerioides. Plant Methods *8*.

Mooney, B.P., Miernyk, J.A., Greenlief, C.M., and Thelen, J.J. (2006). Using quantitative proteomics of Arabidopsis roots and leaves to predict metabolic activity. Physiol Plant *128*, 237-250.

Muchero, W., Ehlers, J.D., Close, T.J., and Roberts, P.A. (2011). Genic SNP markers and legume synteny reveal candidate genes underlying QTL for Macrophomina phaseolina resistance and maturity in cowpea [Vigna unguiculata (L) Walp.]. BMC Genomics *12*.

Mudge, J., Cannon, S.B., Kalo, P., Oldroyd, G.E., Roe, B.A., Town, C.D., and Young, N.D. (2005). Highly syntenic regions in the genomes of soybean, Medicago truncatula, and Arabidopsis thaliana. BMC Plant Biol, 15.

Mun, J.H., Yu, H.J., Park, S., and Park, B.S. (2009). Genome-wide identification of NBS-encoding resistance genes in Brassica rapa. Mol Genet Genomics *282*, 617-631.

Oliver, S.G. (2002). Functional genomics: Lessons from yeast. Philosophical Transactions of the Royal Society of London B Biological Sciences *357*, 17-23.

Poste, G. (2011). Bring on the biomarkers. Nature *469*, 156-157.

Ramachandra, S.B., Sathyanarayana, N.R., Subramonium, S., and Shetty, S.H. (2011). Isolation, cloning and characterization of resistance gene analogues in pearl millet based on conserved nucleotide-binding sites. J Phytopathol *159*, 382-389.

Ratnaparkhe, M.B., Wang, X., Li, J., Compton, R.O., Rainville, L.K., Lemke, C., Kim, C., Tang, H., and Paterson, A.H. (2011). Comparative analysis of peanut NBS-LRR gene clusters suggests evolutionary innovation among duplicated domains and erosion of gene microsynteny. New Phytol *192*, 164-178.

Regan, S., Gustafson, V., Mallubhotla, S., Chakravarty, B., Bagchi, M., Siahbazi, M., Rothwell, C., Sardana, R., Goyer, C., Audy, P., *et al.* (2006). Finding the perfect potato: using functional genomics to improve disease resistance and tuber quality traits. Can J Plant Pathol *28*, S247-S255.

Ribas, A.F., Cenci, A., Combes, M.C., Etienne, H., and Lashermes, P. (2011). Organization and molecular evolution of a disease-resistance gene cluster in coffee trees. BMC Genomics *12*.

Richly, E., Kurth, J., and Leister, D. (2002). Mode of amplification and reorganization of resistance genes during recent Arabidopsis thaliana evolution. Mol Biol Evol *19*, 76-84.

Rico, A., McCraw, S.L., and Preston, G.M. (2011). The metabolic interface between Pseudomonas syringae and plant cells. Curr Opin Microbiol *14*, 31-38.

Robertson, L.D., Singh, K.B., Erskine, W., and El Moneim, A.M.A. (1996). Useful genetic diversity in germplasm collections of food and forage legumes from West Asia and North Africa. Genetic Resources and Crop Evolution *43*, 447-460.

Rubiales, D., Fondevilla, S., Chen, W., Gentzbittel, L., Higgins, T.J.V., Castillejo, M.A., Singh, K.B., and Rispail, N. (2015). Achievements and Challenges in Legume Breeding for Pest and Disease Resistance. Crit Rev Plant Sci *34*, 195-236.

Sabir, J., Mutwakil, M., El-Hanafy, A., Al-Hejin, A., Abdel Sadek, M., Abou-Alsoud, M., Qureshi, M., Saini, K., and Ahmed, M. (2014). Applying molecular tools for improving livestock performance: From DNA markers to next generation sequencing technologies. J Food, Agric Environ *12*, 541-553.

Schwartz, D.A., Freedman, J.H., and Linney, E.A. (2004). Environmental genomics: a key to understanding biology, pathophysiology and disease. Hum Mol Genet *13*, R217-R224.

Sharma, R., Chhokar, V., Jha, S., and Rana, J.S. (2008). Proteomics: Current approaches and future perspectives. Journal of Applied Bioscience *34*, 1-16.

Sharma, S., Upadhyaya, H.D., Varshney, R.K., and Gowda, C.L.L. (2013). Pre-breeding for diversification of primary gene pool and genetic enhancement of grain legumes. Front Plant Sci *4*.

Sharma, T.R., Das, A., Thakur, S., and Jalali, B.L. (2014). Recent understanding on structure, function and evolution of plant disease resistance genes. Proc Indian Natl Sci Acad *80*, 83-93.

Singh, N.K., Gupta, D.K., Jayaswal, P.K., Mahato, A.K., Dutta, S., Singh, S., Bhutani, S., Dogra, V., Singh, B.P., Kumawat, G., *et al.* (2012). The first draft of the pigeonpea genome sequence. J Plant Biochem Biotechnol *21*, 98-112.

Singh, R.J., Chung, G.H., and Nelson, R.L. (2007). Landmark research in legumes. Genome *50*, 525-537.

Sladek, R., and Hudson, T.J. (2006). Elucidating cis- and trans-regulatory variation using genetical genomics. Trends Genet *22*, 245-250.

van Buijtenen, J.P. (2001). Genomics and quantitative genetics. Can J For Res *31*, 617-622.

Van Zee, J.P., Geraci, N.S., Guerrero, F.D., Wikel, S.K., Stuart, J.J., Nene, V.M., and Hill, C.A. (2007). Tick genomics: The Ixodes genome project and beyond. Int J Parasitol *37*, 1297-1305.

Varshney, R.K., Chen, W., Li, Y., Bharti, A.K., Saxena, R.K., Schlueter, J.A., Donoghue, M.T., Azam, S., Fan, G., Whaley, A.M., *et al.* (2012). Draft genome sequence of pigeonpea (Cajanus cajan), an orphan legume crop of resource-poor farmers. Nat Biotechnol *30*, 83-89.

Varshney, R.K., Song, C., Saxena, R.K., Azam, S., Yu, S., Sharpe, A.G., Cannon, S., Baek, J., Rosen, B.D., Tar'an, B., *et al.* (2013). Draft genome sequence of chickpea (Cicer arietinum) provides a resource for trait improvement. Nat Biotechnol *31*, 240-246.

Vo, T.D., and Palsson, B.O. (2007). Building the power house: recent advances in mitochondrial studies through proteomics and systems biology. American Journal of Physiology Cell Physiology *292*, C164-C177.

Vossen, J.H., Dezhsetan, S., Esselink, D., Arens, M., Sanz, M.J., Verweij, W., Verzaux, E., and van der Linden, C.G. (2013). Novel applications of motif-directed profiling to identify disease resistance genes in plants. Plant Methods *9*.

Webb, C.A., and Fellers, J.P. (2006). Cereal rust fungi genomics and the pursuit of virulence and avirulence factors. FEMS Microbiol Lett *264*, 1-7.

Wei, H., Li, W., Sun, X., Zhu, S., and Zhu, J. (2013). Systematic Analysis and Comparison of Nucleotide-

Binding Site Disease Resistance Genes in a Diploid Cotton Gossypium raimondii. PLoS One *8*.
Xu, J.-R., and Xue, C. (2002). Time for a blast: Genomics of Magnaporthe grisea. Mol Plant Pathol *3*, 173-176.
Xu, Q.F., Cheng, W.S., Li, S.S., Li, W., Zhang, Z.X., Xu, Y.P., Zhou, X.P., and Cai, X.Z. (2012). Identification of genes required for Cf-dependent hypersensitive cell death by combined proteomic and RNA interfering analyses. J Exp Bot *63*, 2421-2435.
Xu, W., Yu, Y., Zhou, Q., Ding, J., Dai, L., Xie, X., Xu, Y., Zhang, C., and Wang, Y. (2011). Expression pattern, genomic structure, and promoter analysis of the gene encoding stilbene synthase from Chinese wild Vitis pseudoreticulata. J Exp Bot *62*, 2745-2761.
Young, N.D., Debelle, F., Oldroyd, G.E., Geurts, R., Cannon, S.B., Udvardi, M.K., Benedito, V.A., Mayer, K.F., Gouzy, J., Schoof, H., *et al.* (2011). The Medicago genome provides insight into the evolution of rhizobial symbioses. Nature *480*, 520-524.
Yu, J., Tehrim, S., Zhang, F., Tong, C., Huang, J., Cheng, X., Dong, C., Zhou, Y., Qin, R., Hua, W., *et al.* (2014). Genome-wide comparative analysis of NBS-encoding genes between Brassica species and Arabidopsis thaliana. BMC Genomics *15*.
Zimaro, T., Gottig, N., Garavaglia, B.S., Gehring, C., and Ottado, J. (2011). Unraveling plant responses to bacterial pathogens through proteomics. J Biomed Biotechnol *2011*.

Other Publications of Interest

Xu, J.-P. (2014). Next-generation Sequencing: Current Technologies and Applications. Caister Academic Press, U.K. ISBN: 978-1-908230-33-1
He, Z. (2014). Microarrays: Current Technology, Innovations and Applications. Caister Academic Press, U.K. ISBN: 978-1-908230-49-2
Poptsova, M.S. (2014). Genome Analysis: Current Procedures and Applications. Caister Academic Press, U.K. ISBN: 978-1-908230-29-4
Fuentes, M. and LaBaer, J. (2014). Proteomics: Targeted Technology, Innovations and Applications. Caister Academic Press, U.K. ISBN: 978-1-908230-46-1
Murillo, J., Vinatzer, B.A., Jackson, R.W. and Arnold, D.L. (2015). Bacteria-Plant Interactions: Advanced Research and Future Trends. Caister Academic Press, U.K. ISBN: 978-1-908230-58-4
Caranta, C., Aranda, M.A., Tepfer, M. and Lopez-Moya, J.J. (2011). Recent Advances in Plant Virology. Caister Academic Press, U.K. ISBN: 978-1-904455-75-2
Nannipieri, P., Pietramellara, G. and Renella, G. (2014). Omics in Soil Science. Caister Academic Press, U.K. ISBN: 978-1-908230-32-4
Herold, K.E. and Rasooly, A. (2009). Lab-on-a-Chip Technology: Fabrication and Microfluidics. Caister Academic Press, U.K. ISBN: 978-1-904455-46-2

www.ingramcontent.com/pod-product-compliance
Ingram Content Group UK Ltd.
Pitfield, Milton Keynes, MK11 3LW, UK
UKHW062013041026
15591UKWH00017B/1321

9 781910 190357